THE SCIENCE OF NANOMATERIALS

Basics and Applications

Edited by

Suresh C. Ameta, PhD

Rakshit Ameta, PhD

APPLE ACADEMIC PRESS

First edition published 2023

Apple Academic Press Inc.
1265 Goldenrod Circle, NE,
Palm Bay, FL 32905 USA

760 Laurentian Drive, Unit 19,
Burlington, ON L7N 0A4, CANADA

CRC Press
6000 Broken Sound Parkway NW,
Suite 300, Boca Raton, FL 33487-2742 USA

4 Park Square, Milton Park,
Abingdon, Oxon, OX14 4RN UK

Library and Archives Canada Cataloguing in Publication

Title: The science of nanomaterials / edited by Suresh C. Ameta, PhD, Rakshit Ameta, PhD.

Names: Ameta, Suresh C., editor. | Ameta, Rakshit, editor.

Description: First edition. | Includes bibliographical references and index.

Identifiers: Canadiana (print) 20220284946 | Canadiana (ebook) 20220285004 | ISBN 9781774910726 (hardcover) | ISBN 9781774910733 (softcover) | ISBN 9781003283126 (ebook)

Subjects: LCSH: Nanostructured materials.

Classification: LCC TA418.9.N35 S35 2023 | DDC 620.1/15—dc23

Library of Congress Cataloging-in-Publication Data

Names: Ameta, Suresh C., editor. | Ameta, Rakshit, editor.

Title: The science of nanomaterials : basics and applications / edited by Suresh C. Ameta, PhD, Rakshit Ameta, PhD.

Description: First edition. | Palm Bay, FL : Apple Academic Press, 2022. | Includes bibliographical references and index. | Summary: "Nanoparticles have a smaller size as compared to their micro, macro or bulk counterparts. Reduction in size of these particles provides them with some unique characteristics, such as surface-to-volume ratio, quantum confinement effect, surface plasmon response, widening of band gap, etc. These nanoparticles have attracted attention of scientists all over the globe in last few decades. Nanomaterials are a requirement of the day as their use is pushing a number of bulk materials out of use. These have numerous applications in varied fields like solar cells, food industries, targeted gene and drug delivery, medical imaging, polymeric materials, sensors, treatment of water, soil and air pollution, etc. Written in a convenient and easy-to-read style, this book covers the important aspects of nanomaterials by focusing on the many issues related to food and textile industries, treatment of polluted water, health, energy crises, targeted drug delivery, etc. The editors take an interdisciplinary approach to discussing how the scenario will change on a global level in the future and explore when these nanomaterials will replace almost all micro- and macromaterials. The Science of Nanomaterials is a ready-at-hand guide to the many issues related to the use of nanomaterials in drug and gene delivery, sensors, photosplitting of water, wastewater treatment, microbial diagnosis, textile industries, nanocomposites, food industries (safety, security, packaging and preservation), etc. that will be valuable to students at all levels as well as a refresher for scientists, researchers, policymakers, and others"-- Provided by publisher.

Identifiers: LCCN 2022031832 (print) | LCCN 2022031833 (ebook) | ISBN 9781774910726 (hardcover) | ISBN 9781774910733 (paperback) | ISBN 9781003283126 (ebook)

Subjects: LCSH: Nanostructured materials. | Nanostructured materials--Industrial applications.

Classification: LCC TA418.9.N35 S355 2022 (print) | LCC TA418.9.N35 (ebook) | DDC 620.1/15--dc23/eng/20220711

LC record available at https://lccn.loc.gov/2022031832

LC ebook record available at https://lccn.loc.gov/2022031833

ISBN: 978-1-77491-072-6 (hbk)
ISBN: 978-1-77491-073-3 (pbk)
ISBN: 978-1-00328-312-6 (ebk)

Dedicated to

Mrs. ANITA AMETA

who has helped us at every stage of this work.

Contributors

Jiji Abraham
Vimala College (Autonomous), Thrissur 680009, Kerala, India; E-mail: jijiabraham02@gmail.com

C. O. Adetunji
Applied Microbiology, Biotechnology and Nanotechnology Laboratory, Department of Microbiology, Edo University Iyamho, PMB 04, Auchi, Edo State, Nigeria; E-mail: charliguitar@yahoo.com; adetunjicharles@gmail.com; adetunji.charles@edouniversity.edu.ng

J. B. Adetunji
Nutritional and toxicological Research Laboratory, Department of Biochemistry Sciences, Osun State University, Osogbo, Nigeria

O. O. Ajayi
Department of Biochemistry, Edo University Iyamho, Edo State, Nigeria

Rakshit Ameta
Department of Chemistry, J. R. N. Rajasthan Vidhyapeeth (Deemed to be University), Udaipur, Rajasthan, 313001, India; E-mail: rakshit_ameta@yahoo.in

Suresh C. Ameta
Department of Chemistry, PAHER University, Udaipur 313001, Rajasthan, India

O. A. Anani
Laboratory of Ecotoxicology and Forensic Biology, Department of Biological Science, Faculty of Science, Edo University, Iyamho, Edo State, Nigeria

Anca Armăşelu
Department of Electrical Engineering and Applied Physics, Faculty of Electrical Engineering and Computer Science, Braşov, Romania; E-mail: anca_armaselu@yahoo.com

M. N. Avadhanulu
Department of Physics, Kavikulguru Institute of Technology And Science, Ramtek 441106, Maharashtra, India

Jayesh P. Bhatt
Department of Chemistry, PAHER University, Udaipur 313001, Rajasthan, India; E-mail: jay.bht@gmail.com

Subrata Biswas
Nanoscience Laboratory, Department of Physics, National Institute of Technology, Durgapur 713209, West Bengal, India

R. E. Bodunrinde
Department of Microbiology, Federal University of Technology Akure, Nigeria

Amra Bratovcic
University of Tuzla, Faculty of Technology, Department of Physical Chemistry and Electrochemistry, Urfeta Vejzagica 8, 75000 Tuzla, Bosnia and Herzegovina; E-mail: amra.bratovcic@untz.ba

Neelu Chouhan
Department of Pure and Applied Chemistry, University of Kota, MBS Road, Kota 324005, Rajasthan, India; E-mail: neeluchouhan@uok.ac.in

Soney C. George
Centre for Nano Science & Technology, Amal Jyothi Engineering College, Kanjirappally 686518, Kerala, India; E-mail: soneygeo@gmail.com

D. I. Hefft
Edgbaston Campus, University of Birmingham, School of Chemical Engineering, Birmingham B15 2TT, United Kingdom

A. Inobeme
Department of Chemistry, Edo University Iyamho, Nigeria

Monika Jangid
Department of Chemistry, PAHER University, Udaipur 313001, Rajasthan, India;
E-mail: Monikajangid24@gmail.com

Gajanan Kotharu
Department of Chemistry, Kavikulguru Institute of Technology And Science, Ramtek 441106, Maharashtra, India; E-mail: gajanankotharu1@gmail.com

Pathik Kumbhakar
Nanoscience Laboratory, Department of Physics, National Institute of Technology, Durgapur 713209, West Bengal, India; E-mail: pathik.kumbhakar@phy.nitdgp.ac.in

J. T. Mathew
Department of Chemistry, Ibrahim Badamasi University Lapai, Niger State, Nigeria

O. T. Olaniyan
Laboratory for Reproductive Biology and Developmental Programming, Department of Physiology, Edo University Iyamho, Nigeria

S. Lakshmana Prabu
Department of Pharmaceutical Technology, University College of Engineering (BIT Campus), Anna University, Tiruchirappalli 620024, Tamil Nadu, India; E-mail: slaxmanvel@gmail.com

A. Puratchikody
Department of Pharmaceutical Technology, University College of Engineering (BIT Campus), Anna University, Tiruchirappalli 620024, Tamil Nadu, India

Pedro J. Rivero
Materials Engineering Laboratory, Institute for Advanced Materials and Mathematics (INAMAT2), Department of Engineering, Public University of Navarre, Campus Arrosadía s/n, 31006, Pamplona, Spain; E-mail: pedrojose.rivero@unavarra.es

Shobhana Sharma
Department of Chemistry, S. S. Subodh P. G. (Autonomous) College, Jaipur 302005, Rajasthan, India

M. Dhivya Sree
Department of Biomedical Engineering, PSG College of Technology, Coimbatore 641004, Tamil Nadu, India

Sabu Thomas
International and Inter University Centre for Nano Science & Technology, Mahatma Gandhi University, Kottayam 686560, Kerala, India; E-mail: sabupolymer@yahoo.com

Saliney Thomas
School of Chemical Sciences, Mahatma Gandhi University, Kottayam 686560, Kerala, India
Email: salineythomas@gmail.com

Shaji P. Thomas
Bishop Abraham Memorial College, Thuruthicad 689597, Kerala, India; E-mail: shajiparel@gmail.com

A. Umamaheswari
Department of Pharmaceutical Technology, University College of Engineering (BIT Campus), Anna University, Tiruchirappalli 620024, Tamil Nadu, India

M. Vijayalakshmi
Department of Pharmaceutical Technology, University College of Engineering (BIT Campus), Anna University, Tiruchirappalli 620024, Tamil Nadu, India

Shubang Vyas
Department of Chemistry, PAHER University, Udaipur 313001, Rajasthan, India;
E-mail: shubhangvyas@yahoo.com

Abbreviations

ABS	Acrylonitrile-butadiene-styrene
AR	Acid route
BB	Bismarck brown
BCS	Biopharmaceutics Classification System
BIMS	Brominated poly(isobutylene-*co*-paramethylstyrene)
BMO	Bimetallic oxide
BN	Boron nitride
CA	Cellulose acetate
CCVD	Catalytic chemical vapor disposition
CBNs	Carbon-based nanomaterials
CMC	Carboxylmethyl cellulose
CNCs	Cellulose nanocrystals
CNF	Carbon nanofibers
CNF	Cellulose nanofibrils
CNTA	Carbon nanotubes array
CNTs	Carbon nanotubes
COD	Chemical oxygen demand
CR	Congo red
CS	Chitosan
CTAB	Cetyltrimethylammonium bromide
CV	Crystal violet
CVD	Chemical vapor deposition
CWNs	Carnauba wax nanoparticles
DDS	Drug delivery system
DFT	density functional theory
DLVO	Derjaguin–Landau–Verwey–Overbeek
DOS	Density of states
DPE	Diphenylethane
DRS	Double redox strategy
ECM	Extracellular matrices
EDS	Energy-dispersive x-ray spectrum
EFE	Electron field emission
EMI	Electromagnetic interference
ENMs	Engineered nanomaterials

PMS	Peroxomonosulfate
PMVS	Polymethylvinylsiloxane
PNCs	Polymer nanocomposites
POCT	Point of care testing
POME	Palm oil mill effluent
POSS	Polyhedral oligomeric silsesquioxane
PP	Polypropylene
PPESK	Polyphthalazine ether sulfone ketone
PSS	Poly(sodium 4-styrene sulfonate)
PT	Photothermal
PTOP	Photocatalytic turnover productivity
PV	Photovoltaic
PVD	Physical vapor deposition
PVDF	Polyvinylidene fluoride
PVX	Potato virus X
QDs	Quantum dots
QT	Quercetin
RBM	Radial breathing modes
rGO	Reduced graphene oxides
RHE	Reversible hydrogen electrode
RI	Refractive index
RL	Random laser
ROS	Reactive oxygen species
RRDE	Rotating ring-disk electrode
SAOS	Small-amplitude oscillatory shear
SAVR	Surface area to volume ratio
SBR	Styrene butadiene rubber
SBT	Sea buckthorn fruit
SEM	Scanning electron microscopy
SERLs	Surface emitting random lasers
SLN	Solid nanoparticles
SLS	Sodium lignin sulphonate
SOG	Soap oil and grease
SPR	Surface plasmon resonance
SR	Surfactant route
STH	Solar-to-hydrogen
SWCNT	Single wall carbon nanotube
TCE	Trichloroethene
TEM	Transmission electron microscopy

TNBC	Triple-negative breast cancer
TOC	Total organic compound
TOF	Turn over frequency
TON	Turnover number
TPP	Tripolyphosphate
TRG	Thermally reduced graphene
UA	Ursolic acid
UHD	Ultra-high definition
VBB	Victoria blue B
VC	Vinyl chloride
VCNFs	Vapor-grown carbon nanofibers
VLS	Vapor–liquid–solid
VNPs	Viral nanoparticles
VSS	Vapor–solid–solid
WVP	Water vapor permeability
XLPE	Cross-linked polyethylene
XRD	X-ray diffraction

Preface

Nanoparticles have a smaller size as compared to their micro, macro, or bulk counterparts. Reduction in size of these particles provides them some unique characteristics, such as surface-to-volume ratio, quantum confinement effect, surface plasma response, widening of band gap, etc. These nanoparticles have attracted the attention of scientists all over the globe in the last few decades.

A nanomaterial is a requirement of the day as their use is pushing a number of bulk materials out of use. These have numerous applications in varied fields such as solar cells, food industries, targeted gene delivery, medical imaging, polymeric materials, sensors, treatment of water, soil, and air pollution.

In this book, water pollution, food preservation, food packaging, detection of foodborne pathogens, targeted drug delivery, hydrogen generation from photosplitting of water, nanomedical and nanobiotechnological applications, gene delivery, nanocomposites, antimicrobial surface in textile industry, optical properties, and sensing applications are discussed.

—Suresh C. Ameta
Rakshit Ameta

CHAPTER 1

Nanomaterials: An Introduction

SURESH C. AMETA*

Department of Chemistry, PAHER University, Udaipur 313001, India

**E-mail: ameta_sc@yahoo.com*

ABSTRACT

Nanostructured material is one billionth of a meter. It is 10^{-5} times smaller than the diameter of a human hair. These materials are classified as zero-dimensional, one-dimensional, two-dimensional, and three-dimensional nanomaterials, which are represented by 0D, 1D, 2D, and 3D, respectively. Examples of nanomaterials are quantum dots, nanowires, nanosheets, and nanocrystals. These nanomaterials are unique in nature because they have a high surface-area-to-volume ratio, surface plasmon resonance, and quantum confinement effect. These can be prepared by many methods but they can be categorized either in top-down or bottom-up approaches, that is, either converting bulk or macroparticles to nanoparticles or from atoms/molecules or clusters to build up nanoparticles. These materials show interesting properties like optical, electrical, magnetic, or mechanical properties making them quite useful with potential applications in sensors, water purification, targeted drug delivery, scratch resistant, data storage, displays, food packaging, catalytic convertor, antimicrobial, smart solar panel, coating, self-cleaning glass, magnetic resonance imaging, lighter sporting goods, and many more. This is all possible due to change in conductivity, catalytic behavior, lighter weight, high strength (stronger), elastic modulus, rigidity, toughness, hardness, plasticity, etc.

The Science of Nanomaterials: Basics and Applications. Suresh C. Ameta & Rakshit Ameta (Eds.)

1.1 INTRODUCTION

Whenever any material has at least one of its dimensions in the nanometer range, then it is called a nanomaterial. A nanometer (nm) is 1 billionth part of a meter, that is, 10^{-9} m or 1 nm. It is an extremely small value, which is equal to approximately the length, when 10 hydrogen or 5 silicon atoms are placed in a line. Nanomaterials show significantly different optical, magnetic, electrical, and mechanical properties than macromaterials and bulk materials, and therefore, these are used in many energy applications, such as information technology, bioengineering, and environmental applications.

Some of the chemical as well as physical properties of a material may differ significantly as compared to their bulk counterpart. Many properties of material must now be reinvestigated because there is a large increase in its surface-area-to-volume ratio (SAVR) on reducing the size of any material to the nanoscale, which may have a prominent effect on the performance of that material.

Nanotechnology and nanoscience are not new areas of research as nanomaterials are being used for centuries. The Chinese used to introduce gold nanoparticle to give red color to ceramic porcelains more than thousand years ago while Romans used metal nanoparticles to give beautiful colors to glass artifacts. Nanoparticles were also used for the decoration of cathedral windows. The synthesis, characterization, and uses of nanostructured materials give a broad frame to nanoscience.

These nanosized materials are now commonly used in many industries. Nano-ZnO and TiO_2 are used for providing protection from ultraviolet rays. Oxide of iron is used in audio–video tapes and disk drives as it is a magnetic material. Carbon black particles are used to create wear resistance in rubber tires. The reinforcement and insulation of composites can be done by using some nanofibers. The use of these nanoparticles can make any product more conducting in nature, stronger, and lighter in weight. Some of the products are already available in the market and these are computer hard drives, automobile catalytic converters, magnetic recording tapes, car bumpers, dental bonding agents, solid-state compasses, metal-cutting tools, eyeglasses and windows with glare-reducing/protective coatings, longer-lasting tennis ball, inks, burn and wound dressing, and so on. Nanotechnology finds numerous potential applications in the field of medicine and/or biology and also in nanomedicine, development of nanoscale devices for improving therapy, targeted drug delivery, and quick and correct diagnostics. It has been observed that such nanoscale devices or nanorobots help in detecting or providing protection/prevention against disease in early stages as vehicles for delivering therapeutic agents, and may be helpful in repairing metabolic or genetic defects. A major challenge in their application in medical fields is preparing instruments of such a small size, that they can act at a molecular level; for example, it may require sensors of cellular size or even smaller so as to observe functions at that level.

The applications of nanomaterials in varied fields have been excellently reviewed by Liu and Bashir (2015), Bhagyaraj (2018), Kumar and Kumbhat (2016), Liao et al. (2017), Hou and Sellmyer (2017), Acharya (2020), Shi (2014), Khan (2017), Zenkina (2019), Giersig and Khomutov (2008), Grumezescu (2019), Amparo et al. (2018), Aguilar (2012), Kane et al. (2016), De Oliveira Jr. et al. (2016), Grassian (2008), Hu and Apblett (2014), Wang and Chen (2018).

There are three major factors, which are responsible for some unique properties in nanomaterials, which are totally different from those of bulk materials. These are as follows:

- Increase in SAVR,
- Surface plasmon resonance (SPR), and
- Quantum confinement effect.

These three factors can change the characteristic properties of the material like strength, reactivity, optical, magnetic, electrical properties, and so on.

1.2 CLASSIFICATION OF NANOMATERIALS

Nanomaterials have extremely small sizes having at least one dimension of less than 100 nm. These can exist in different forms such as single, fused, agglomerated, or aggregated forms having different shapes, such as spherical, tubular, irregular, and so forth.

Nanomaterials are classified based on the number of dimensions in the nanoscale, which are less than 100 nm. These are as follows:

- Zero-dimensional (0D) nanomaterials,
- One-dimensional (1D) nanomaterials,
- Two-dimensional (2D) nanomaterials, and
- Three-dimensional (3D) nanomaterials (Figure 1.1).

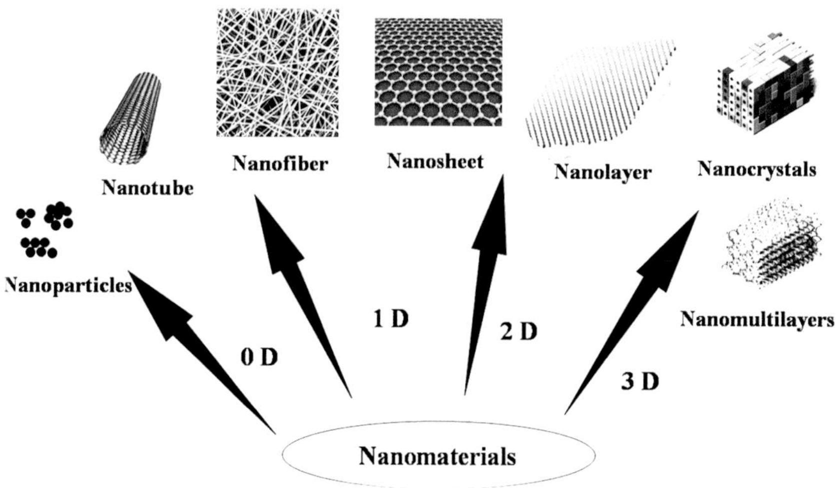

FIGURE 1.1 Schematic illustration of structural dimensionality of nanomaterial.

1.2.1 ZERO-DIMENSIONAL NANOMATERIALS

In zero-dimensional (0D) nanomaterial, all the dimensions of the material are in the nanoscale. It means that no dimensions are larger than 100 nm and therefore, these are called zero-dimensional. Nanoparticles are the most common 0D nanomaterials. Examples are nanoparticles, nanograins, nanoshells, nanocapsules, nanorings, quantum dots, and fullerene. Two 0D nanomaterials (quantum dots and fullerene) are presented in Figure 1.2.

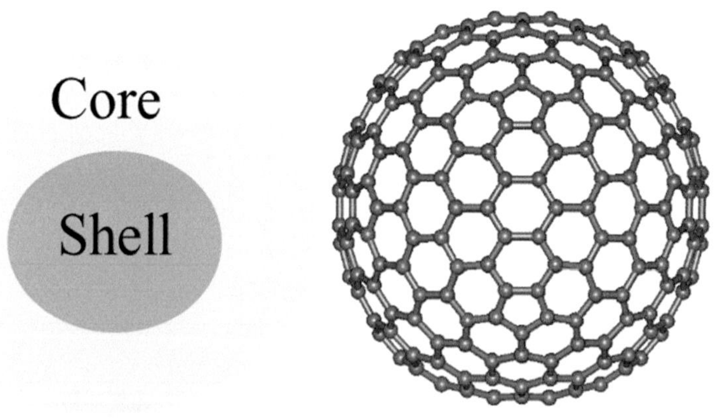

FIGURE 1.2 Zero-dimensional nanomaterial—Quantum dots, fullerene.

1.2.2 ONE-DIMENSIONAL NANOMATERIALS

When one dimension is outside the range of nanoscale, then such materials are called one-dimensional (1D) nanomaterial and represented as 1D. This leads to a needle-shaped nanomaterials. Examples include nanowires, nanofibers, nanotubes, etc. 1D nanomaterials—nanorod and nanotube are shown in Figure 1.3.

FIGURE 1.3 One-dimensional nanomaterial—nanorods, nanotube.

1.2.3 TWO-DIMENSIONAL NANOMATERIALS

When two dimensions (2D) of any material are not confined to the nanoscale, then these are termed as 2D nanomaterials. They exhibit plate-like shapes. Examples are nanofilms, graphene, nanolayers, and nanocoating, are good examples of 2D materials. 2D nanomaterial—nanosheet and nanolayer are shown in Figure 1.4.

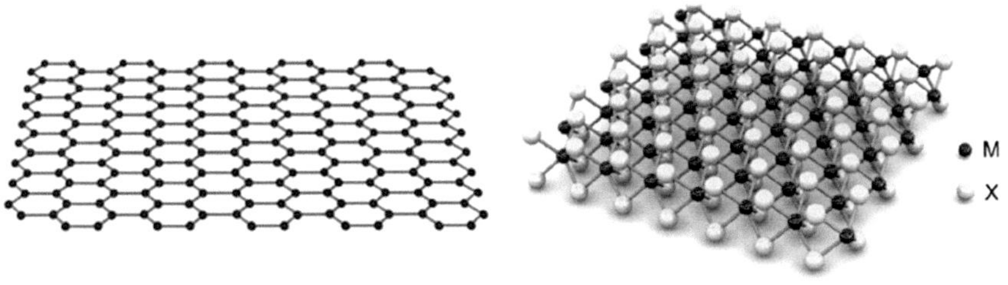

FIGURE 1.4 Two-dimensional nanomaterial—nanosheet, nanolayer.

1.2.4 THREE-DIMENSIONAL NANOMATERIALS

Bulk nanomaterials are known as three-dimensional (3D) nanomaterials. These nanomaterials are not confined to the nanoscale in any dimension and have the dimension in all directions above 100 nm. Nanomaterials in bulk can be in the form of a multiple arrangements of crystals of nanosize, mostly with different orientations. This may be in the form of nanotubes, bundles of nanowires, powders, dispersions of nanoparticles, as well as multinanolayers.

3D nanomaterial—nanocrystals are shown in Figure 1.5.

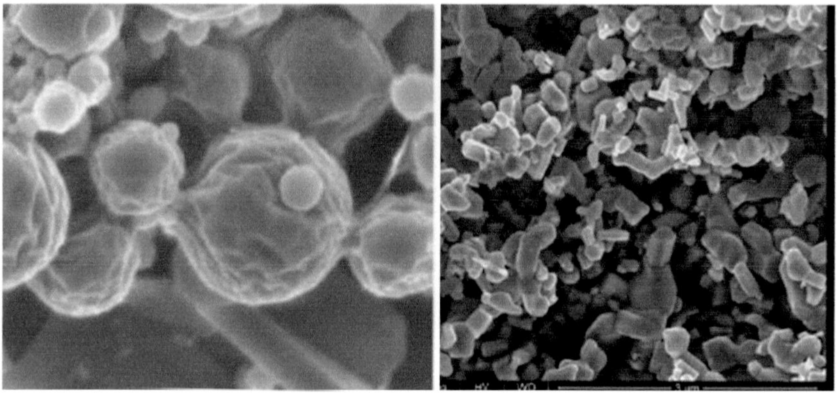

FIGURE 1.5 Three-dimensional nanomaterial—nanocrystal.

Nanomaterials are available in different shapes and these are named according to their resemblance to particularly structures like nanoarrows, nanobelts, nanobowls, nanobricks, nanocapsules, nanocombs, nanocones, nanoclusters, nanocoreshells, nanocubes, nanodisks, nanodots, nanodumbbells, nanofilms, nanoflakes, nanoflowers, nanonails, nanoneedles, nanopencils, nanopillars, nanopins, nanoplatelet, nanopowders, nanoribbons, nanorice, nanorings, nanorods, nanosheets, nanospheres, nanotripods, nanotubes, and nanowires. Some of the nanomaterials having different shapes are shown in Figure 1.6.

1.3 UNIQUE NATURE OF NANOPARTICLES

1.3.1 INCREASE IN THE SURFACE-AREA-TO-VOLUME RATIO

Both SAVR as well as surface-to-bulk-atom ratio increase dramatically as the size of any particle is reduced to a few nm. Accordingly, properties like color solubility, conductivity, catalytic behavior, absorption or emission wavelength, transparency, and melting point are changed due to the variations in the particle size.

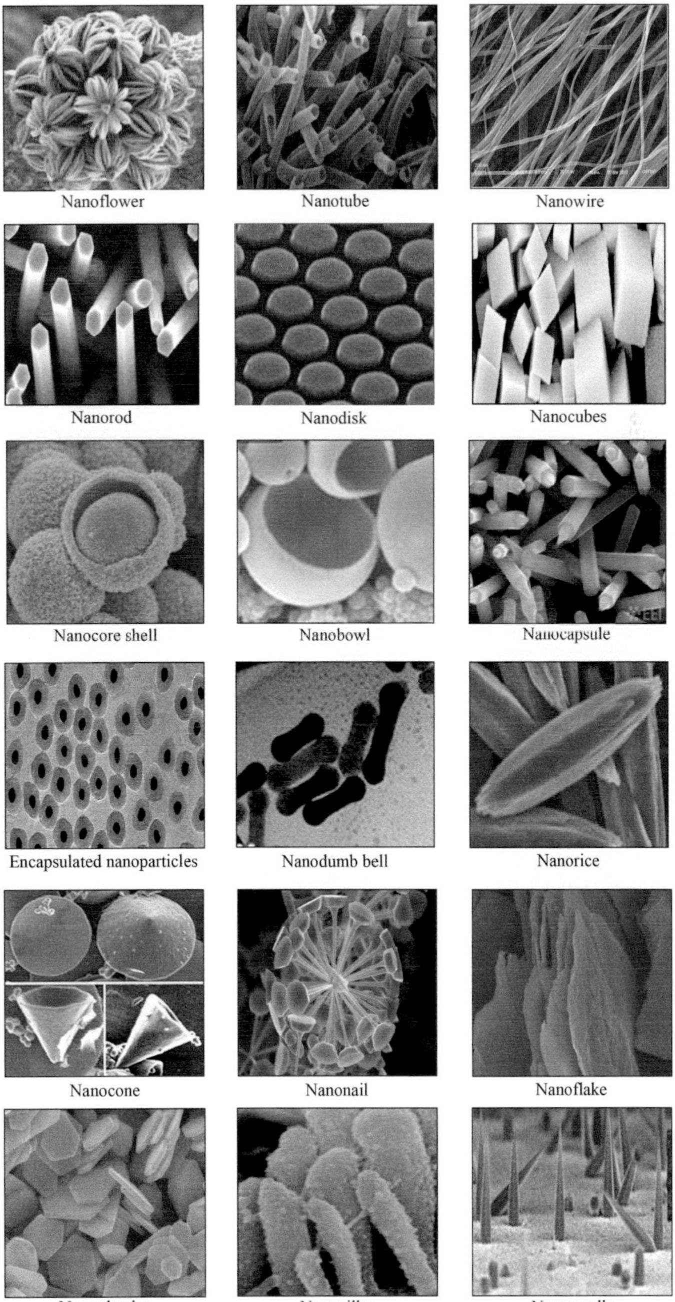

FIGURE 1.6 Nanomaterials in different shapes.

Source: Adapted from Solanki et al. (2018). With permission.

A larger surface area is there in the case of materials made up of nanoparticles as compared to the same materials with larger particles. It means that the same material has a larger surface area in nanoscale, when compared to its bulk counterpart. There is an increase in SAVR of the material on decreasing its particle size and vice versa. SAVR is generally also called surface-to-volume ratio. The generation of porosity in materials can produce surface area much more than produced by only particle size reduction. It may be especially due to small pores.

The larger surface-to-volume ratio of these nanoparticles opens new avenues for fabricating newer materials and is also responsible for facilitating various chemical processes. Most of the atoms are not at surface in bulk of many conventional materials. This bulk does not exist in the case of nanomaterials. Therefore, materials with this property are unique. These nanomaterials can serve as very efficient catalysts or these may be used in thin films so as to serve as thermal barriers or for improving the wear resistance of materials.

A larger portion of the atoms is present at the surface as compared to those present inside on decreasing particle size. As catalysis in any chemical reaction occurs on surface and therefore, a particular mass of a nanomaterial will be more reactive as compared to the equivalent mass of that material, which is made up of larger particles. It was interesting to note that those materials become much more reactive in their nanoscale form, which was almost inert in nature in their counter bulk form melting point of any substance also with increasing decreases in surface energy.

1.3.2 SURFACE PLASMON RESONANCE

There is a resonance, whenever a polarized light strikes an electrically conducting surface. Due to this, some electron-change density waves are obtained, which are called plasmons. SPR is a collective oscillation of valence electrons in a solid simulated by light. This type of resonance is obtained, whenever the frequency of light photons (incident) becomes equal to the natural frequency of surface electrons, which are oscillating against the restoring force of the positive nuclei. Localized surface plasmon resonance(LSPR) is nothing but SPR observed in structures within nanometric size. These are collective oscillations of electron charge in metallic nanoparticles on exposure to light. Solutions having nanoparticles of metals like silver and gold are colored due to the SPR of nanoparticles. Properties like dispersion, optical, conductivity, and catalytic behavior change with different surface properties of the particle.

1.3.3 QUANTUM CONFINEMENT EFFECT

When the size of any particle is so small that it can be compared with the wavelength of the electron, then the quantum confinement effect is observed. This effect can be observed when the diameter of any material is of the same magnitude as that of de Broglie wavelength of the electron wave function. When materials are so small, their properties deviate substantially from the properties of bulk materials.

If this effect is considered in two words such as quantum and confinement, it becomes clearer. The word confinement means to confine the motion of any randomly moving electron to restrict its motion in a specific direction, decreasing energy levels, while quantum reflects the atomic realm of size of a particle to a nanoparticle dimension. As a result, a decrease in confining dimension will make the energy levels discrete. Hence, an increase or widening of bandgap is observed.

1.4 PROPERTIES OF NANOMATERIALS

Microstructure materials have similar properties to their bulk counterparts, but the nanomaterials are having the larger ratio of surface area to volume and this makes them unique. As this ratio is too large in surface or interfacial atoms, this results in more surface-dependent material properties. SAVR has an important feature in the development of nanoscience and nanotechnology. As different shapes of the nanomaterials have different surface-to-volume ratios, they show diversity in different properties. Significant differences between the nanomaterials and its bulk materials are as follows:

- Large fraction of surface atoms.
- High surface energy.
- Spatial confinement.
- Reduced imperfections.

1.4.1 OPTICAL PROPERTIES

One of the most important properties of any nanomaterial is its optical properties. The nanomaterials have considerably different electronic configurations from their bulk materials. Much more changes in the optical properties of the nanomaterials are based on their size, which changes the density of electronic energy levels. The main contributing factors for its optical property include quantum confinement of electrical carriers within nanomaterials, charge transfer, and efficiency energy. Other major factors for change in the optical property of the nanomaterials are dimensions of the crystals, its surface chemistry, and fabrication technology.

Optical properties of nanomaterials depend upon different parameters such as size, shape, feature, surface characteristics, doping, and its interaction with other nanostructures or the surrounding environment. These include absorption, reflection, diffraction, refraction, polarization, dispersion, color, photosensitivity, photoluminescence (fluorescence), transmittance, dichroism, scattering, birefringence, and so on. The nanomaterials are used in laser, sensor, optical detector, solar cell, imaging, photocatalysis, phosphor, display, photoelectrochemistry, and biomedicine due to its optical properties.

1.4.2 ELECTRICAL PROPERTIES

The electrical properties of nanomaterials are based on electrical conductivity. These nanostructured materials can hold larger energy because of their more grain boundary area as compared to its bulk materials. The electrical conductivity of the nanomaterials is a little bit complex and it has distinct mechanisms based on their size. The electrical conductivity mechanism is generally classified into these four categories:

- Surface scattering including grain boundary
- Quantized conduction including ballistic conduction
- Coulomb charging
- Tunneling

The electrical conductivity of nanomaterials is affected by major factors like structural defects (dislocations), reduced impurity, widening and discrete bandgap, and changes in the microstructure. Single molecule is able to control the electron transport in molecular electronics. This electron transport property plays an important role in the functioning of electronic devices and working circuits.

Properties like conductivity or resistivity are types of electrical properties. These properties are also changed at the level of nanosize. These changes in electrical properties of nanomaterials include the following:

- It was observed that the conductivity of a bulk or large material does not depend upon its area of cross-section, diameter, twist in the conducting wire, and so on, but it changes with variation in area of cross-section in carbon nanotubes.

- It was also observed that it also changes, when some shear force (like twisting) is applied on the nanotube.
- Graphite is well-known for its electrical conductivity, while on the other hand, carbon nanotubes can act as either conducting or semi-conducting in behavior.

1.4.3 MECHANICAL PROPERTIES

Materials are either polycrystalline, nanocrystalline, or amorphous in structure, but most of the materials are polycrystalline (ordered crystals). Mechanical properties of nanomaterials deal with influence of porosity and grain size, particle-filled polymers, filled polymer composites, polymer-based nanocomposites filled with platelets, superplasticity, carbon nanotube based composites, bulk metallic, and ceramic materials.

Mechanical properties of any material are decided on the basis of its mechanical characteristics of under various environments and different external loads. Therefore, different materials exhibited different mechanical properties. Examples of some mechanical properties are plasticity, tensile strength, hardness, modulus of elasticity, brittleness, rigidity, elongation, toughness, fatigue strength, ductility, and yield stress. On the contrary, nanomaterials have appreciable mechanical properties because of their high surface-to-volume ratio and quantum confinement effects.

- Carbon nanotubes are cylindrical structures made of a graphene sheet, and these are expected to show outstanding mechanical properties such as elastic modulus and high tensile strength. These nanotubes can be used as ideal filler material for preparing composites used in sporting goods and aerospace materials.
- The application of nanomaterials can also be used in solving problems in oil well cementing in the oil and gas industries as potential accelerators due to permeability/porosity.

1.4.4 MAGNETIC PROPERTIES

Saturation of magnetization and coercivity are the measures of strength of a magnet. Surface atoms of the nanomaterials are different from its bulk atoms and these can be modified by interaction with the chemicals. Physical properties of the nanoparticles can also be modified by capping these with some appropriate molecule. The magnetic property of a nanomaterial increases with the decrease in grain size, but increases with surface area and as a result, nanomaterials have good magnetic properties due to their small particle size and high surface area. This magnetic property of the nanomaterials is utilized in ultra-sensitive analytical instruments and also in medical diagnostics for magnetic resonance imaging (MRI).

Iron, nickel, cobalt, and so on, are known magnetic materials and magnetic properties of nanoparticles with a chemical component having functionality can be changed in the presence of a magnetic field. Such magnetic nanoparticles are called nanobeads, which are clusters composed of a number of individual nanoparticles. These clusters form magnetic nanochains by further magnetic assembly. These days, magnetic nanoparticles have become a field of research interest, because of their certain properties such as catalysis, specific targeting of tissues, microfluidics, MRI, magnetic particle imaging, optical filters, nanofluids, environmental remediation, magnetically tunable colloidal photonic crystals, cation sensors, data storage, magnetic cooling, and defect sensor.

1.5 SYNTHESIS OF NANOMATERIALS

A nanometer is 1 billionth of a meter; that is approximately 10^5 times smaller than the human hair's diameter. Different methods are used to prepare a variety of nanostructured products. Nanomaterials can be prepared by these two approaches:

- *Top-down approach:* It involves disassembling, breaking, dissociating, or reducing the large-sized materials into nanostructured materials.
- *Bottom-up approach:* It involves changing individual atoms, molecules, assemble atoms together into nanostructures.

Schematic representation for the preparation of nanomaterials by bottom-up and top-down is shown in Figure 1.7.

1.5.1 TOP-DOWN APPROACH

In this approach, bulk material is successively cut (sliced) to achieve nanoparticles. It means that nanomaterials may be prepared by etching outer crystal planes already present on the bulk material that is building blocks are removed from the substrate slowly to obtain the final nanostructure. It does not mean that top-down approach involves no chemistry; it does. Top-down approach provides a better control. This approach has major advantages such as low cost, scalability as well as the formation of better and uniform product, but there is a big problem in this approach that is the development of imperfections in the structure of surface supported by some crystallographic damage to processed patterns but even then, this process is still important for the synthesis of nanostructures. Solid state is the starting material in these methods. Some of these methods are as follows:

- Mechanical methods: cutting, etching, grinding (ball milling).
- Lithographic methods: photolithography-electron beam lithography.
- Thermal decomposition.
- Pulsed laser ablation.
- Templating.

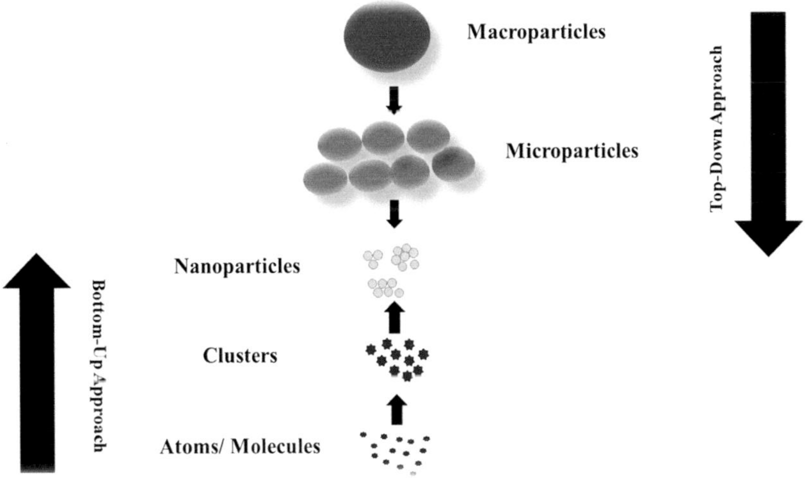

FIGURE 1.7 Schematic representation of nanomaterials preparation.

1.5.2 BOTTOM-UP APPROACH

In this approach, stacking of atoms is there on each other, which may result into crystal planes; these planes are again stack onto each other to form nanoparticles. Here, the building blocks are added, ultimately providing nanostructures. In this method, starting point may be some homogeneous solution or gas, which slowly builds up nanoparticles with different sizes and shapes. In other words, one starts assembling from atom to atom, molecule to molecule, or cluster to cluster to reach a nanomaterial. This approach is more beneficial as in this case, nanostructures with less defects and more homogeneous chemical composition are obtained

Particles having a very small number of atoms (about a few hundred atoms or smaller) are termed as clusters. A nanocrystallite possesses crystalline order in addition to nanoscale size. If one of the dimensions of the three-dimensional nanostructure is at nanoscale, then it is termed as quantum well, but if two dimensions of the three-dimensional nanostructure are at nanoscale, then it is termed as a quantum wire. If all the three dimensions of the nanostructure are at nanoscale, then it is called a quantum dot.

Clusters come closer to the surface of substrate to form self-assembled monolayers. Here, either the gaseous or the liquid state of matter is the starting material.

Some of these methods are:

- Chemical vapor deposition: Deposition of vapor phase of reaction species:
- Sputtering.
- Pyrolysis.
- Sol–gel method.
- Microemusion route.
- Physical vapor deposition: condensation of vapor phase species.

1.6 DISADVANTAGES OF NANOMATERIALS

Although there are many advantages of nanomaterials, these are also associated with some major disadvantages. These include the following:

- Instability of particles.
- Impurity.
- Biologically harmful (toxic).
- Difficult synthesis, isolation, and application.
- Recycling.
- Disposal.

1.7 APPLICATION OF NANOMATERIALS

Unique structures and materials can be synthesized/prepared using nanoparticles, which are relatively low cost, stronger, stable for the long term, lighter in weight, and so on, and these may be used in a wide variety of products and various fields, for example, computers made by nanomaterials may be a million times smaller in size and a billion times faster in speed.

Manufacturing can be at a very low cost. The automatic pollution cleanup is also possible. Hence, there are numerous applications of nanomaterials, such as:

- Specific site of action in the body is targeted by the drug (targeted drug delivery).
- These can be made lighter, stronger, durable, reactive, highly electrical conducting, more sieve-like, and so forth. Many commercial products using nanomaterials are presently available in the market.
- Surface treatment of any fabric with nanoparticles can resist its wrinkling, staining, and bacterial growth.
- Nanofilms used in displays of computer and cameras, windows, glasses, and eyeglasses can make these surfaces water- and dirt-repellent, resistant to ultraviolet, antifogging, antimicrobial, scratch-resistant, self-cleaning, antireflective, electrically conductive, and so forth.
- Nanomaterials can be used to make cars, buses, trucks, boats, airplanes, and spacecraft, which are lighter in weight and their use will result in savings of fuel.
- Nanoscale additives are added in polymer composite materials so as to make helmets, tennis rackets, baseball bats, bicycles, automobile parts, and it makes them, stiff, durable, and lighter in weight.
- Nanomaterials based on cellulose may find promising applications in a wide range of industrial sectors including packaging, food, energy, electronics, healthcare, automotive, construction, and defence forces because these cellulosic nanomaterials have a significant strength-to-weight ratio.
- Nanomaterials can also be used in automobile sector with significant advantages such high-power rechargeable battery systems, tires with lower rolling resistance, temperature control, sensors, smart solar panels, and as an additive for clean exhaust.
- Nanoparticles find potential application in catalyzing chemical reactions. Their use will reduce the quantity of catalyst required and with a lower cost. They also reduce pollutants and therefore can be used in refining of petroleum and also as catalytic converters.
- The properties of nanomaterials are also used in many household products, such as degreasers, stain removers, air purifiers, environmental sensors, self-cleaning paints, and antibacterial uses that resist dirt and marks.

- A number of personal care products also contain nanoparticles. Nanoscale ZnO and TiO_2 are being used in sunscreen so as to provide protection from harmful rays of Sun and along with this, these are also invisible on the skin.
- Nanomaterials have greatly contributed in the field of electronics, leading to smaller and more portable systems. These are faster, which can manage and store much larger amounts of information.
- Ultra-high-definition displays and televisions using quantum dots are presently available in the market, which produces more contrast colors with more energy efficiency.
- As nanomaterials are stretchable, flexible, foldable, bendable, and rollable in nature, these can used in smartphone and e-reader display electronics.
- Nanomaterials find numerous applications particularly in the field of medicine area such as prevention, diagnosis, and treatment of many diseases; even some deadly disease.
- Gold nanoparticles are used clinically in the treatment for cancer as well as other diseases.
- A nanoparticle can help to deliver a particular medicine directly to the affected part of the body due to malignant cells, which will minimize the risk of damaging to healthy tissues as it involves targeted drug delivery.
- Graphene nanoribbons are also studied for repairing spinal cord injury.
- Nanoparticles are also used for drug delivery in the treatment of tuberculosis.
- Society is facing two major problems of scarcity of energy sources and pollution causing global warming. Therefore, the whole world is looking for some alternative source of energy, which is clean, affordable, and renewable.
- Wires containing carbon nanotubes can be used in the electric grid as they have lower resistance than presently used high-tension wires, so there will be a reduction in power loss as heat during transmission.
- Nanostructure materials are being incorporated into solar panels. Such cells will be low-cost and more efficient. Apart from this, they can be easily installed.
- Nanomaterial can be used in fabricating batteries, which are quickly charged, more efficient, lighter in weight, with higher power density, and discharge slowly.
- Nanomaterials are also used in treatment of impurities in water and provide clean drinking water.

- Membrane containing molybdenum disulfide thin film with nanopores can filter water 2–5 times faster than current conventional filters used so far.
- Nanoparticles (carbon nanotubes and nanofibers) are used to increase the strength, durability, and control of pollution of different cementitious composites.
- Various coats and thin films are produced with self-cleansing ability as well as self-color change.
- Some of these composites have sensing and self-repairing abilities.
- Photocatalytic capacity of nanomaterials. Self-cleaning properties of catalysts materials and photocatalytic capacity can reduce indoor air pollution using P25 TiO_2.
- Photocatalytic properties of nanomaterials also help in the destruction of fungi and bacteria. The development of indoor fungi and bacteria is one of the main causes, which is responsible for degradation of construction materials as well as some health problems.

There are limitless benefits of nanomaterials in varied fields making them useful in the majority aspects of life. Various nanostructured materials have been developed in past and this is continuing today also. One can anticipate newer and newer nanomaterials in future with excellent optical, electrical, magnetic, and mechanical properties. This will result into many more fascinating materials with astonishing properties and applications. They will find still better future in years to come. Coming few decades will develop as a new era of nano-dimensional materials. Apart from their advantages, these nanomaterials are also associated with some limitations, some of which are not fully explored, and therefore, these must be used with caution at present.

KEYWORDS

- **nanomaterials**
- **nanoparticles**
- **classification**
- **top-down**
- **bottom-up**
- **application**
- **disadvantage**

REFERENCES

Acharya, A., Ed. *Nanomaterial-Based Biomedical Applications in Molecular Imaging, Diagnostics and Therapy*; Springer Nature, 2020.

Aguilar, Z. *Nanomaterials for Medical Applications*; Newnes, 2012.

Amparo, L. R., Rovira, M. J. F., Sanz, M. M.; Gomez-Mascaraque, L. G. *Nanomaterials for Food Applications*; Elsevier, 2018.

Bhagyaraj, S. M.; Oluwafemi, O. S.; Kalarikkal, N.; Thomas, S., Eds. *Applications of Nanomaterials: Advances and Key Technologies*, 2018.

De Oliveira, Jr, O.; Ferreira, L.; Marystela, G.; de Lima Leite F.; Da Róz A. L. *Nanoscience and Its Applications*; William Andrew, 2016.

Giersig, M.; Khomutov, G. B., Eds. *Nanomaterials for Application in Medicine and Biology*; Springer: Dordrecht, 2008.

Grassian, V. H., Ed. *Nanoscience and Nanotechnology: Environmental and Health Impacts*; John Wiley & Sons, 2008.

Grumezescu, A. M. *Nanomaterials for Drug Delivery and Therapy*; William Andrew, 2019.

Hou, Y.; Sellmyer, D. J., Eds. *Magnetic Nanomaterials: Fundamentals, Synthesis and Applications*; John Wiley & Sons, 2017.

Hu, A.; Apblett, A., Eds. *Nanotechnology for Water Treatment and Purification*; Springer International Publishing: Switzerland, 2014.

Kane, D. M.; Micolich, A.; Roger, P., Eds. *Nanomaterials: Science and Applications*; CRC Press, 2016.

Khan, Z. H., Ed. *Nanomaterials and their Applications*, Vol. 84; Springer, 2017.

Kumar, N.; Kumbhat, S. *Essentials in Nanoscience and Nanotechnology*, Vol. 486; Wiley: Hoboken, 2016.

Liao, J. W.; Zhang, H. W.; Lai, C. H. Magnetic Nanomaterials for Data Storage. Magnetic Nanomaterials: Fundamentals, Synthesis and Applications, 2017.

Liu, J. L.; Bashir, S. *Advanced Nanomaterials and Their Applications in Renewable Energy*, 2015.

Shi, D. *Nanomaterials and Devices*; Elsevier, 2014.

Solanki, M. S.; Benjamin, S.; Ameta, S. C. *Chapter 6: Nanomaterials, in Nanomaterials in Sonochemistry: An Emerging Green Technology*; Edited by Ameta, S. C., Ameta, R., Ameta, G.; CRC Press, 2018.

Wang, X.; Chen, X., Eds. *Novel Nanomaterials for Biomedical, Environmental and Energy Applications*; Elsevier, 2018.

Zenkina, O. V., Ed. *Nanomaterials Design for Sensing Applications*; Elsevier, 2019.

CHAPTER 2

Optical Properties of Nanostructured Materials and Their Sensing Applications

SUBRATA BISWAS and PATHIK KUMBHAKAR*

Nanoscience Laboratory, Department of Physics, National Institute of Technology Durgapur 713209, India

**Corresponding author. E-mail: pathik.kumbhakar@phy.nitdgp.ac.in*

ABSTRACT

In recent times, nanomaterials have drawn immense attention of technologists and research communities due to their interesting and superior physicohemical properties over their bulk counterparts. Here we provide a brief introduction about nanomaterials and classification of nanomaterials in terms of dimensionality, and discuss about the optical properties of the nanomaterials and its mathematical modeling. A brief review on the interesting optical sensing application of metal nanoparticles and its composites based on previous reports has also been included. In the last section, syntheses and optical properties of some plasmonic-semiconductor nanocomposites of ZnO/Ag and ZnS/Ag via some innovative approaches are reported. It has been found that the photocatalytic performances of hydrothermally synthesized ZnO/Ag and ZnS/Ag plasmonic photocatalysts are very interesting. Therefore, the performances of prepared composites in UV–vis light driven photocatalytic degradation of a typical test dye of methylene blue (MB) are compared with those of rod-like ZnO (a few μm in length) and spherical ZnS (a few μm in diameter) samples. An improved photocatalytic degradation efficiency (η) of 90 (80)% has been achieved in ZnS/Ag (ZnO/Ag) in 60min as compared with that of η = 39 (48)% in ZnO (ZnS), under the similar experimental conditions. These results clearly opened up a way to improve

The Science of Nanomaterials: Basics and Applications. Suresh C. Ameta & Rakshit Ameta (Eds.)

the photocatalytic performance of ZnO and ZnS nanoparticles by making plasmonic-nanoparticle composites with Ag.

2.1 INTRODUCTION

Nanomaterial can be a particle, an aggregate, or an agglomerate with at least one or more of their dimensions in the range of 1–100 nm (Jeevanandam et al., 2018; Barros and Casey, 2020). The nanomaterials demonstrate drastic improvements in their optical (Coronado et al., 2011; Mie and Beirtrage, 1908), electrical, magnetic (Yurkov et al., 2007), mechanical (Wu et al., 2020), structural, and biological properties (Hajipour et al., 2012) due to their reduced size and shape. There are several naturally occurring nanomaterials like viruses, protein, lotus leaf, spider-mite silk, butterfly wings, and so on (Griffin et al., 2018). However, more advanced nanomaterials such as Au NPs (Biswas et al., 2019), Ag NPs (Biswas and Kumbhakar, 2018), CNT (Brozena et al., 2019), graphene (Biswas et al., 2016a), and MoS_2 (Karmakar et al., 2017) have been synthesized in the last few decades with top-down and bottom-up approaches. Nanoscience and nanotechnology are mainly concerned with the synthesis, characterization, investigation, and use of nanostructured materials (Jeevanandam et al., 2018).

Nanostructures are creating a connection between molecules and bulk materials. Nanostructures include clusters (Suber et al., 2018), quantum dots (Bera et al., 2010), nanocrystals (Jarvis et al., 2019), nanowires (Jagadish, 2010), nanotubes (Brozena et al., 2019), arrays, assemblies (Huang et al., 2020a), and superlatives of particular nanostructures. Physical and chemical properties of nanomaterials can differ significantly from the bulk materials of the same structure (Barros and Casey, 2020). The uniqueness of the structural characteristics, capacity, reaction, dynamics, and chemistry of nanostructures forms the basis of nanoscience (Brozena et al., 2019; Biswas et al., 2016b; Karmakar et al., 2017; Jeevanandam et al., 2018; Suber et al., 2018; Bera et al., 2010; Jarvis et al., 2019; Jagadish, 2010; Huang et al., 2020a, 2020b). The electronic states of nanostructures contribute to the appearances of new electrical, thermal, magnetic, optical, and mechanical properties at the nanoscale (Yurkov et al., 2007; Wu et al., 2020; Hajipour et al., 2012; Griffin et al., 2018; Biswas et al., 2019; Biswas and Kumbhakar, 2018; Brozena et al., 2019; Biswas et al., 2016a, 2016b; Karmakar et al., 2017; Jeevanandam et al., 2018; Suber et al., 2018; Bera et al., 2010; Jarvis et al., 2019; Jagadish, 2010; Huang et al., 2020b). Physicochemical nanostructures are tailored to their size and forms at the nanoscale, leading to novel applications (Yurkov et al., 2007; Wu et al., 2020; Hajipour et al., 2012; Griffin et al., 2018; Biswas et al., 2019; Biswas and Kumbhakar, 2018; Brozena et al., 2019; Biswas et al., 2016a, 2016b; Karmakar et al., 2017; Jeevanandam et al., 2018; Suber et al., 2018; Bera et al., 2010; Jarvis et al., 2019; Jagadish, 2010; Huang et al., 2020b; Flory et al., 2011; Son et al., 2014; Gleiter, 2000; Nirmal and Brus; 1999; Amendola and Meneghetti, 2009). The physicochemical properties of nanostructures are, however, restricted only by the accuracy of the experimental methods used in the manufacturing process of the low-dimensional structures (Gatoo et al., 2014).

To date, nanomaterials' research has been gaining traction and gradually growing with a multitude of practical applications due to their specific physical, chemical, mechanical, and optoelectronic properties, which can be tuned in terms of size and dimensions (Barros and Casey, 2020; Jeevanandam et al., 2018). Among these, the optical properties of nanomaterials have gained a great deal of interests having different advanced optical applications (Flory et al., 2011). The optical properties of nanomaterials are dependent on the internal electronic structures, which can be modified by adjusting their size and thus, the size of nanomaterials can be tailored depending on the need of the application (Flory et al., 2011; Son et al., 2014). However, before going into the depth, one has to understand the fundamentals of the optical material and the quantum containment effect (Son et al., 2014) of nanomaterials, which are responsible for the exceptional optical properties needed for optical applications.

Material optics usually deals with the interaction of light with matter resulting in reflection, refraction, absorbance, fluorescence, dispersion, frequency change, and focusing or splitting of the optical beam (Gleiter, 2000; Nirmal and Brus, 1999; Amendola and Meneghetti, 2009). The characteristics of the optical substance are a strict function of the wavelength of light used. As the dimensions of nanomaterials are represented in the nanoscale, which is often smaller than the wavelength of light, indicates that understanding the interaction of light matter becomes a key step in the optical application of nanomaterials. Therefore, the optical properties of the materials can be divided into two key sets of linear and nonlinear properties (Zhang and Wang, 2017). The interaction of materials with optical waves and photons relies heavily on the structure that can then be used to track the distribution of light field and propagation of light (Nirmal and Brus, 1999; Amendola and Meneghetti, 2009). This allows the development of a wide range of key components for optical systems and is now a significant field of photonics.

Among a variety of nanomaterials noble metal nanoparticles (NPs) (Sreeprasad and Pradeep, 2013; Zhang and Wang, 2017), especially those made of Ag and Au, have attracted a great deal of significance due to their size and shape-dependent optical properties resulting from their surface plasmon resonance (SPR) (Zhang and Wang, 2017). SPRs are originated from coherent oscillations of metal electrons in the vicinity of the metal–dielectric interface. SPR can be generally divided into two categories, that is, localized SPR (LSPR) and propagating SP polaritons. In the case of noble metals, the resonant frequencies lie at the middle of the optical spectrum and hence, the surface plasmons interact strongly with the light. In recent times, it has been noticed that there is a remarkable increase in interest in the field of plasmonics (Eustis and El-Sayed, 2006; Barnes et al., 2003) caused by (1) Creation of new preparation methods for a wide range of metallic NPs and nanostructures, (2) Availability of methods and techniques for assessing their morphology and architecture with unprecedented resolution, and (3) Creation of new quantitative analytical methods for modeling and predicting optical behavior. Taking advantage of the greatly enhanced local electrical field, metal nanoparticles will remarkably improve the power of interaction between photons and materials, stimulate fast-growing developments in plasma-enhanced fluorescence (Jeong et al., 2018), Raman spectroscopy (Almohammed et al., 2018), heat generation, photoacoustics (Li and Chen; 2015), photocatalysis (Han et al., 2018), non-linear optics (Zhang and Wang, 2017), solar energy conversion (Guo et al., 2014), and so on. Plasmonics can be considered an interdisciplinary area of research with considerable impact due to its potential applications in miniaturized optical instruments (Mody et al., 2010), sensors (Malekzad et al., 2017), photonic circuits (Yetisen et al., 2015), medical diagnostics, and therapeutics (Doria et al., 2012).

Noble metals are eclectic nontoxic agents in the sense that they have a wide range of medicinal applications, such as use in highly sensitive diagnostic assays (Doria et al., 2012), thermal ablation as radiotherapy enhancers (Ashikbayeva et al., 2019), and drug- and gene-delivery vehicles (Patra et al., 2018). The combination of laser dyes as medium gain with noble metal NPs as scattering centers is of particular interest in the case of random laser (RL) generation (Wiersma, 2008). Noble metal NPs display localized plasma resonances within the visible spectrum (Sreeprasad and Pradeep, 2013), which allow incoming lights to be more efficiently distributed than their similar dimensional dielectric counterparts, made of high refractive index dielectrics such as TiO_2 or ZnO. The spectral position of the plasmon resonance peak depends heavily on the material, shape, and surrounding media of the NPs (Sreeprasad and Pradeep, 2013; Zhang and Wang, 2017), so that the variance of one or more of these parameters allows the spectral alignment of the plasmon resonance to overlap the emission spectrum of the desired active medium (Wang et al., 2017).

On the other hand, the properties of semiconductor photocatalysts (Mills and Hunte, 1997) are comparatively ineffective for practical use due to the recombination of fast charge carriers and reduced light absorption. When combined with metals, especially the noble metals, the properties of semiconductor photocatalysts can be dramatically improved due to increased charge separation at the metal semiconductor interface and increased visible light absorption due to the SPR of metals (Fu et al., 2019).

The creation of a novel, quick, reversible, inexpensive, enzyme-free, and colorimetric paper assay for the determination of glucose levels has recently been documented by Pinheiro et al. (2020). This sensing strategy is based on the synthesis of gold NPs (Au NPs) by reduction of a gold salt precursor in which glucose acts simultaneously as reducing and capping agent. The novel coronavirus disease (COVID-19) emerged in Wuhan in 2019 and has spread rapidly in too many other countries and regions. COVID-19 has a high human-to-human transmission infectivity and it can cause acute respiratory diseases. Thus, efforts are underway to devise a diagnostic tool for cost effective and rapid detection of COVID-19. Huang et al. (2020b) have developed colloidal Au NP-based lateral-flow assay to achieve rapid diagnosis and on-site detection of IgM antibody against the SARS-CoV-2 virus through the indirect immunochromatography method. Cassano et al. (2019) have reported both the straightforward production of narrow-NIR-absorbing gold ultrasmall-in-nano architectures (tNAs) and their suitability as platforms for photothermal (PT) upon CW-irradiation at 808 nm. Li et al. (2017) introduced a simple and novel colorimetric sensor array for rapid identification of microorganisms. Interactions between Au NPs and microorganisms have been found to contribute to apparent color changes, which could be seen with the naked eye (Li et al., 2017).

Multicolor random lasing has a wide range of uses in the area of imaging, sensing, and optoelectronics. Chang et al. (2017) synthesized silver nanoflowers (Ag NFs) with abundant nanogaps using a quick one-step solution-phase synthesis method and proposed as effective broadband plasmon dispersers to achieve different color random lasing. Redding et al. (2012) used the low spatial coherence of specially engineered RLs to show off specular-free full-field imaging in the intense optical scattering environment. Quantitatively, they have shown that the images generated by RL illumination are of superior quality than images provided by spatially coherent illumination. Song et al. (2010) have shown that the peculiar characteristics of random bone lasing can be used to test nanoscale unique alterations as a mechanical or structural biosensor, provided that bone is a partially disrupted biological nanostructure.

Photocatalyst is considered to be one of the clean technologies used in a number of medical and environmental applications. Photocatalytic reactions most popularly involve the use of efficient semiconductive photocatalysts that could absorb light energy and convert it into chemical forms (Ziashahabi et al., 2019). The new synergetic hybrid Ag/ZnO nanostructure has been manufactured and shown that these are capable of causing photocatalytic degradation (in high yields) of methylene blue (MB) dye both under visible light and in the dark, the heterostructure SnO_2/ZnO nanocomposite photocatalyst has recently been prepared by Uddin et al. (2020) and its photocatalytic activity was studied under UV light irradiation.

2.2 CLASSIFICATION OF NANOMATERIALS

Nanomaterials can be categorized by source, scale, constituent material, and possible toxicity (Moon et al., 2011). Nanomaterials can also be distinguished by origin: Natural or anthropogenic. The latter group can be divided into incidental and engineered nanomaterials, depending on whether their creation is intentional or not (Pokropivny and Skorohod, 2007). Natural and accidental nanoparticles are commonly referred to as ultrafine particles. Nanomaterials are now manufactured using a wide variety of chemical elements, such as metals, semiconductor, metal oxides, carbon, and polymers. They come in a wide variety of shapes such as balls, wires, fibers, needles, sticks, tubes, shells, rings, plates, and coatings, as well as in more rare, flower-like designs (Figure 2.1).

The classification of nanomaterials can also be rendered on the basis of their dimensionality (Wang et al., 2020). Quantum dots can be described as zero-dimensional, semiconductor nanocrystals with dimensions <10 nm. One-dimensional (1D) nanomaterials (Benelmekki, 2019) have two external dimensions at the nanoscale, of which the third is usually at the microscale. These include nanofibers, nanotubes, nanowires, and nanotubes. Nanowires have the largest aspect ratio of 1D nanomaterials with a length to width ratio > 1000. This allows them to confine electrons sideways, which is of great importance for applications in electronics. With just one external dimension at the nanoscale, 2D nanomaterials are made up of thin films, nanosheets, nanocoatings, and nanoplates (Yang et al., 2019). The last dimensional category of nanomaterials, 3D nanomaterials (Gough et al., 2009), has internal nanoscale features but no external dimension at the nanoscale level. This includes nanocomposites and nanostructured materials, which are multi-component solid materials having at least one nanoscale external dimension. Nanocomposites and nanostructured materials are very important materials in terms of their real-life applications.

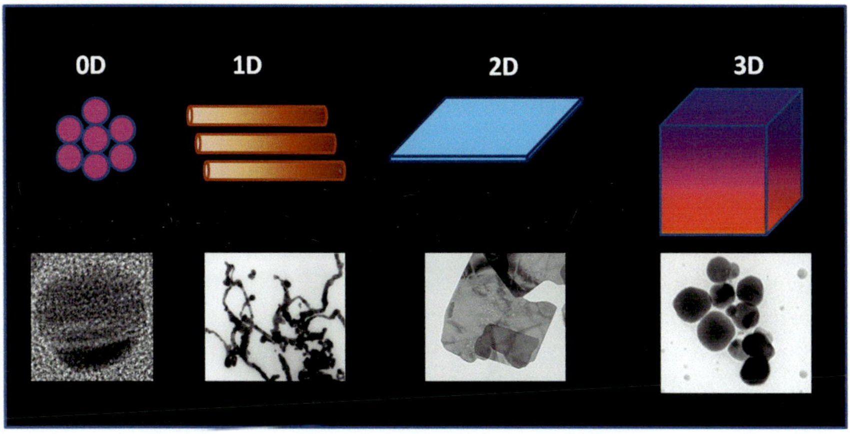

FIGURE 2.1 Classification of the nanomaterials according to the dimensionality is shown in the upper panel (schematically) and lower panel shows their representative TEM images.

2.3 OPTICAL PROPERTIES OF NANOSTRUCTURED MATERIALS

Nanostructured materials and NPs have fascinating optical properties (Becker, 2012; Van Dijk et al., 2006; Ahn et al., 2006; Zayats et al., 2005; Koprucki et al., 2006; Yang and Jiang, 2006; Smith and Nie, 2010). Whenever there is an incidence of light of appropriate wavelength on NP, it may be scattered or absorbed. The sum of scattered and absorption is referred to as extinction (Becker, 2012). In case of NPs, the fraction of light is scattered or absorbed vary greatly depending on the particle diameter and almost all of the extinction at a diameter of less than 20 nm is due to absorption and above 100 nm, the extinction is mostly due to scattering (Becker, 2012; Van Dijk et al., 2006). Thus, an optimal amount of dispersion and absorption can be obtained by preparing a NP with a larger or smaller diameter. However, the interaction between light and matter in nanoscale dimension can be categorized into three ways (Becker, 2012): (1) Interaction of light with nanometer-sized matter, (2) Containment of light to nanoscale dimensions that are far smaller than the wavelength of light, and (3) Nanoscale containment of a photo-process that can cause a light-induced phase change.

Prominent materials for photonic and optoelectronic components operating in the visible and near-infrared sections of the electromagnetic spectrum include metals and semiconductors (Van Dijk et al., 2006). Metals such as gold, silver, aluminum, and copper are historically not products of choice due to their large absorption losses caused by both, intraband and interband transitions (Sreeprasad and Pradeep, 2013; Zhang and Wang, 2017). Metals, however, give a special opportunity for photonics in this part of the spectrum, namely the presence of extremely small surface waves at interfaces with dielectrics (Sreeprasad and Pradeep, 2013). Here, under sufficient conditions, the electromagnetic field can be coupled with the electron plasma conduction to set up a hybrid mode called the plasmon surface (Sreeprasad and Pradeep, 2013; Zhang and Wang, 2017). For extended interfaces, modes with mode zones below the diffraction limit (Plasmon polaritons surface) can be propagated, whereas metal nanostructures function as optical nanoresonators with minute subdiffraction-limit mode volumes (Zayats et al., 2005). Usually, the models used to detect the optical properties of nanostructures are based on electromagnetic theory, but when the dimension of the nanocrystal semiconductor is smaller than that of the de Broglie wavelength, quantum phenomena (Zayats et al., 2005) must be considered. The de Broglie wavelength can be seen as the length of the electron wave function.

2.3.1 OPTICAL ABSORPTION PROPERTIES OF SEMICONDUCTOR NANOMATERIALS

Electronic level of energy and density of states (DOS) (Koprucki et al., 2006) defines the properties of materials, including optical and electronic properties, as well as their functionality. For nanoscale materials, the levels of energy and DOS vary depending on the size, resulting in dramatic changes in the properties of the materials. Energy level spacing increases with a decreasing dimension, and this is known as the quantum size containment effect (Yang and Jiang, 2006). The classical particle-in-a-box model can be used to consider the effect of quantum size containment. If the particle size decreases, the energy level spacing increases quadratically as the length of the box decreases. The quantum containment effect can be explained qualitatively by the means of an accurate estimate of the mass. In the case of a spherical particle with radius *r*, the effective bandgap, *Eg* (*r*), is given as (Smith and Nie, 2010):

$$E_g(r) = E_g(\infty) + \frac{\hbar^2 \pi^2}{2r^2}\left(\frac{1}{m_e^*} + \frac{1}{m_h^*}\right) - \frac{1.8e^2}{\varepsilon r}. \tag{2.1}$$

where, $Eg(\infty)$ is the bulk band gap, me^* and mh^* are the effective masses of the electron and hole, and ε is the bulk optical dielectric constant or relative permittivity. The second term to the right shows that the effective band gap is inversely proportional to r^2 and increases as the size decreases. On the other hand, the third term suggests that band gap energy decreases with decreased *r* due to increased Coulombic interactions (Smith and Nie, 2010). However, as the second term becomes dominant for small *r*, effective band gap is expected to increase with declining r, especially, when *r* is small.

In general, the consequences of containment must be taken into account as the size of the particles decreases below the Bohr exciton diameter of the same substance (Cassidy and Zamko, 2020). In the NP semiconductor, certain physical properties depend on size, when the size of the NP is comparable to the Bohr diameter of its bulk counterpart. The key change in optical properties is the blue shift of the excitonic peak in the absorption spectrum that occurred due to the containment effect as the particle size decreased below the Bohr exciton diameter (Cassidy and Zamko, 2020). There is also a spectral change in the fluorescence emission of light, when the particle size is reduced (Yao and Fukuia, 2016). Since NPs are characterized by a broad surface-to-volume ratio, certain surface-related effects are observed and thus, corresponding changes in emission properties are observed.

The absorption of light is a crucial property of semiconductor materials. Light absorption can be expressed in terms of the coefficient α (*hv*) (Pankove, 1971; Parkash and Kulkarni, 2011), which is defined as the relative rate of light intensity *I* (*hv*) decreases along its path of propagation (*x*):

$$\alpha(hv) = \frac{1}{hv}\frac{dI}{dx}, \tag{2.2}$$

In case of direct band gap transitions, if electrical dipoles are taken into consideration, it can be shown that:

$$\alpha(hv) = A(hv - E_g)^{\frac{1}{2}}. \tag{2.3}$$

In case of indirect band gap transition, which are only allowed at $k \neq 0$ but forbidden at $k = 0$, the absorption coefficient can be expressed as (Pankove, 1971; Parkash and Kulkarni, 2011):

$$\alpha(h\nu) = B(h\nu - E_g)^{\frac{3}{2}}. \tag{2.4}$$

where, A and B are constant determined by the index of refraction and electron and hole effective masses. For electric dipole forbidden or indirect band gap transitions that require phonon assistance, the electronic absorption coefficient can also be given as (Pankove, 1971):

$$\alpha(h\nu) = \frac{C(h\nu - E_g + E_p)^2}{\exp\frac{E_p}{kT} - 1}. \tag{2.5}$$

where, C is a constant similar to A and B, and Ep is the phonon energy.

2.3.2 SURFACE PLASMON RESONANCES IN METAL NANOPARTICLES: INFLUENCE OF SHAPE, SIZES, AND ENVIRONMENT

The theoretical basis of the SPR was provided by Mie (1908) in the beginning of the nineteenth century and determined the exact solution of the Maxwell's equations for the light interaction with an isotropic spherical NP surrounded by a nonabsorbing homogenous medium. The effects of the size, shape, and dielectric function of the surrounding medium on the optical properties of the NPs were then studied in details by various researchers (Garcia, 2011; Noguez, 2007; Alabastri et al., 2013; Kreibig and Vollmer, 1995; Gans, 1912; Link and El-Sayed, 2005; Borja et al., 2009; Yeshchenko et al., 2014; Mérai et al., 2020; Nobbmann and Morfesis, 2009). In an approximation called the quasi-static approximation, particles sizes are taken to be smaller than the wavelength of the incident light so that the spatial distribution of the incident electric field remains almost homogeneous over the surface of the NPs. In the limit of the quasi-static approximation, the molecular polarizability (αp) of the spherical metal NPs having dielectric constant ε embedded in a medium of dielectric constant εr can be written as:

$$\alpha_p = 3V\varepsilon_0\left(\varepsilon - \varepsilon_m / \varepsilon + 2\varepsilon_m\right). \tag{2.6}$$

where, $V = 4/3\pi a^3$ is the volume of the NP of radius a. The dielectric function, $\varepsilon = \varepsilon 1 + i\varepsilon_2$, where ε_1 and ε_2 are the real and imaginary parts of the dielectric function, respectively. However, in case of noble metals, the imaginary part of the dielectric function varies slowly with the incident wave frequency. Thus, the major contribution to the dielectric function comes from the real part and resonance frequency, called SPR frequency, can be obtained by solving the equation $\varepsilon_1 + 2\varepsilon m = 0$ and consequently polarizability of the NPs reaches its maximum value. The extinction and scattering efficiencies in terms of the dipolar polarizability can be written as:

$$E_{\text{ext}} = \left(\frac{8\pi^2 a^3 \varepsilon_m^{1/2}}{\lambda}\right) \text{Im}\left(\frac{\varepsilon - \varepsilon_m}{\varepsilon + 2\varepsilon_m}\right), \tag{2.7}$$

and

$$E_{\text{scatt}} = \left(\frac{128\pi^4 a^4 \varepsilon_m^2}{\lambda^4}\right)\left(\frac{\varepsilon - \varepsilon_m}{\varepsilon + 2\varepsilon_m}\right). \tag{2.8}$$

However, for the larger particle size, higher multipoles, especially the quadrupole term need to be taken into consideration for the calculation of the extinction and scattering spectra. Following the same analogy, as that of the quasi-static approximation, the extinction and scattering efficiencies can be written as:

$$E_{\text{ext}} = \left(\frac{8\pi\ a\ \varepsilon_m^{1/2}}{\lambda}\right)$$

$$\text{Im}\left[\left(\frac{\varepsilon - \varepsilon_m}{\varepsilon + 2\varepsilon_m}\right) + \left(\frac{\pi^2 a^2 \varepsilon_m}{3\lambda^2}\right)\left(\frac{\varepsilon - \varepsilon_m}{\varepsilon + 1.5\varepsilon_m}\right) + \left(\frac{2\pi a \varepsilon_m}{15\lambda^2}\right)(\varepsilon - 1)\right], \tag{2.9}$$

$$E_{\text{scatt}} = \left(\frac{128\pi^4 a^4 \varepsilon_m^2}{3\lambda^4}\right)$$

$$\left[\left(\frac{\varepsilon - \varepsilon_m}{\varepsilon + 2\varepsilon_m}\right)^2 + \left(\frac{2\pi^4 a^4 \varepsilon_m^2}{15\lambda^2}\right)\left(\frac{\varepsilon - \varepsilon_m}{\varepsilon + 1.5\varepsilon_m}\right)^2 + \left(\frac{4\pi^4 a^4 \varepsilon_m^2}{225\lambda^2}\right)(\varepsilon - 1)\right]. \tag{2.10}$$

In bulk metal, the contribution to the dielectric function comes either from transition between electronic states within the band called interband transition or from the band-to-band transition called intraband electron transitions. Thus, interband transition involves the bound electrons whereas the intraband transition involves transition of the free electrons. The free electron can be described by Drude model (Garcia, 2011), where equation of motion of the electron of mass *m* and in an external electromagnetic field of frequency υ can be written as:

$$m\frac{d^2x}{dt} + \gamma_0\frac{dx}{dt} + Kx = eE_0 e^{-i\upsilon t}. \tag{2.11}$$

where, γ_0 is the bulk damping constant, *e* is the electronic charge and the *K* is the force constant. *K* is reduced to zero for the bulk metal with free electron model. Under these considerations, the solution of the eq 2.11 takes the following form:

$$x = \frac{eEe^{-i\nu t}}{m\left(i\gamma_o\nu + \nu^2\right)}. \tag{2.12}$$

Now, if we consider that the volume concentration of the electrons is *N*, then the polarizability per unit volume can be expressed as:

$$P = \text{Nex} = \frac{Ne^2E_0e^{-i\nu t}}{m\left(i\gamma_0\nu + \nu^2\right)}. \tag{2.13}$$

Using eq 2.13, ε can be extracted from the relation $P = (\varepsilon_{\text{intra}} + 1)\,\varepsilon_0 E$ and can be expressed in the following simplified form, that is,

$$\varepsilon_{\text{int ra}} = 1 - \frac{\nu_p^2}{\left(\nu^2 + i\gamma_0\nu\right)}. \tag{2.14}$$

where, υ_{p} is the bulk Plasmon frequency. As discussed earlier that $\varepsilon = \varepsilon_1 + \text{i}\,\varepsilon_2$, so the real and imaginary part of the ε can be found out from the eq 2.14 and are expressed as:

$$\varepsilon_1 = 1 - \frac{\nu_p^2}{\left(\nu^2 + \gamma_0^2\right)},\ \varepsilon_2 = \frac{\nu_p^2\gamma_0}{\left(\nu^3 + \gamma_0^2\nu\right)}. \tag{2.15}$$

In terms of the expression of the real part of the dielectric function given by eq 2.15, the resonant condition $\varepsilon_1 + 2\varepsilon m = 0$ can now be modified as:

$$1 - \frac{\nu_p^2}{\left(\nu_{\text{SPR}}^2 + \gamma_0^2\right)} + 2\varepsilon_m = 0 \tag{2.16}$$

At high frequency $\upsilon^2 >> \gamma 0^2$ and from eq 2.16, one can obtain the expression for the SPR,

$$\nu_{\text{SPR}} = \frac{\nu_p}{\left(1 + 2\varepsilon_m\right)^{1/2}}. \tag{2.17}$$

The SPR frequency as obtained theoretically from eq 2.17 does not meet the experimental results, many a times, as it does not include the contribution from the bound electrons. As the bound electrons create a positive background that screens the free electrons, the bulk plasma frequency υ_{SPR} generally decreases. For example, the bulk plasma frequency of Ag, as predicted by the Drude's theory, corresponds to the energy value of 9.2 eV. However, the measured value was found to be 3.9 eV, showing a detectable variation from the Drude model. Thus, total dielectric constant for the NPs must include the contribution from the bound electrons and the dielectric function and it can be rewritten as (Noguez, 2007):

$$\varepsilon_{\text{total}} = \varepsilon_{\text{inter}} + \varepsilon_{\text{intra}},$$

$$\varepsilon_{\text{total}} = \varepsilon_{\text{inter}} + 1 - \frac{v_p^2}{\left(v^2 + i\gamma v\right)}. \tag{2.18}$$

where, γ is the surface damping term at the NPs surface and it is related to the bulk damping constant by the relation,

$$\gamma = \gamma 0 + VF/R. \tag{2.19}$$

where, V_F is the Fermi velocity at the metal surface and R is the radius. In the framework of Mie theory, the far-field spectrum of the spherically shaped NPs in response to the external electric field can be analytically evaluated by solving Maxwell's equations using a series expansion of spherical harmonic wave equations. The solution contains a series of plasmonic modes starting from the lowest order dipolar modes to multipolar modes. Now for the NPs of radius a having the dielectric function are of the form $\varepsilon = \varepsilon 1 + i\varepsilon_{2,}$ as discussed earlier and embedded in the medium of dielectric function ε_m, the extinction and scattering cross-section can be written as (Alabastri et al., 2013):

$$\sigma_{\text{ext}} = \left(\frac{2\pi R^2}{t^2}\right)\sum(2l+1)\text{Re}\left[a_l^2 + b_l^2\right], \tag{2.20}$$

$$\sigma_{\text{scatt}} = \left(\frac{2\pi R^2}{t^2}\right)\sum(2l+1)\left[a_l^2 + b_l^2\right]. \tag{2.21}$$

where, $t = (2\pi/\lambda)\,\varepsilon_m^{1/2}$ and al, bl are computed from the Riccati–Bessel functions:

$$a_l = \left[p\psi_l(pt)\psi_l'(t) - \psi_l(t)\psi_l'(pt)\right] / \left[p\psi_l(pt)\xi_l'(t) - \xi_l(t)\psi_l'(pt)\right], \tag{2.22}$$

$$b_l = \left[\psi_l(pt)\psi_l'(t) - p\psi_l(t)\psi_l'(pt)\right] / \left[\psi_l(pt)\xi_l'(t) - p\xi_l(t)\psi_l'(pt)\right] \tag{2.23}$$

where, $p = (\varepsilon/\varepsilon_m)^{1/2}$ and $\psi l(t) = (\pi t/2)^{1/2}\ Jl_{+1/2}$ (t) and $\xi l(t) = (\pi t/2)^{1/2}\ [Jl_{+1/2}$ (t)+i$Yl_{+1/2}$ (t)]. Jl and Yl are the first- and second-order Bessel functions, respectively. When the particles size is smaller than the wavelength of the light used, that is, under the quasistatic approximation and only dipolar field are taken into consideration the scattering and absorption cross-section cab be approximated as (Alabastri et al., 2013):

$$\sigma_{\text{scatt}} = \left(\frac{24\pi^3 V^2 \varepsilon_m^2}{\lambda^2}\right)\frac{\left[(\varepsilon_1 - \varepsilon_m)^2 + \varepsilon_2^2\right]}{\left[(\varepsilon_1 + 2\varepsilon_m)^2 + \varepsilon_2^2\right]}, \quad (2.24)$$

$$\sigma_{\text{scatt}} = \left(\frac{18\pi\ V\ \varepsilon_m^{3/2}}{\lambda}\right)\left[\frac{\varepsilon_2}{(\varepsilon_1 + 2\varepsilon_m)^2 + \varepsilon_2^2}\right]. \quad (2.25)$$

Although, optical properties of solid spherical metal particles have been discussed using the Mie theory, but it can be extended for other cases also, where spherical symmetry is preserved like in core shell NPs. The Mie theory has been used to evaluate the properties of anisotropic particles like oblate and ellipsoidal NPs under the dipolar approximation and these solutions are termed as Gans theory (Alabastri et al., 2013). Gans theory is proven to be very useful in explaining the observed plasmonic scattering and absorption of the nanorods where it is assumed that the nanorods, are special case of the ellipsoidal NPs. In the Gans theory, the absorption cross-section of the ellipsoidal NPs can be written as (Kreibig and Vollmer, 1995; Gans, 1912):

$$\sigma_{\text{scatt}} = \left(\frac{8\pi^3 V^2}{9\lambda^4}\right)\varepsilon_m^2 \sum \left(1/P_j^2\right)$$

$$\left[\left\{(\varepsilon_1 - \varepsilon_m)^2 + \varepsilon_2^2\right\} / \left\{(\varepsilon_1 + (1-P_j)\varepsilon_m / P_j)^2 + \varepsilon_2^2\right\}\right] \quad (2.26)$$

$$\sigma_{\text{scatt}} = \left(\frac{2\pi\ V}{3\lambda}\right)\varepsilon_m^{3/2} \sum \left(1/P_j^2\right)\left[\left\{(\varepsilon_1 + (1-P_j)\varepsilon_m / P_j)^2 + \varepsilon_2^2\right\}\right] \quad (2.27)$$

where, P_j is the depolarization factor along the major and minor axes of the ellipsoid. Along the major axis, $P_A = \{(1-e^2)/e^2\}[(1/2e)\ln\{(1+e)/(1-e)\}-1]$ and along the minor axes $P_B = P_C = (1-P_A)/2$. Here, e is the eccentricity, and $e = (1-AR2)^{1/2}$, where AR is the aspect ratio.

Now let us study the effect of the surrounding environment on the SPR properties of the metal NPs. In the visible and infrared region $\gamma << \upsilon_p$, so from eq 2.15, we get $\varepsilon_1 = 1 - \upsilon_p^2/\upsilon^2$. Now, when the resonance occurs, $\varepsilon_1 = -2\varepsilon_m$ and $\upsilon = \upsilon SPR$. Thus, at resonance, one can write as:

$$v_{\text{SPR}} = \frac{v_p}{(1+2\varepsilon_m)^{1/2}} \quad (2.28)$$

In terms of the wavelength, the eq 2.28 has been reduced to (Link and El-Sayed, 2005),

$$\lambda_{\text{SPR}} = \lambda_P \left(1 + 2\varepsilon_m\right)^{1/2} = \lambda_p \left(1 + 2n_m^2\right)^{1/2}. \tag{2.29}$$

where, $\varepsilon_m = n_m^{\ 2}$, and n_m is the refractive index (RI) of the medium. From eq 2.29, it is clearly evident that with increasing RI of the medium, the peak position of the LSPR can be shifted to the longer wavelength side. Further, it can be noted here that the variation of the absorption peak position of metal NPs with the RI of the surrounding medium as stated in eq 2.29 may form a basis to high-sensitivity detection of any change in RI or any change at the molecular level and it can be used in the LSPR-based sensing.

The intensity and peak position of the SPR band also depend on the temperature. It has been reported earlier that with the increasing temperature, the SPR band gets broadened (Borja et al., 2009) and hence, intensity drops with a redshift in the SPR peak position. The observed phenomenon can be explained in terms of the electron–phonon scattering, thermal expansion of the NPs, and temperature dependence of the dielectric permittivity of the surrounding medium. Now if we consider the electron scattering at the NPs' surface, then eq 2.16 can be written as (Yeshchenko et al., 2014):

$$v_{\text{SPR}} = \left[\frac{v_p^2}{\left(1 + 2\varepsilon_m + \varepsilon_{\text{inter}}\right)} - \gamma^2\right]^{1/2} \tag{2.30}$$

where, $\gamma = \gamma_0 + VF/R$. Now, as the surface temperature of the metal NPs increases, there is the expansion of the particle radius and consequently, γ will decrease, which in turn decrease the overall SPR energy. As a result, SPR band gets broadened, its intensity drops and also SPR peak showed slight blue shift. The decrease in the SPR peak intensity as well as the shift in SPR peak position with temperature are found to be almost linear as reported by several researchers. This property of the NP can be suitably used in temperature sensing activity in research as well as in industry.

2.3.3 LIGHT-SCATTERING APPLICATIONS IN NANOSTRUCTURED MATERIALS

Light-scattering methods are used in a wide variety of fundamental and applied optical problems, including the regulation of material properties in the chemical and material sciences (Flory et al., 2011; Mérai et al., 2020; Nobbmann and Morfesis, 2009; Bussche et al., 2004; Stetefeld et al., 2016; Santra et al., 2017; Lorber et al., 2012; Chavel and Lowenthal, 1978; Laing et al., 2017). RLs (Cao, 2005) are of special type of optical devices, which rely on multiple scattering of light on nanoparticles or microparticles embedded in some suitable gain media. RLs are different from normal lasers in which a well-defined feedback system, that is, optical cavity is required (Consoli and López, 2015). Multiple light scattering can be considered unfavorable in a normal laser; however, in RL, multiple light scattering is used beneficially to increase the length of the light path in the active medium. RLs can be used in many fields due to their unique characteristics; low production costs, specific operating wavelength, compact form and substrate compatibility (Hoang et al., 2010). Redding et al. (2012) have shown that RLs are capable of generating high-intensity and specular-free light leading to the development of a compact on-chip random spectrometer. RLs also tend to be the best in the image quality test overpowering all other sources of light. RL plays an important role in the medical field in the identification of tumors and photodynamic therapy (Noginov 2005).

A recent development in the field of laser technology is the manufacturing of the disturbed laser systems, where the built-in gain medium scatterers cause multiple scattering, which plays the role of feedback mechanism, causing amplification and stimulated emission in the medium to produce light emission. RL, and surface plasmon amplification by induced radiation (SPASER) systems are working on this principle (Chen et al., 2016; Bergman and Stockmanm (2003). Appropriately structured nanomaterials are important for these devices to serve as scatters and sometimes to replace the optical cavity. The striking point is that the standard emission characteristics of RL are highly dependent on the dispersing properties of the medium as well as on the geometry of the active zone and the pump profile. It has been observed that RL emission is extremely sensitive to small-scale alterations in the dimension, distribution, and optical response of the scatterers. The RL concept has, therefore, been recognized as a great opportunity to develop a new form of the active optical sensor (Tommasi et al., 2018) for different dispersion media. Since the stimulated emission mechanism changes the shape of the emission spectrum, the effect of the disorder can be identified by the measurement of variations in spectral properties rather than by energy measurement or light path reconstruction as in passive systems. Moreover, in a RL, scattering does not play the function of a loss factor, as in many passive methods, but actually feeds the signal, leading to more effective strategies for detecting heavy scattering samples. Recently, Shi et al. (2020) reported low-cost biosensors on fiber facet for label-free detecting biomolecules based on a plasmonic RL. A RL cytometer to detect malignant cell is demonstrated by He et al. (2019) in an optofluidic system loaded with medium gain and normal/cancer human breast cells. Liu et al. (2019) have demonstrated perovskite-based surface-emitting RLs (SERLs) and explore their applications in speckle-free imaging.

As discussed earlier that RLs are special lasers that rely on the multiple light scatter and optical gain and in RL, multiple light scattering is used to provide feedback. A schematic diagram has been given in Figure 2.2, which shows the comparison of the conventional laser and RL system. RL systems were first theoretically proposed by Letokhov (1968) using the diffusion equation.

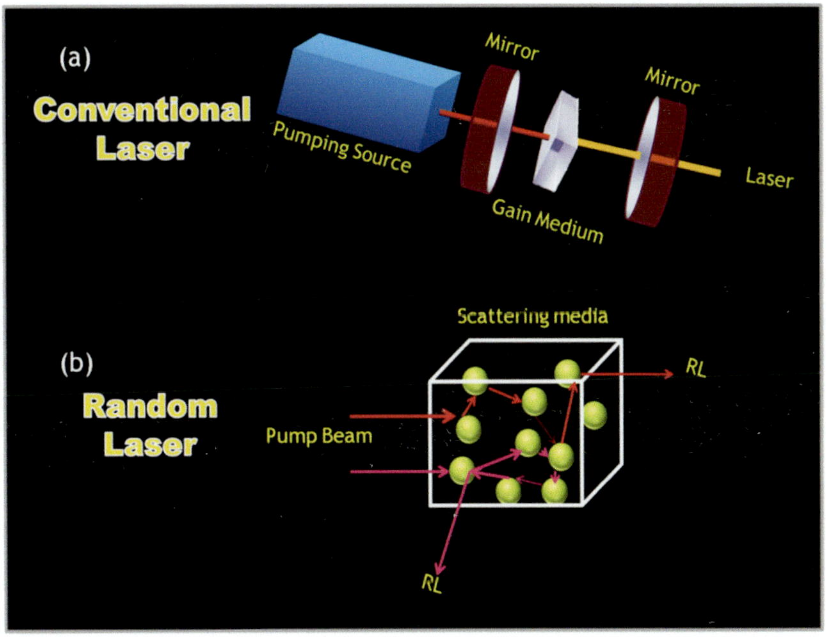

FIGURE 2.2 Schematic of the (a) Conventional laser and (b) RL system.

However, it has long been assumed that the dispersion in random systems is harmful to the lasing process, as the dispersion decreases the consistency of light input in typical laser systems. Thus, RL systems had been relatively unknown for about 35 years, until the group of Lawandy et al., (1994) experimented and clarified. Cao et al. (2003) divided the RL into two groups based on the lasing feedback: coherent RL and an inconsistent RL. They first indicated that the RL should have consistent behavior. Consistent properties of random lasing in ZnO powder have been experimentally demonstrated.

They proposed that light be restricted through recurrent processes in a solid and gain-driven medium, and provided coherent feedback. There are two distinct spectral signatures of RLs, such as incoherent and coherent emissions. Coherent/resonant emission is characterized by several narrow emission peaks of sub-nanometer line width, whereas incoherent/non-resonant emission has a single-peak range of a few nm, and both are randomly distributed in frequency. Redding et al. (2012) have shown that despite their brightness, lasers, and super-luminescent diodes are not well suited for full-field imaging applications because of their high spatial coherence which contributes to coherent artifacts, such as speckles that corrupt image formation. They have shown that RL can be designed to provide low spatial coherence and that the low spatial coherence of the explicitly designed RL can be used to demonstrate speckle-free full-field imagery in the sense of extreme optical scattering.

Liu et al. (2019) recently demonstrated that multiple scattering induced the RL generation in a perovskite polycrystalline film sandwiched by two Bragg-distributed reflectors. Due to the close containment in the vertical direction and a large number of random resonances, the wave vectors of the RL are dominated by their vertical components, resulting in multimode SERLs with a divergence angle of 3–5° and low spatial coherence. By directly illuminating the patterns with the SERLs, the speckle noise of standard optical images has been significantly suppressed. Recently, Biswas and Kumbhakar (2017) have also demonstrated the RL action ca. 674 nm in *Hibiscus rosasinensis* leaves extract (HRLe), by using triangular-shaped Ag NPs as plasmonic scatterer particles, pumped by a very low power CW He–He laser radiation of 632.8 nm. As the gain media and scatterer used in this case are biocompatible, this may lead to a new initiative to look for biocompatible gain media as well as an effective scatter for RL action.

2.3.4 ABSORPTION AND LUMINESCENCE CHARACTERISTICS OF METAL NANOSTRUCTURED MATERIALS

The UV–visible (UV–vis) absorption and fluorescence spectrophotometric techniques can be employed for various sensing applications of metal NPs (Tauc and Menth, 1972; Behzadi et al., 2015; Mourdikoudis et al., 2018; Wei et al., 2010; Kole et al., 2016; Biswas et al., 2016a, 2016b; Calcerrada et al., 2015; Saha et al., 2012; McGrath and Scanaill, 2013; Soares et al., 2014; Martinsson et al., 2014). The detection, analysis, and sensing of chemicals and biochemicals are very much important for a wide range of industrial and technological applications, ranging from medicine to food, electronics, safety, and the chemical industry (Calcerrada et al., 2015; Saha et al., 2012). The target analytes may be used to identify the target. The signal is produced, when the analyzer or sensor interacts with the target analyte and the interaction normally needs to be activated by a stimulus or energy source. Source stimuli and the signal detected can come in several ways, for example, light, heat, electrical current or voltage, magnetic, mechanical, and sound (McGrath and Scanaill, 2013). Among these, optical stimulation and detection are the most common and have the benefit of low cost and noninvasiveness.

Noble metal nanostructures demonstrate LSPR in near infrared and visible regions, resulting in sharp absorption peaks in their absorption spectra (Wei et al., 2010). The properties (such as positioning and widening) of the LSPR peaks are very sensitive to the dielectric medium covering the surface of the metal nanostructures and can be used effectively to detect biomolecules (Soares et al., 2014). Ag nanostructures also demonstrate some beneficial properties during sensing applications. Some of them are (1) No labeling is required, (2) Real-time glucose concentration monitoring is possible, (3) Silver nanostructures are more susceptible to the RI (Martinsson et al., 2014) than Au nanostructures (AUNs) of the same size and shape: 161 nm/RUI *vs.* 44 nm/RUI for silver and gold, and (4) Low cost and simple instrumental configuration.

2.3.4.1 *INTRODUCTION TO UV–VIS ABSORPTION SPECTROPHOTOMETRY*

Here, a brief introduction to UV–vis absorption spectrophotometry has been developed, which can be used as a tool for various metal NP sensing applications as discussed earlier. According to the Beer's law, the absorption of light in a material is proportional to the amount of absorbing molecules that depend on the concentration of absorbing molecules, and Lambert's law states that the fraction of absorbed radiation is independent of the intensity of the radiation. The fusion of these two rules resulted in the well-known Beer-Lambert's law (Tauc and Menth, 1972).

$$A = \log_{10}\left(\frac{I_0}{I}\right) = \varepsilon cl. \tag{2.31}$$

where, A is the absorbance, I_0 is the intensity of the incident radiation, I is the intensity of the transmitted radiation, ε is the molar absorption coefficient (mol/dm^3 cm), l is the path length of the absorbing solution in centimeters, and c is the concentration of the absorbing species (mol/dm^3).

In UV–vis absorption spectroscopy, the absorption of ultraviolet or visible radiation by a molecule leads to a transition between the electronic energy levels of the molecule (Behzadi et al., 2015; Mourdikoudis et al., 2018). The peculiar optical properties of metal NPs are the result of the electron conduction oscillations called SPR caused by electromagnetic radiation and can be described by the use of UV–vis spectroscopy for metal NPs (Wei et al., 2010). The UV–vis absorption spectroscopy is a fast and easy-to-operate NP characterization technique, particularly for colloidal suspensions.

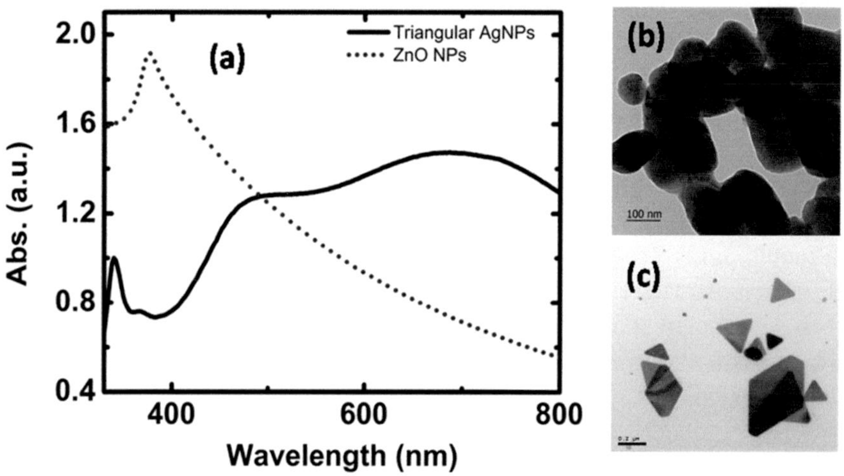

FIGURE 2.3 (a) UV–vis absorption spectra of triangular AgNPs (solid line) and ZnO NPs (short dashed line); (b) and (c) TEM images of ZnO and triangular AgNPs.
Source: Adapted from Biswas, et al. (2019). With permission. © 2019 Springer.

However, in Figure 2.3a, the typical UV–vis absorption spectra of the triangular-shaped AgNPs (solid line) and ZnO NPs (dashed line) are shown together and Figure 2.3b and c showed the corresponding TEM images of the ZnO and triangular-shaped AgNPs, respectively (Kole et al., 2016; Biswas et al., 2016a, 2016b). In the case of triangular-shaped AgNPs, it is observed that multiple absorption peaks are aroused from different dipolar and multipolar surface plasmon oscillations (Wei et al., 2010), whereas single absorption peak in the UV region is seen in ZnO NPs due to band edge absorption. Therefore, by observing the features of UV–vis absorption, it is possible to predict the particular types of NPs, when the effects of this technique are combined with other microscopic characterization techniques.

2.3.4.2 FLUORESCENCE PROPERTIES OF METAL NANOSTRUCTURES FOR SENSING APPLICATIONS

The basic mechanism behind many optical sensors is the identification of changes in optical signatures of nanomaterials, when communicating with target analyte molecules (Jackson et al., 2014). The optical signature may be absorption, luminescence, or Raman scattering. Among these, luminescence is the most common because it is highly sensitive, ubiquitous, and easy to detect (Jackson et al., 2014; Strianese et al., 2012; Demchenko, 2009). Another most widely used technique is Förster resonance energy transfer or FRET (Chen et al., 2020). In the FRET process, an excited donor passes energy (not an electron) to an acceptor group through a nonradiative process. This method is highly distance-dependent, allowing one to examine biological structures. One typical application is simply to calculate the distance between two positions of interest on a large molecule, generally a biological macromolecule, by adding suitable donor–acceptor groups to a large molecule (Sobakinskaya et al., 2018). If only one donor and one acceptor group are involved in a large molecule, the distance between the donor and the acceptor can be easily determined, if there is no conformational change in this process. Figure 2.4 shows the schematic of the FRET mechanism (Förster, 1948).

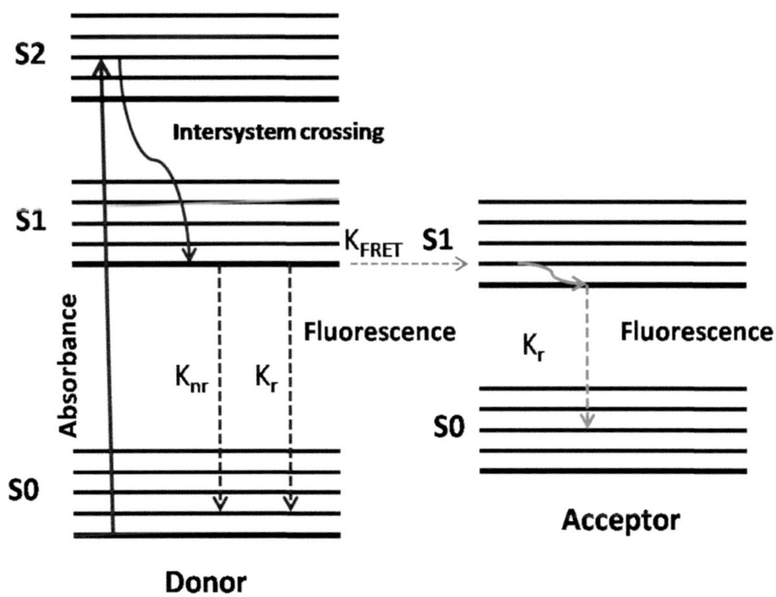

FIGURE 2.4 The schematic of the FRET mechanism.

FRET results from the excited state of the donor and provides an alternate pathway for the donor to relax back to the ground state in addition to relaxation by nonradiative or radiative processes. FRET, therefore, competes kinetically with other relaxation mechanisms (Figure 2.4) and its efficiency *EFRET* is calculated by the ratio between the constant rate for FRET, $K_{\mathrm{FRET},}$ and the sum of the constant rate for all relaxation processes (Förster, 1948, 1959; Jones and Bradshaw, 2019):

$$E_{FRET} = \frac{K_{FRET}}{K_{FRET} + K_{nr} + K_r}. \tag{2.32}$$

where, *Kr* is the rate constant for radiative relaxation of donor molecules in the excited state, and *Knr* is the rate constant for nonradiative relaxation processes. The distance dependence of the FRET efficiency E_{FRET} is obtained as (Förster, 1959):

$$E_{FRET} = \frac{R_0^6}{r^6 + R_0^6}, \tag{2.33}$$

where, r is the distance between the donor and the acceptor fluorophores and R_0 is the Förster radius or characteristic transfer distance at which the energy transfer efficiency is 50%. The Förster radius R_0 depends on the spectral properties of the donor and the acceptor pair more specifically on the quantum yield of the donor ϕ_{D} (Förster, 1959). The integral overlap *J*, which measures the spectral overlap between the emission of the donor and the absorption spectrum of the acceptor, and the orientation factor k^2 is a test for the relative orientation of the transition dipoles of the donor and the acceptor. It can be shown that:

$$R_0^6 = \frac{9 \ln 10 \varphi_D k^2}{128 \pi^2 N_A n^2} J. \tag{2.34}$$

where, $\mathrm{N_A}$ is the Avogadro's number, n is the RI of the buffer. The overlap integral J can be evaluated from the normalized emission of the donor *FD* and the extinction coefficient of the acceptor εA in the region of spectral overlap (λ_{min} to λ_{max}) according to the relation:

$$J = \int_{\min}^{\max} F_D(\lambda) \varepsilon_A(\lambda) \lambda^4 d\lambda. \tag{2.35}$$

Recently, Biswas et al. (2019) have also reported syntheses of different AUNs (sphere, nanorice, rod shaped, etc.) of different shapes and sizes by using simple wet chemical techniques and demonstrated FRET process in rhodamine 6G (Rh 6G)-AUNs system. Different optical properties are presented in Figure 2.5 such as UV–vis absorption and PL emission characteristics of AUNs of different shapes. In the inset upper panel, the sample color sunder sunlight are also shown.

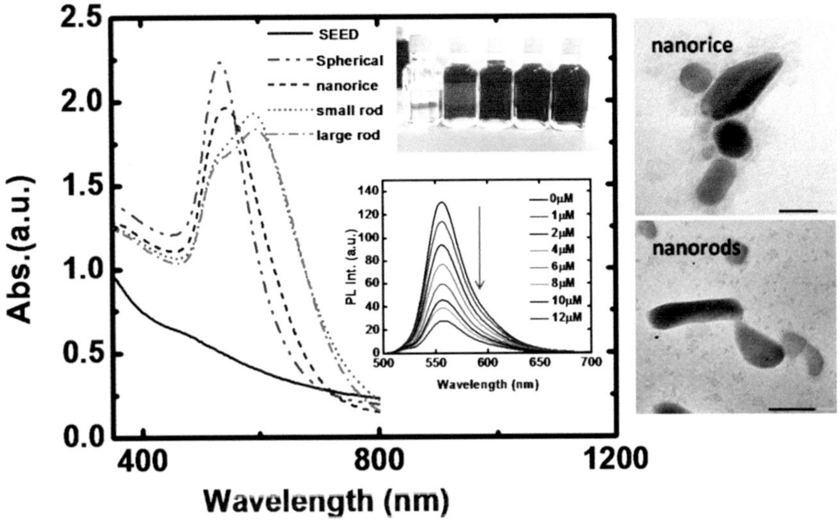

FIGURE 2.5 UV–vis absorption spectra of different shaped AUNs. In the inset, upper panel shows the sample color under sunlight. Lower panel of the inset shows the PL quenching of Rh-6G when loaded with different concentration of AUNs. Side panel shows the typical TEM images of the nanorice- and nanorod-shaped AUNs.

Source: Adapted from Biswas et al. (2019). With permission. © 2019 Springer.

The lower panel of the inset of the Figure 2.5 shows the PL quenching properties of Rh-6G, when loaded with different concentration of AUNs. The typical TEM images of the nanorice- and nanorod-shaped AUNs are shown in the side panel of Figure 2.5. The advantageous properties of AUNs can further be used in the future to study different bimolecular activities within the human body. It was found that the PL of Rh 6G had been extinguished, when it was kept in the vicinity of the synthesized AUNs, which was explained in terms of the Förster energy transfer mechanism (Biswas et al., 2019). They have also investigated the energy transfer process between different formed AUNs with Rh-6G and observed that with an increase in the value of the overlapping integral (*J*) between the absorption spectrum of the receiver (AUNs) and the fluorescence spectrum of the donor (Rh-6G), the energy transfer efficiency has reached a limit and almost 80% PL quenching in the case of rod-shaped AUNs was observed.

The FRET-based fluorescent systems also find applications in various other fields, some of them are given below. A supramolecular host–guest FRET pair based on carboxyfluorescein-labeled cucurbituril (CB7-CF, as an acceptor) and fluorescent dye 4′,6-diamidino-2-phenylindole (DAPI, as a donor) has been developed for DNA sensing by Zhang et al. (2019). Niu et al. (2020) have developed a two-photon fluorescent probe based on a dual-site response for SO_2/HOCl. Zhang et al. (2015a) developed and synthesized a ratiometric FRET-based probe to detect HNO in living cells. Koa and Lee (2019) developed a FRET protein sensor for quantitative analysis of L-Met using unnatural fluorescent amino acids and YFP as a FRET pair.

2.3.4.3 SURFACE PLASMON RESONANCE BASED SENSING

As discussed above, the SPR properties of metal nanostructures are very interesting and can be used in different sensing applications (Canovi et al., 2012). The absorption/reflection properties of metal nanostructures depend on a number of variables, such as the dielectric constant of the imbedding material and even minor variations in the dielectric constant, which may have an observable effect on the wavelength of the resonance (Daghestani and Day, 2010). This can be used in RI sensing. Moreover, it is even more efficient in functional sensing in which the receptor monolayer is sensitive to the desired analyte applied to the film surface. When a relatively small number of analyte molecules are bound to these surfaces, the change in the dielectric constant is sufficient to cause a significant change in the plasmon resonance. This has been seen in a clinical diagnostic procedure and has the potential to become an efficient and highly sensitive biomedical detection technique (Daghestani and Day, 2010; Homola, 2008). Due to the high degree of tunability in the SPR position of the metal nanostructures in the visible region of the electromagnetic spectrum, which can be easily achieved by adjusting their shape and scale, they are sensitive to detect minor changes in the RI of the surrounding medium (Szunerits and Boukherroub, 2012). The change in the SPR properties of metal NPs may also lead to the corresponding color change in the sample, viz. the spherical mono-dispersed Au NPs, which are usually pink in color, when aggregated due to a change in any dispersive medium properties (such as pH and temperature), the SPR location shifts to a longer wavelength and the solution color changes accordingly from pink to blue-purple (Biswas et al., 2019). Since the assemblies of Au nanostructured materials are often accompanied by distinct color shifts, colorimetric sensors using Au nanostructured have been extensively explored. Thus, Au nanostructured can be extensively used in the detection of pH (Biswas et al., 2019), DNA, proteins (Teengam et al., 2017), and metallic ions (Priyadarshini and Pradhan, 2017), and so on.

Wang (2018) has theoretically examined the nonlocal effect on the evolution of different plasmon resonance modes in closely coupled NP dimer and trimer antennas with a distance of less than 1 nm. Then RI sensing has been demonstrated in these nonlocal systems and these results were compared with classical calculations. Shiohara et al. (2014) reported widespread use of Au nanostars for ultrasensitive molecular identification, based on both LSPR and surface-enhanced Raman scattering. They have prepared Au nanostar monolayers and immobilized it on transparent, flexible polydimethylsiloxane substrates, and their RI sensitivity and SERS performance have been studied. Recently, spiky Au NPs have been prepared using different biocompatible buffers through a seed-mediated synthesis by Pallares et al. (2019). Li et al. (2019) recently introduced a way to endow Au NPs with aggregation behavior induced by pH tuning.

2.3.4.4 GLUCOSE SENSING ACTIVITY THROUGH SPR IN ANISOTROPIC SILVER NANOSTRUCTURES

Different nanostructures of Ag can be synthesized with different morphologies and sizes, such as quasi-spherical, hexagonal, triangular, and so forth (Biswas et al., 2014). Anisotropic nanostructures, that is, nanobars, nanowires of a few micrometers in length, can also be synthesized by a simple and low-cost chemical pathway. The synthesis processes reported do not require any complex laboratory setup and the chemicals used are also non-toxic and nonhazardous. The synthesized anisotropic silver nanostructures are found to be capable of glucose sensing and the RI sensitivity of 402 nm/RUI has been reported. The demonstrated method is easy and it is possible to track glucose in real-time. A minimum glucose concentration of 0.011 mM has also been shown to cause a significant shift in the position of LSPR.

As shown in eq 2.29, $\lambda_{SPR} = \lambda_P \sqrt{(1+2nm^2)}$, (Biswas et al., 2014), where, λ_{SPR} is the wavelength at the LSPR peak and λ_p is the wavelength corresponding to the plasma frequency of the bulk metal. From this equation, a modification of the LSPR peak position was found with RI of the surrounding medium is almost linear and this formed the basis of the glucose as well as RI-sensing platform. For the glucose-sensing application, two of our synthesized samples were picked, namely, nanowire-shaped and ellipsoid-shaped silver nanostructures, which are anisotropic in shape. Changes in the absorption spectra of the samples were tracked, when they were subjected to an increase in glucose concentrations from 0.011 to 1.110 mM. Figure 2.6 shows the schematic of the glucose sensing experiment with nanowire-shaped Ag nanostructures and also the shift in the LSPR peak position, when the sample is loaded with the highest concentration of the glucose. Standard glucose levels in adult humans are between 70 and 130 mg/dl before meals and less than 180 mg/dl 2 h after the start of a meal (Hong et al., 2012; Nayak et al., 2005). Now in this case, Authors were able to trace the lowest concentration of 198 mg/dL of glucose in the blood of a diabetic patient. Biswas et al. (2014) contrasted the observed LSPR change for wire-shaped and ellipsoid-shaped Ag nanostructures. Ellipsoidal formed Ag nanostructure has been shown to have a greater change in the LSPR peak position for a specific glucose concentration; thus, providing a higher RI sensitivity and being 402 nm/RUI.

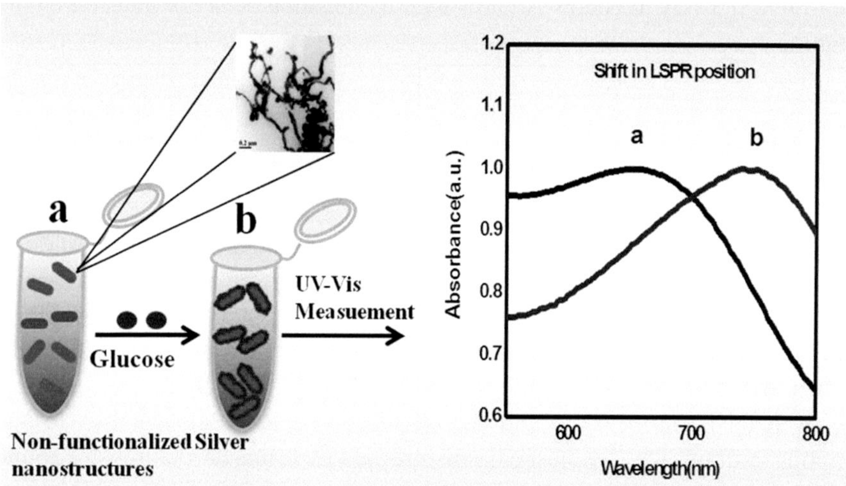

FIGURE 2.6 Schematic of the glucose sensing experiment with nanowire-shaped silver nanostructures and also the shift in the LSPR position when the samples loaded with highest concentration of the glucose.

Source: Adapted from Biswas et al. (2014). With permission. © IOP Publishing.

2.3.4.5 REFRACTIVE INDEX AND PH SENSING THROUGH SPR IN METAL NANOSTRUCTURES

In the recent past, Biswas and Kumbhakar (2018) have demonstrated the syntheses of triangular AgNPs by solvothermal method in DMF with PVP as capping agent. Very interestingly, it has been found that the samples remained stable even more than 3 years after synthesis. The stability of the samples was monitored by a UV–vis absorption and transmission electron micrograph (TEM) data analyses. The TEM images of the freshly synthesized sample and those taken after 3 years of synthesis, respectively, have been depicted in Figure 2.7a and b. It clearly indicates the unprecedented stability of the synthesized samples.

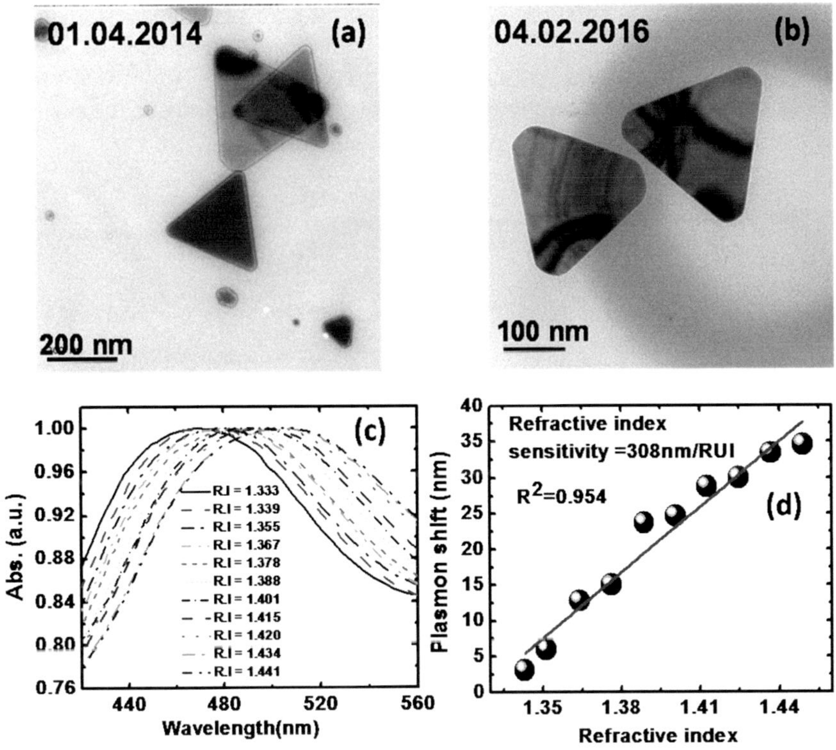

FIGURE 2.7 (a) TEM image of the as synthesized samples, (b) TEM image of the sample taken after 3 years, (c) Shift in plasmon peak position in different RI medium and (d) Scattered points are the plot of plasmon shift *vs.* RI and solid line is the linear fit.

Source: Adapted from Biswas and Kumbhakar (2018). With permission. © Springer Nature.

Figure 2.7c shows the change in the SPR peak position of the Ag NPs in media with different refractive indices, prepared by a mixture of water–glycerine. In Figure 2.7d, Plasmon shift with RI is plotted and the slope of the linear fit of the experimental points gives the RI sensitivity, which was found to be 308 nm/RUI. This investigation shows that the synthesized silver nanostructures are worth enough even after 3 years of synthesis and can be used for RI sensing in industry and science.

It has also been demonstrated that the syntheses of various architectures of AUNs, demonstrated quick, simple and IoT-based calorimetric sensing of pH using a smartphone (Biswas et al., 2019). The change in the color of the samples due to the change in pH (2–8) of the solution was digitally recorded in the Android mobile app and then the images were analyzed based on the CIELab 1931 color space to study the output of the perceptual colorimetric sensor. The UV–vis absorption spectra of the AUNs in different pH media have been presented in Figure 2.8a. The inset at the bottom shows the color change of AUNs due to agglomeration and Figure 2.8b shows the schematic of the smart phone-based pH sensing mechanism.

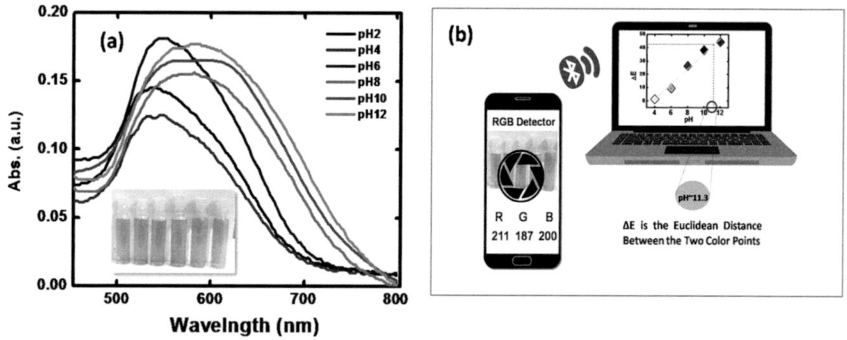

FIGURE 2.8 (a) The UV–vis absorption spectra of the AUNs in different pH media. Inset at the bottom shows the color change of AUNs due to agglomeration and (b) The schematic of the smart phone-based pH sensing mechanism.

Source: Adapted from Biswas et al. (2019). With permission. © Royal Society of Chemistry.

The quantitative color difference (*ΔE*) between the two images taken at different pH levels was estimated (Biswas et al., 2019). It was observed that *ΔE* depends on the form of the synthesized nanostructures and in the highly acidic medium (pH 2–4), the spherical AUN specimen showed the highest *ΔE* value and for the highly alkaline medium (pH 10–12), the small AUNs showed the highest *ΔE* value. Ratiometric pH sensing was also done in the same pH range in all samples. The absorption spectra of the samples at different pH levels were deconvoluted in terms of the absorption peak at 540and 600 nm and the ratiometric plot (*I540/I600*) was obtained. The experimental points were fitted with a sigmoid curve and the pH of the unknown solution was determined using both; calorimetric and ratiometric methods, which were found to align reasonably well with each other. Since the current pH detection method does not need any sophisticated equipment, it has potential application in fieldwork in remote areas, where pH detection can only be performed using a smartphone.

2.4 PHOTOCATALYTIC PROPERTIES OF SOME PLASMONIC PHOTOCATALYSTS

Different nanomaterials are reported to show the photocatalytic properties but semiconductor photocatalysts are very often used for photodegradation of different toxic dyes from water (Lu et al., 2008a; Fang et al., 2015; Wang et al., 2015). However, in the recent years composites of noble metals with semiconductor materials have been widely studied as photocatalytic materials for wastewater decontamination due to their high photocatalytic activity (Lu et al., 2008b; 2008c), low cost (Fang et al., 2015), and nonhazardous nature (Wang et al., 2015). The semiconductor materials having wide bandgap like ZnO and ZnS have a unique capability to convert the light energy to chemical energy and produce charge carriers upon light irradiation and due to the relatively negative reduction potential of excited electrons, it offers an excellent photocatalytic activity (Sharma et al., 2010).

However, the photocatalytic potential of ZnO and ZnS greatly suffers by the drawbacks of fast recombination of photoexcited charge carriers. Moreover, the ZnO and ZnS possesses large bandgap, which limits their ability to efficiently absorb solar light and cannot used as solar light driven photocatalytic activity. Therefore, the optical and electronic properties of their semiconductors need to be tuned effectively by synthesizing the composites with different noble metal specially the noble metal (Kim et al., 2005; Lee et al., 2004) which have absorption in visible region of the spectrum due to SPR (Lu et al., 2008b). Among the different noble metals, silver has attracted significant attraction due to its nontoxic nature, high-catalytic property, low cost, and antibacterial property (Koga et al., 2009). (Lu *et al.*, 2008c) have demonstrated the high photocatalytic activity of Ag/ZnO self-assembled 3D hollow microspheres synthesized by hydrothermal synthesis route. ZnO/Ag composite nanosphere with an average diameter of 400 nm have been synthesized by one-pot sovothermal synthesis route and these are shown to exhibit enhanced photocatalytic activity over pure ZnO (Xie and Wu, 2010).

Zhang et al. (2015b) have investigated the photocatalytic properties of Ag/ZnO composites prepared by microwave-assisted combustion method.

Sivakumar et al. (2014) have synthesized ZnS and ZnS-Ag nanobelts by chemical precipitation method and observed the excellent photocatalytic performance of ZnS and ZnS-Ag nanobelts by irradiating the solution of MB in the presence of visible light. It has been also reported by several other researchers that the ZnO–ZnS–Ag ternary structures may also improve the photocatalytic performance under the UV–vis light irradiation. Subash et al. (2014) have synthesized ZnS-loaded Ag–ZnO by a precipitation–thermal decomposition method and used it for degradation of four toxic azo dyes. Zamiri et al. (2014) have demonstrated fabrication of ZnO–ZnS–Ag_2S ternary nanostructures with efficient solar-light photocatalytic activity. However, all the techniques for the synthesis of ZnO or ZnS composites with the Ag are complex and required a precise control over the synthesis procedure.

Here, the syntheses of some freshly prepared ZnO/Ag and ZnS/Ag nanocomposites by some easy technique have been described and their photocatalytic degradation capabilities have also been studied. The mechanism responsible for the photocatalytic degradation of the MB dye has also been provided. The formations of the nanocomposites were confirmed by XRD, SEM, and TEM studies. The syntheses of these plasmonic nanocomposites of ZnO or ZnS with Ag surely pave the way to synthesize a new class of composites materials by simple hydrothermal route to enhance the photocatalytic performance.

2.4.1 SYNTHESES OF ZnO/Ag AND ZnS/Ag COMPOSITES

First of all, ZnS and ZnO are prepared. ZnO has been synthesized freshly by the hydrothermal synthesis route. In a typical synthesis process, 8 mM of Zn $(NO_3)_2$, $6H_2O$ was dissolved in 30 ml of Deionized water (DI) and then 10 ml of KOH solution containing 40 mM of KOH was added drop-wise under magnetic stirring. After a few minutes, the clear solution turned into milky white, and the whole solution is then transferred to an autoclave and maintained at 140°C for 8 h. After that autoclave was left to cool at room temperature and precipitates were collected and washed with DI water several times to remove the impurities and the ZnO was collected by centrifugation. The collected sample was then dried at room temperature.

ZnS has been synthesized by the solvothermal route. In a typical synthesis procedure, the requisite amount of Zn $(NO_3)_2$, $2H_2O$, and thioacetamide (TTA) were dissolved in ethanol, and then it is stirred for 30 min. After that solution was transferred to an autoclave and maintained at 120°C for 6 h. Then the autoclave was left for cooling at room temperature and precipitates were collected by centrifugation. The collected ZnS nanoparticles were dried at room temperature.

For the synthesis of composites, 10 ml solution of $AgNO_3$ containing 0.4 mM $AgNO_3$ was prepared after that 0.5 mM of $NaBH_4$ was added to it under rigorous magnetic stirring. This solution was added to the precursor solution of ZnO and ZnS before hydrothermal process, stirred for 30 min and then transferred to an autoclave maintained at the same temperature and time stated earlier. The composites were then collected by a similar procedure. Then the dry samples were dispersed in water for further characterization by using different standard techniques (i.e., XRD, SEM, and TEM).

2.4.2 RESULTS AND DISCUSSION

Figure 2.9a exhibits UV–vis absorption spectra of ZnO and ZnO/Ag nanocomposites measured with a double beam Hitachi U-3010 spectrophotometer. The absorption spectra of ZnS and ZnS/Ag nanocomposites are depicted in Figure 2.9b. A well-defined excitonic absorption peak centered at approximately 373 nm was clearly observed in both the samples, which is very close to the bulk excitonic absorption peak of ZnO at approximately 373 nm (Haase et al., 1988) and an absorption shoulder extending up to 800 nm. The absorption in the visible region has been greatly improved in ZnO/Ag nanocomposites in comparison with the pure ZnO, which is helpful for the better photocatalytic performance of ZnO/Ag composites over ZnO on using UV–vis light. The excitonic absorption peak of ZnS appeared at approximately 338 nm and a detectable blue shift of 7 nm has been observed.

This blue shift may have arisen from the quantum size effect (Jiang et al., 2007) of ZnS nanocrystallines, which served as building blocks of the ZnS mesospheres formed during the synthesis by self-assembly (Bai et al., 2014) as evident from TEM images discussed later. In the absorption spectra of the ZnS/Ag, the excitonic absorption peak of ZnS remained almost in same position but the absorption in the visible region is highly enhanced due to the formation of nanoparticle composites with Ag. From the absorption spectra of ZnO/Ag and ZnS/Ag nanoparticle composites, it was observed that absorption in the visible region is highly enhanced in ZnS/Ag as compared to ZnO/Ag nanocomposites. As the photocatalytic experiment has been performed with UV–vis light irradiation, the enhancement in absorption in the visible region may be the possible reason of the better photocatalytic performance of ZnS/Ag nanoparticle composites in comparison with ZnO/Ag nanoparticle composites.

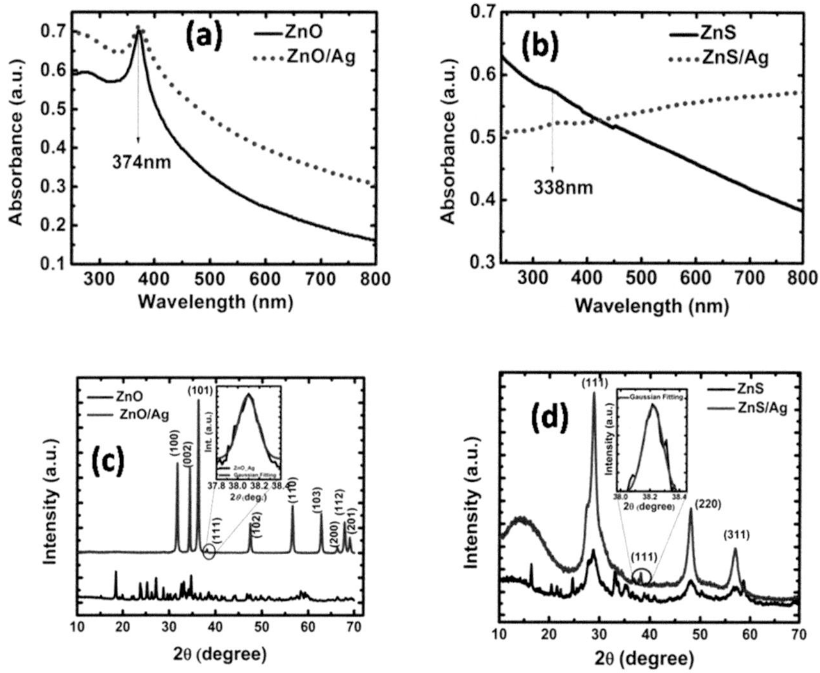

FIGURE 2.9 a Solid line represents UV–vis absorption spectra of ZnO and dashed line ZnO/Ag composites, (b) Solid line represents the UV–vis absorption spectra of ZnS and dashed line ZnS/Ag. (c) XRD pattern of the ZnO and ZnO/Ag composites and inset shows the Gaussian fitting of the (*111*) peak of silver, and (d) XRD pattern of ZnS and ZnS/Ag composites and inset shows the Gaussian fitting of the (*111*) peak of silver.

The crystal structures and the compositions of pure ZnO and ZnS and its composites with Ag have been investigated by the XRD analysis (X-ray diffractometer [PAN ANLYTICAL] with Cu K_α radiation of wavelength 0.15406 nm). The Figure 2.9c depicts the XRD pattern of the pure ZnO and ZnO/Ag nanocomposites. The reflections from the ZnO crystal planes have been indexed and the position corresponds to the wurtzite structure of ZnO (JCPDS Card No. 80-0075). A very feeble peak at 38.1° has been identified, which corresponds to the (*111*) refection of the Ag. This observation also confirms the presence of Ag in the ZnO/Ag nanocomposites. The peak corresponds to the Ag has been Gaussian fitted shown in the inset of the Figure 2.9c. The crystallites size of Ag has been calculated using the Debye–Scherrer equation (Zak et al., 2011) and was estimated to be 35 nm, which are very close to the size of the Ag nanoparticle embedded on ZnO as evident from the TEM images discussed later. Similarly the Figure 2.9d shows the XRD pattern of ZnS and ZnS/Ag nanocomposites together and the reflection from ZnS crystal planes matches well with the (*111*), (*220*), and (*311*) crystallographic planes of the cubic zinc-blende phase of ZnS (Kole and Kumbhakar, 2012) (JCPDS Nos. 00-80-0020). The peak at 38.1° in the XRD pattern of ZnS/Ag nanocomposites is identified as the reflection from the (*111*) peak of Ag. The Gaussian fitting of the peak corresponds to (*111*) plane of Ag was exhibited in the inset of the Figure 2.9d and from fitting using Debye–Scherrer equation, the crystallites size has been calculated to be 43 nm.

The SEM images (measured in FEG-SEM of SIRION, FEI) of the synthesized ZnO/Ag and ZnS/Ag are shown in the Figure 2.10a and b. The SEM micrographs of ZnO/Ag composites demonstrated that the rod-like ZnO with semispherical Ag nanoparticles attached at the surface grown densely and uniformly over large areas. The average diameter of the nanorods was estimated to be approximately 80 nm and the average length was estimated to be approximately 1000 nm. The SEM of ZnS/Ag composites revealed that Ag NPs are well distributed on the surface of large ZnS microspheres with an average diameter of 2.4 μm. The geometrical surface area of ZnS microspheres are 4 times larger than the Zn nanorods and may provide the greater surface area of dye adsorption.

TEM images (measured in Tecnai T-20 microscope operating at 300 kV) of the ZnO/Ag and ZnS/Ag composites have been shown in the Figure 2.10c and e. The TEM images of the ZnO/Ag showed that the Ag NPs are well anchored on the surface of the ZnO nanorods. The HRTEM image of the ZnO surface exhibited in the Figure 2.10d revealed the highly crystalline nature of the ZnO surface and inset shows the SAED pattern of the crystallographic planes of ZnO/Ag composites. The TEM images of ZnS/Ag nanocomposites showed in Figure 2.10e confirmed the porous structures of the ZnS microspheres. TEM analysis has revealed that the ZnS microspheres are not solid and it is formed due to the self-assembly of much smaller ZnS nanospheres (Jiang et al., 2007), which are formed at the initial stage of hydrothermal reaction. Such porous structures will provide the large surface area for the effective degradation of the dye. Figure 2.10f showed the SAED pattern of ZnS/Ag composites and the ring pattern indicated the polycrystalline nature of the ZnS/Ag composites.

The PL emission spectra of the synthesized pure ZnO, ZnS, and nanoparticles composites ZnO/Ag, ZnS/Ag were also taken (Perkin Elmer LS-55 spectrofluorimeter) at an excitation wavelength of 225 nm, which is depicted in the Figure 2.11a. The PL emission spectra can be divided into two regions; namely, UV region and visible region. For all the samples, a deep level UV emission band is centered at approximately 310 nm. The UV emission for all the samples may be attributed to the radiative recombination at the near band edge excitons (Yu and Yoshimura, 2002).

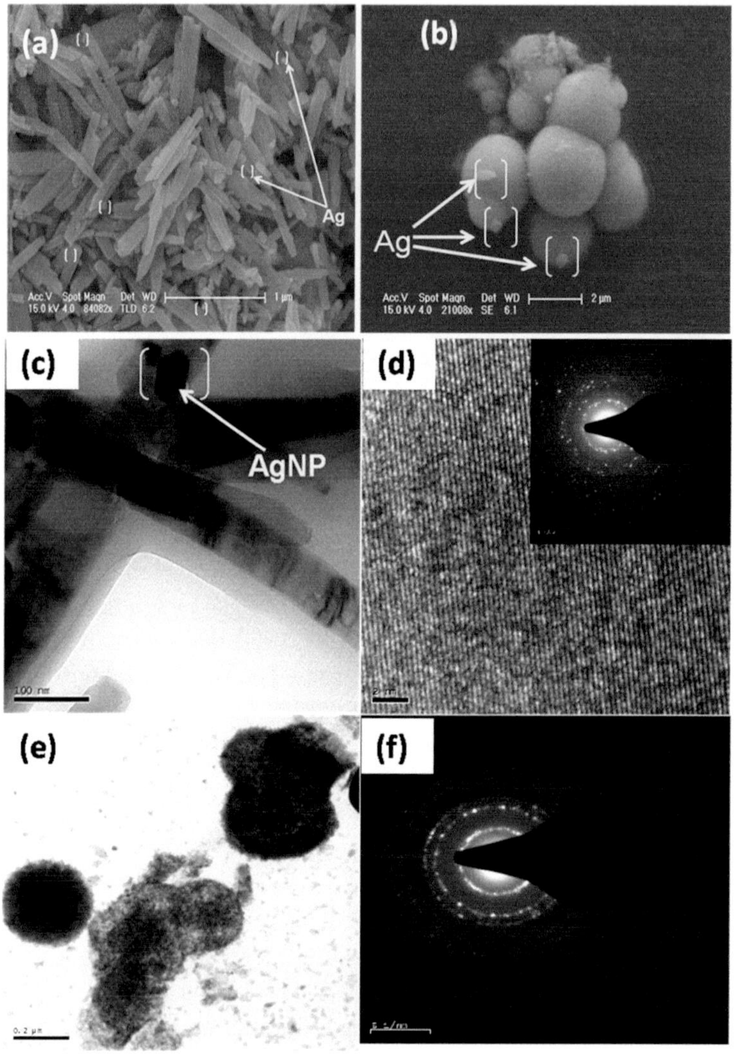

FIGURE 2.10 (a) SEM images of the ZnO/Ag composites and yellow parenthesis exhibited the attached AgNPs at the surface of the ZnO, (b) SEM images of the ZnS/Ag composites and yellow parenthesis exhibited the attached AgNPs at the surface of the ZnS, (c) TEM image of the ZnO/Ag composites with Ag embedded on the ZnO surface is highlighted by the yellow parenthesis, (d) HRTEM image of the crystallographic planes of ZnO/Ag, an inset shows the SAED pattern, (e) TEM image of the ZnS/Ag composites, and (f) SAED pattern of crystallographic plane of ZnS/Ag crystallographic surface.

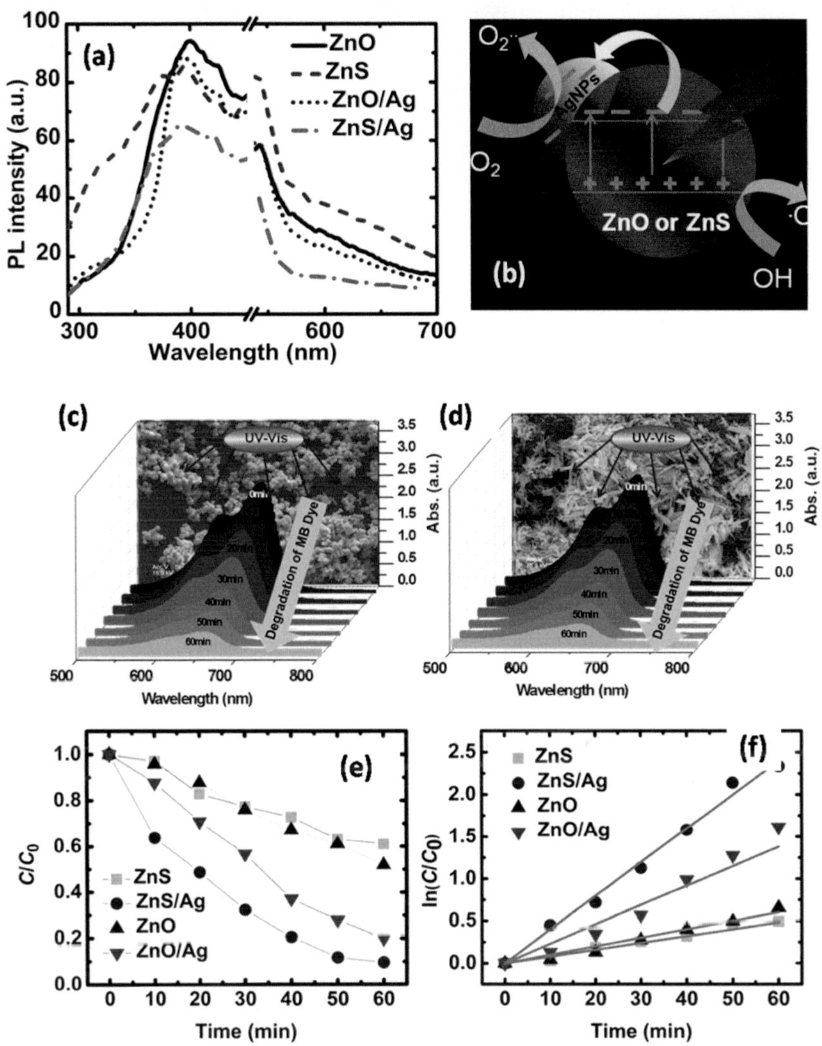

FIGURE 2.11 (a) Solid line represents PL emission spectra of the ZnO; the short-dashed lines represent PL emission spectra of ZnO/Ag short dot line represent PL emission spectra of ZnS and short-dashed dot line represent the PL emission spectra of ZnS/Ag at an excitation wavelength of 275 nm, (b) Schematic of the charge transfer mechanism, (c) UV-Vis absorption spectra of the MB in the presence of ZnS/Ag composites under visible light irradiation show the degradation of MB dye with irradiation time, (d) UV–vis absorption spectra of the MB in the presence of ZnO/Ag composites under visible light irradiation show the degradation of MB dye with irradiation time, (e) Plot of MB dye concentration with irradiation time, and (f) The concentration data of MB dye for different irradiation time has been fitted with first order kinetic equation the scattered points are the experimental points and the solid lines are the theoretical fitting.

Another intense blue emission band at approximately390 nm has been observed in all the samples. In case of ZnO and ZnO/Ag, the blue emission may arise from the radiative recombination of zinc interstitial energy level to zinc vacancy energy level in ZnO (Li et al., 2009). However, in case of ZnS and ZnS/Ag, the emission might have occurred from the transitions involving interstitial zinc or sulfur atoms. In the visible region, a green emission band at approximately 500 nm was easily identified for all the samples and the origin of such emission is the zinc and oxygen vacancies (Kar and Chaudhuri, 2005). Another orange emission band at approximately 600 nm was also present in the PL emission spectra. In case of ZnO, the red–orange emission comes from the oxygen interstitial defects (Djurišić et al., 2006) and for ZnS, orange emission is originated from the recombination between interstitial zinc states and zinc vacancies (Wang et al., 2011). It has been observed that intensity of PL emission band of pure ZnO and ZnS has significantly reduced after the formation of the composites with Ag. In case of ZnS/Ag nanoparticle composites, reduction in PL emission intensity is maximum. It has been reported earlier that the incorporation of metal nanoparticle at the surface of ZnO and ZnS nanoparticles may improve the separation of photo-induced electron and holes (Lu et al., 2008b; 2008c; Choi et al., 2016), and the reduction in PL emission intensity was also observed by the authors. As the Fermi level of Ag lies above the Fermi level of the ZnO and ZnS and transfer of electrons from Ag to semiconductor nanoparticles continues until equilibrium has been achieved. At this equilibrium condition, the conduction band of ZnO and ZnS lies close to the newly formed Fermi level and as a result, the photoexcited electron in the conduction band of ZnO and ZnS first transferred to the Ag NPs. Therefore, the overall charge separation decreases and such a decrease in PL emission intensity was observed. The delayed electron-hole recombination enhances the probability of capturing the electron in Ag NPs by chemisorbed O_2 and hole in the semiconductor by hydroxyl ion and better photocatalytic performance in ZnO/Ag and ZnS/Ag nanocomposites was observed. A schematic of the charge transfer mechanism has been depicted in Figure 2.11b.

The photocatalytic activities of the pure ZnO and ZnS and their composites have been evaluated by measuring the degradation of MB dye under UV–vis light illumination. The degradation MB dye was measured by monitoring the decrease in absorbance of the characteristic absorption peak of approximately 660 nm for each sample under UV–vis light irradiation and for ZnO/Ag and ZnS/Ag composites are presented in Figure 2.11c and d. 100 mL solution of MB dye was taken having concentration of 200 mL/L t and

0.006 g of sample was added to it. The solution was continuously stirred in dark before irradiating it to UV–vis light to obtain the adsorption equilibrium. A particular amount of the above solution during irradiation was called and absorption spectra were taken to measure the degradation of MB dye.

From Figure 2.11c and d, it can be observed that the absorbance characteristic peak of MB dye has been decreased gradually with irradiation time, indicating the degradation of the MB dye with irradiation time, which is also evident from the digital images of the MB dye solution taken in different time interval given in the inset. The degradation rate of the individual sample has been plotted in Figure 2.11e considering that the ratio of absorption of initial dye solution and the solution after particular irradiation time is equivalent to the ratio of initial dye concentration (C_0) and the dye concentration of dye after particular irradiation time (C). From the graph, it is clear that the degradation efficiency of the composites has been improved considerably in comparison to the pure ZnS and ZnO and for ZnS/Ag nanoparticle composites; it is highest and could reach almost 90% within 60 min of irradiation time. The kinetic curve of degradation of MB dye has been plotted in the Figure 2.11f, which is fitted with pseudo first-order equation given by Langmuir–Hinshelwood (Liang et al., 2010),

$$\ln(C_0/C) = K \times t. \tag{2.36}$$

where K is the rate constant and the value of the rate constant for ZnS/Ag nanoparticles composites is 0.04 min^{-1}, which is highest among all the synthesized samples and explains its ability for better photocatalytic degradation of MB dye.

2.5 SUMMARY

In summary, the optical properties of nanostructured materials and their sensing applications have been briefly discussed. It starts with a short introduction about nanomaterials, and then the significance of nanomaterials in research and industry has been discussed. Then the classification of nanomaterials is also described. Optical properties of nanomaterials especially the optical absorption and luminescence properties of metal nanoparticles are presented in details along with the theoretical models of SPR in metal NPs. Sensing applications in terms of the SPR such as glucose, RI, and pH using Ag and Au NPs are discussed. The results on the investigations on photocatalytic activities of Ag/ZnO and Ag/ZnS nanocomposites have been presented. This chapter will provide a concise understanding about the optical properties of nanomaterials and create enough curiosity among well-established researchers as well as in beginners.

KEYWORDS

- **nanomaterials**
- **sensing applications**
- **optical properties**
- **surface plasmon resonance**
- **nanocomposite**

REFERENCES

Ahn, J. H.; Kim, H. S.; Lee, K. J.; Jeon, S.; Kang, S. J.; Sun, Y.et al. Heterogeneous Three-Dimensional Electronics by Use of Printed Semiconductor Nanomaterials. *Science* **2006,** *314*, 1754–1757.

Alabastri, A.; Tuccio, S.; Giugni, A.; Toma, A.; Liberale, C.; Das, G. et al. Molding of Plasmonic Resonances in Metallic Nanostructures: Dependence of the Non-Linear Electric Permittivity on System Size and Temperature. *Materials* **2013,** *6*, 4879–4910.

Almohammed, S.; Zhang, F.; Rodriguez, B. J.; Rice, J. H. Photo-Induced Surface-Enhanced Raman spectroscopy from a Diphenylalanine Peptide Nanotube-Metal Nanoparticle Template., *Sci. Rep.*, **2018,** *8* (1), DOI: 10.1038/s41598-018-22269-x.

Amendola, V.; Meneghetti, M. Size Evaluation of Gold Nanoparticles by UV− vis Spectroscopy. *J. Phys. Chem. C* **2009,** *113*, 4277–4285.

Ashikbayeva, Z.; Tosi, D.; Balmassov, D.; Schena, E.; Saccomandi, P.; Inglezakis, V. Application of Nanoparticles and Nanomaterials in Thermal Ablation Therapy of Cancer. *Nanomaterials (Basel).***2019,** *9* (9), DOI: 10.3390/nano9091195.

Bai, W.; Cai, L.; Wu, C.; Xiao, X.; Fan, X.; Chen, K. Alcohothermal Synthesis of Flower-Like ZnS Nano-Microstructures with High Visible Light Photocatalytic Activity. *Mater. Lett.* **2014,** *124*, 177–180.

Barnes, W. L.; Dereux, A.; Ebbesen, T. W. Surface Plasmon Subwavelength Optics. *Nature*, **2003,** *424* (6950), 824–830.

Barros, C. H. N.; Casey, E. A Review of Nanomaterials and Technologies for Enhancing the Antibiofilm Activity of Natural Products and Phytochemicals. *ACS Appl. Nano Mater*. **2020,** *3* (9), 8537–8556.

Becker, J. Light-Scattering and -Absorption of Nanoparticles. In *Plasmons as Sensors. Springer Theses*; Recognizing Outstanding Ph.D. Research; Springer: Berlin, Heidelberg, 2012.

Behzadi, S.; Ghasemi, F.; Ghalkhani, M.; Ashkarran, A. A.; Akbari, S. M.; Pakpour, S. Determination of Nanoparticles Using UV-Vis Spectra. *Nanoscale* **2015,** *7*, 5134–5139.

Benelmekki, M. *One-Dimensional Nanomaterials*; Morgan & Claypool Publishers, Bristol, 2019.

Bera, D.; Qian, L.; Tseng, T.; Holloway, P. H. Quantum Dots and Their Multimodal Applications: A Review. *Material (Basel)* **2010,** *3* (4), 2260–2345.

Bergman, D. J.; Stockmanm, M. I. Surface Plasmon Amplification by Stimulated Emission of Radiation: Quantum Generation of Coherent Surface Plasmons in Nanosystems. *Phys. Rev. Lett.*, **2003,** *90*. DOI: 10.1103/PhysRevLett.90.027402.

Biswas, P. S.; Chakraborty, J.; Agarwal, A.; Kumbhakar, P. Gold Nanostructures for the Sensing of pH Using a Smartphone. *RSC Adv.* **2019,** *9*, 34144–34151.

Biswas, S.; Kole, A. K.; Sarkar, R.; Kumbhakar, P. Synthesis of Anisotropic Nanostructures of Silver for Its Possible Applications in Glucose and Temperature Sensing. *Mater. Res. Exp.*, **2014,** *1*. DOI: 10.1088/2053-1591/1/4/045043.

Biswas, S.; Kole, A. K.; Tiwary, C. S.; Kumbhakar, P. Enhanced Nonlinear Optical Properties of Graphene Oxide–Silver Nanocomposites Measured by Z-Scan Technique. *RSC Adv.* **2016a,** *6*, 10319–10325.

Biswas, S.; Kole, A. K.; Tiwary, C. S.; Kumbhakar, P. Observation of Size-Dependent Electron–Phonon Scattering and Temperature-Dependent Photoluminescence Quenching in Triangular-Shaped Silver Nanoparticles. *Plasmonics* **2016b,** *11*, 593–600.

Biswas, S.; Kumbhakar, P. Continuous Wave Random Lasing in Naturally Occurring Biocompatible Pigments and Reduction of Lasing Threshold Using Triangular Silver Nanostructures as Scattering Media. *Nanoscale* **2017,** *9*, 18812–18818.

Biswas, S.; Kumbhakar, P. Refractive Index and Temperature Sensing in Anisotropic Silver Nanostructures with Stable Photo-Physical Properties. *Appl. Phys. A*. **2018,** *124*, 6–16.

Borja, S.; Paula, C. A.; Laura, M. L.; Luis, M. L. LSPR-Based Nanobiosensors. *Nano Today* **2009,** *4*, 244–251.

Brozena, A. H.; Kim, M.; Powell, L. R.; Wang, Y. Controlling the Optical Properties of Carbon Nanotubes with Organic Colour-Centre Quantum Defects. *Nat. Rev. Chem.* **2019,** *3*, 375–392.

Bussche, E. V.; Deene, Y. D.; Dubruel, P.; Vergote, K.; Schacht, E.; Wagter, C. D. The Use of Static Light Scattering for the Structure Analysis of Radiosensitive Polymer Gels: A Literature Survey. *J. Phys.: Conf. Ser.* **2004,** *3*. DOI: 10.1088/1742-6596/3/1/022.

Calcerrada, M.; García-Ruiz, C.; González-Herraez, M. Chemical and Biochemical Sensing Applications of Microstructured Optical Fiber-Based Systems. *Laser Photon. Rev.*, **2015,** *9*, 604–627.

Canovi, M.; Jacopo, L.; Matteo, S.; Francesca, R.; Davide, M.; Paolo, B. et al. Applications of Surface Plasmon Resonance (SPR) for the Characterization of Nanoparticles Developed for Biomedical Purposes. *Sensors (Basel)*. **2012,** *12* (12), 16420–16432.

Cao, H. Random Lasers: Development, Features and Applications. *Opt. Photonics News* **2005,** *16*, 24–29.

Cao, H.; Xu, J. Y.; Ling, Y.; Burin, A. L.; Seeling, E. W.; Liu, X. Random Lasers with Coherent Feedback. *IEEE JSTQE* **2003,** *9*, 111–119.

Cassano, D.; Santi, M.; D'Autilia, F.; Mapanao, A. K.; Luin, S. Voliani, V. Photothermal Effect by NIR-Responsive Excretable Ultrasmall-in-Nano Architectures. *Mater. Horiz.* **2019,** *6*, 531–537.

Cassidy, J.; Zamko, M. Nanoshell Quantum Dots: Quantum Confinement beyond the Exciton Bohr Radius. *J. Chem. Phys*. **2020,** 152, DOI: 10.1063/1.5126423.

Chang, Q.; Shi, X.; Liu, X.; Tong, J.; Liu, D.; Wang, Z. Broadband Plasmonic Silver Nanoflowers for High-Performance Random Lasing Covering Visible Region. *Nanophoton* **2017,** *6*, 1151–1160.

Chavel, P.; Lowenthal, S. Noise and Coherence in Optical Image Processing. I. The Callier Effect and Its Influence on Image Contrast. *JOSA*, **1978,** *68*, 559–568.

Chen, H.; Ding, F.; Zhou, Z.; He, X.; Shen, J. FRET-Based Sensor for Visualizing pH Variation with Colorimetric/Ratiometric and Application for Bioimaging in Living Cells, Bacteria and Zebrafish. *Analyst* **2020,** *145*, 4283–4294.

Chen, Y.; Chena, Q.; Fan, X. Optofluidic Chlorophyll Lasers. *Lab Chip* **2016,** *16*, 2228–2235.

Choi, Y. I.; Lee, S.; Kim, S. K.; Kim, Y.; Cho, D. W.; Khan, M. M. et al. Fabrication of ZnO, ZnS, Ag-ZnS, and Au-ZnS Microspheres for Photocatalytic Activities, CO Oxidation and 2-hydroxyterephthalic Acid Synthesis. *J. Alloys. Compd.* **2016,** *675*, 46–56.

Consoli, A.; López, C. Decoupling Gain and Feedback in Coherent Random Lasers: Experiments and Simulations. *Sci. Rep.* **2015,** *5*. DOI: 10.1038/srep16848.

Coronado, E. A.; Encinaa, E. R.; Stefani, F. D. Optical Properties of Metallic Nanoparticles: Manipulating Light, Heat and Forces at the Nanoscale. *Nanoscale* **2011,** *3*, 4042–4059.

Daghestani, H. N.; Day, B. W. Theory and Applications of Surface Plasmon Resonance, Resonant Mirror, Resonant Waveguide Grating, and Dual Polarization Interferometry Biosensors. *Sensors* **2010,** *10* 9630–9646.

Demchenko, A. P. Fluorescence Detection Techniques. In *Introduction to Fluorescence Sensing*; Demchenko, A. P., Ed.; Springer: Dordrecht, 2009.

Djurišić, A. B.; Leung, Y. H.; Tam, K. H.; Ding, L.; Ge, W. K.; Chen, H. Y. et al. Green, Yellow, and Orange Defect Emission from ZnO Nanostructures: Influence of Excitation Wavelength. *Appl. Phys. Lett.* **2006,** *88*, 103107–103110.

Doria, G., Conde, J.; Veigas, B.; Giestas, L., Almeida, C.; Assunção, M. et al. Noble Metal Nanoparticles for Biosensing Applications. *Sensors* **2012,** *12*, 1658–1687.

Eustis, S.; El-Sayed, M. A. Why Gold Nanoparticles Are More Precious Than Pretty Gold: Noble Metal Surface Plasmon Resonance and Its Enhancement of the Radiative and Nonradiative Properties of Nanocrystals of Different Shapes. *Chem. Soc. Rev.* **2006,** *35* (3), 209–217.

Fang, Z.; Weng, S.; Ye, X.; Feng, W.; Zheng, Z.; Lu, M. et al. Defect Engineering and Phase Junction Architecture of Wide-Bandgap ZnS for Conflicting Visible Light Activity in Photocatalytic H_2 Evolution. *ACS Appl. Mater. Interf.* **2015,** *7*, 13915–13924.

Flory, F.; Escoubas, L.; Berginc, G. Optical Properties of Nanostructured Materials: A Review. *J. Nanophoton.* **2011,** *5*, 052502–052522.

Förster, T. Intermolecular Energy Migration and Fluorescence. *Ann. Phys.* **1948,** *2*, 55–75.

Förster, T. 10th Spiers Memorial Lecture. Transfer Mechanisms of Electronic Excitation *Disc. Faraday Soc.* **1959,** *27*, 7–17.

Fu, Y.; Li, J.; Li, J. Fundamentals, Structures, Applications and Properties. *Nanomaterials* **2019,** *9* (3). DOI: 10.3390/nano9030359.

Gans, R. Über die form ultramikroskopischer goldteilchen. *Ann. Phys.* **1912,** *342*, 881–900.

Garcia, M. A. Surface Plasmons in Metallic Nanoparticles: Fundamentals and Applications. *J. Phys. D: Appl. Phys.* **2011,** *44*, 283001–283021.

Gatoo, M. A.; Naseem, S.; Arfat, M. Y.; Mahmood Dar, A.; Qasim, K.; Zubair, S. Physicochemical Properties of Nanomaterials: Implication in Associated Toxic Manifestations. *Biomed. Res. Int.* **2014,** *2014*. DOI: 10.1155/2014/498420.

Gleiter, H. Nanostructured Materials: Basic Concepts and Microstructure. *Acta Mater.* **2000,** *48*, 1–29.

Gough, D. V.; Juhl, A. T.; Braun, P. V. Programming Structure into 3D Nanomaterials. *Mater. Today* **2009,** *12*, 28–35.

Griffin, S.; Masood, M. I., Nasim, M. J., Sarfraz, M., Ebokaiwe, A. P., Schäfer, K. et al. Natural Nanoparticles: A Particular Matter Inspired by Nature. *Antioxidants* **2018,** *7*, 2–21.

Guo, C. F.; Sun, T.; Cao, F.; Liu, Q.; Ren, Z. Metallic Nanostructures for Light Trapping in Energy-Harvesting Devices. *Light Sci. Appl.* ***2014,*** *3*. DOI: 10.1038/lsa.2014.42.

Haase, M.; Weller, H.; Henglein, A. Photochemistry and Radiation Chemistry of Colloidal Semiconductors. 23. Electron Storage on Zinc Oxide Particles and Size Quantization. *J. Phys. Chem.* **1988,** *92*, 482–487.

Hajipour, M. J.; Fromm, K. M.; Ashkarran, A. A.; de Aberasturi, D. J.; de Larramendi, I. R.; Rojo, T. et al. Inorganic Nanoparticles and the Microbiome. *Trends Biotechnol.* **2012,** *30*, 499–511.

Han, P.; Martens, W.; Waclawik, E. R.; Sarina, S.; Zhu, H.; Metal Nanoparticle Photocatalysts: Synthesis, Characterization, and Application. *Part. Part. Syst. Charact.* **2018.** DOI: 10.1002/ppsc.201700489.

He, J.; Hu, S.; Ren, J.; Cheng, X.; Hu, Z. Wang, N. Biofluidic Random Laser Cytometer for Biophysical Phenotyping of Cell Suspensions. *ACS Sens.* **2019,** *4* (4), 832–840.

Hoang, D. V.; Phuong, N. T.; Phu, N. V. Random Lasers: Characteristics, Applications and Some Research Results. *CMST* **2010,** *2*, 47–51.

Homola, J. Surface Plasmon Resonance Sensors for Detection of Chemical and Biological Species. *Chem. Rev.* **2008,** *108* (2), 462–493.

Hong, Y.; Huh, Y. M.; Yoon, D. S.; Yang, J. Nanobiosensors Based on Localized Surface Plasmon Resonance for Biomarker Detection. *J. Nanomater.* **2012,** *2012*. DOI: 10.1155/2012/759830

Huang, C.; Chen, X.; Xue, Z.; Wang, T. Effect of Structure: A New Insight Into Nanoparticle Assemblies from Inanimate to Animate. *Sci. Adv.* **2020a,** *6*. DOI: 10.1126/sciadv.aba1321

Huang, C.; Wen, T.; Shi, F.; Zeng, X.; Jiao, Y. Rapid Detection of IgM Antibodies against the SARS-CoV-2 Virus via Colloidal Gold Nanoparticle-Based Lateral-Flow Assay. *ACS Omega* **2020b,** *5* (21), 12550–12556.

Jackson, R.; Oda, R. P.; Bhandari, R. K.; Mahon, S. B.; Brenner, M.; Rockwood et al. Development of a Fluorescence-Based Sensor for Rapid Diagnosis of Cyanide Exposure. *Anal. Chem.* **2014,** *86* (3), 1845–1852.

Jagadish, C.; Nanowires. *Semicond. Sci. Technol.* **2010,** *25*. DOI: 10.1088/0268-1242/25/2/020301.

Jarvis, M.; Krishnan, V.; Mitragotri, S. Nanocrystals: A Perspective on Translational Research and Clinical Studies. *Bioeng. Transl Med.* **2019,** *4* (1), 5–16.

Jeevanandam, J.; Barhoum, A.; Chan, Y. S.; Dufresne, A.; Danquah, M. K. Review on Nanoparticles and Nanostructured Materials: History, Sources, Toxicity and Regulations *Beilstein J. Nanotechnol.* **2018,** *9*, 1050–1074.

Jeong, Y.; Kook, Y.; Lee, K.; Koh, W. Metal Enhanced Fluorescence (MEF) for Biosensors: General Approaches and a Review of Recent Developments. *Biosens. Bioelectron.* **2018,** *111*, 102–116.

Jiang, C.; Zhang, W.; Zou, G.; Yu, W.; Qian, Y. Hydrothermal Synthesis and Characterization of ZnS Microspheres and Hollow Nanospheres. *Mater. Chem. Phys.* **2007,** *103*, 24–27.

Jones, G. A.; Bradshaw, D. S. Resonance Energy Transfer: From Fundamental Theory to Recent Applications. *Front. Phys.* **2019,** *7*. DOI: 10.3389/fphy.2019.00100.

Kar, S.; Chaudhuri, S. Controlled Synthesis and Photoluminescence Properties of ZnS Nanowires and Nanoribbons. *J. Phys. Chem. B* **2005,** *109*, 3298–3302.

Karmakar, S.; Biswas, S.; Kumbhakar, P. Low Power Continuous-Wave Nonlinear Optical Effects in MoS_2 Nanosheets Synthesized by Simple Bath Ultrasonication. *Opt. Mater.* **2017,** *73*, 585–594.

Kim, H.; Achermann, M.; Balet, L. P.; Hollingsworth, J. A.; Klimov, V. I. J. Synthesis and Characterization of Co/CdSe Core/Shell Nanocomposites: Bifunctional Magnetic-Optical Nanocrystals. *J. Am. Chem. Soc.* **2005,** *127*, 544–546.

Koa, W.; Lee, H. S. Development of Specific l-Methionine Sensors by FRET-Based Protein Engineering. *RSC Adv*. **2019,** *9*, 15648–15656.

Koga, H.; Kitaoka, T.; Wariishi, H. In Situ Synthesis of Silver Nanoparticles on Zinc Oxide Whiskers Incorporated in a Paper Matrix for Antibacterial Applications. *J. Mater. Chem.* **2009,** *19*, 2135–2140.

Kole, A. K.; Biswas, S.; Tiwary, C. S.; Kumbhakar, P. A Facile Synthesis of Graphene Oxide–ZnS/ZnO Nanocomposites and Observations of Thermal Quenching of Visible Photoluminescence Emission and Nonlinear Optical Properties. *J. Lumin.* **2016,** *179*, 211–221.

Kole, A. K.; Kumbhakar, P. Cubic-to-Hexagonal Phase Transition and Optical Properties of Chemically Synthesized ZnS Nanocrystals. *Results Phys.* **2012,** *2*, 150–155.

Koprucki, T.; Kaiser, H. C.; Fuhrmann, J. Electronic States in Semiconductor Nanostructures and Upscaling to Semi-Classical Models. In *Analysis, Modeling and Simulation of Multiscale Problems*; Mielke, A., Eds.; Springer: Berlin, Heidelberg, 2006.

Kreibig, U.; Vollmer, M. *Optical Properties of Metal Clusters*; Springer: Berlin, 1995.

Laing, S.; Jamieson, L. E.; Faulds, K.; Graham, D. Surface-Enhanced Raman Spectroscopy for in Vivo Biosensing. *Nature Rev. Chem.* **2017,** *1*. DOI: 10.1038/s41570-017-0060.

Lawandy, N. M.; Balachandra, R. M.; Gomes, A. S. L.; Sauvain, E. Laser Action in Strongly Scattering Media. *Nature* **1994,** *368*, 436–438.

Lee, J.; Govorov, A. O.; Dulka, J.; Kotov, N. A. Bioconjugates of CdTe Nanowires and Au Nanoparticles: Plasmon-Exciton Interactions, Luminescence Enhancement, and Collective Effects. *Nano Lett.* **2004,** *4*, 2323–2330.

Letokhov, V. S. Generation of Light by a Scattering Medium with Negative Resonance Absorption *JETP* **1968,** *26*, 835–840.

Li, B.; Li, X.; Dong, Y.; Wang, B.; Li, D.; Shi, Y. et al. Colorimetric Sensor Array Based on Gold Nanoparticles with Diverse Surface Charges for Microorganisms Identification. *Anal. Chem.* **2017,** *89* (20), 10639–10643.

Li, F.; Jiang, Y.; Hua, L.; Liua, L.; Li, Z.; Huang, X. J. Structural and Luminescent Properties of ZnO Nanorods and ZnO/ZnS Nanocomposites. *J. Alloys. Compd.* **2009,** *474*, 531–535.

Li, S.; Shiying, L.; Kwok-Ho, L.; Tik-Hung, T.; Wai-Sum, L.; Xin, L.; Xuesen, H. et al. pH-Responsive Targeted Gold Nanoparticles for in Vivo Photoacoustic Imaging of Tumor Microenvironments. *Nanoscale Adv.* **2019,** *1*, 554–564.

Li, W.; Chen, X. Gold Nanoparticles for Photoacoustic Imaging. *Nanomedicine* **2015,** *10* (2), 299–320.

Liang, Y.; Wang, H.; Casalongue, H.; Chen, Z.; Dai, H. TiO_2 Nanocrystals Grown on Graphene as Advanced Photocatalytic Hybrid Materials. *Nano Res.* **2010,** *3*, 701–705.

Link, S.; El-Sayed, M., Additions and Corrections to Simulation of the Optical Absorption Spectra of Gold Nanorods as a Function of Their Aspect Ratio and the Effect of the Medium Dielectric Constant. *J. Phys. Chem. B* **2005,** *109*, 10531–10532.

Liu, Y.; Yang, W.; Xiao, S.; Zhang, N.; Fan, Y.; Qu, G.; et al., Surface-Emitting Perovskite Random Lasers for Speckle-Free Imaging. *ACS Nano* **2019,** *13*, 10653–10661.

Lorber, B.; Fischer, F.; Bailly, M.; Kern, D. Protein Analysis By Dynamic Light Scattering: Methods and Techniques for Students *Biochem. Mol. Biol. Educ.* **2012,** *40* (6), 372–382.

Lu, F.; Cai, W.; Zhang, Y. ZnO Hierarchical Micro/Nanoarchitectures: Solvothermal Synthesis and Structurally Enhanced Photocatalytic Performance. *Adv. Funct. Mater.* **2008a,** *18*, 1047–1056.

Lu, W.; Gao, S.; Wan, J. One-Pot Synthesis of Ag/ZnO Self-Assembled 3D Hollow Microspheres with Enhanced Photocatalytic Performance. *J. Phys. Chem. C*, **2008b,** *112*, 16792–16800.

Lu, W.; Liu, G.; Gao, S.; Xing, S.; Wang, J. Tyrosine-Assisted Preparation of Ag/ZnO Nanocomposites with Enhanced Photocatalytic Performance and Synergistic Antibacterial Activities. *Nanotech.* **2008c,** *19*, 445711–445721.

Malekzad, H.; Zangabad, P. S.; Mirshekari, H.; Karimi, M.; Hamblin, M. R. Noble Metal Nanoparticles in Biosensors: Recent Studies and Applications. *Nanotechnol. Rev.* **2017,** *6* (3), 301–329.

Martinsson, E.; Otte, M. A.; Shahjamali, M. M.; Sepulveda, B.; Aili, D. Substrate Effect on the Refractive Index Sensitivity of Silver Nanoparticles. *J. Phys. Chem. C* **2014,** *118* (42), 24680–24687.

McGrath, M. J.; Scanaill, C. N.; Sensing and Sensor Fundamentals. In *Sensor Technologies*; McGrath, M. J., Scanaill, C. N., Eds.; Apress: Berkeley, CA, 2013.

Mérai, L.; Janovák, L.; Kovács, D. S.; Szenti, I.; Vásárhelyi, L.; Kukovecz, Á. et al. Fast Optical Method for Characterizing Plasmonic Nanoparticle Adhesion on Functionalized Surfaces. *Anal. Bioanal. Chem.* **2020,** *412*, 3395–3404.

Mie, G. Articles on the Optical Characteristics of Turbid Tubes, Especially Colloidal Metal Solutions. *Ann. Phys.* **1908,** *25*, 377–445.

Mills, A.; Hunte, S. An Overview of Semiconductor Photocatalysis. *J. Photochem. Photobiol. A* **1997,** *108*, 1–35.

Mody, V. V.; Siwale, R.; Singh, A.; Mody, H. R. Introduction to Metallic Nanoparticles. *J. Pharm. Bioallied. Sci.* **2010,** *2* (4), 282–289.

Moon, R. J.; Martini, A.; Nairn, J.; Simonsen, J.; Youngblood, J. Cellulose Nanomaterials Review: Structure, Properties and Nanocomposites. *Chem. Soc. Rev.* **2011,** *40*, 3941–3994.

Mourdikoudis, S.; Pallares, R. M.; Thanh, N. T. K. Characterization Techniques for Nanoparticles: Comparison and Complementarity upon Studying Nanoparticle Properties. *Nanoscale* **2018,** *10*, 12871–12934.

Nayak, S.; Maiya, A.; Hande, M. Influence of Aerobic Treadmill Exercise on Blood Glucose Homeostasis in Noninsulin Dependent Diabetes Mellitus Patients. *Indian J. Clin. Biochem.* **2005,** *20*, 47–51.

Nirmal, M.; Brus, L. Luminescence Photophysics in Semiconductor Nanocrystals. *Acc. Chem. Res.* **1999,** *32*, 407–414.

Niu, H.; Tang, J.; Zhu, X.; Li, Z.; Zhang, Y.; Ye, Y.; Zhao, Y. A Three-Channel Fluorescent Probe to Image Mitochondrial Stress. *Chem. Commun.* **2020,** *56*, 7710–7713.

Nobbmann, U.; Morfesis, A. Light Scattering and Nanoparticles. *Mater. Today* **2009,** *12*, 52–54.

Noginov, M. *Solid-State Random Lasers. Springer Series in Optical Sciences*, Vol. 105; Springer: New York, 2005.

Noguez, C. Surface Plasmons on Metal Nanoparticles: The Influence of Shape and Physical Environment. *J. Phys. Chem. C* **2007,** *111*, 3806–3819.

Pallares, R. M.; Stilson, T.; Choo, P.; Hu, J.; Odom, T. W. Using Good's Buffers to Control the Anisotropic Structure and Optical Properties of Spiky Gold Nanoparticles for Refractive Index Sensing. *ACS Appl. Nano Mater.* **2019,** *2* (8), 5266–5271.

Pankove, J. I. *Optical Properties in Semiconductors*; Dover Publications, Inc.: New York, 1971; p 422.

Parkash, V.; Kulkarni, A. K. Optical Absorption Characteristics of Silicon Nanowires for Photovoltaic Applications. *IEEE Trans. Nanotechnol.* **2011,** *10*, 1293–1297.

Patra, J. K.; Das, G.; Fraceto, L. F.; Campos, E. V. R.; del Pilar Rodriguez-Torres, M. et al. Nano Based Drug Delivery Systems: Recent Developments and Future Prospects. *J. Nanobiotechnol.* **2018,** *16*. DOI: 10.1186/s12951-018-0392-8.

Pinheiro, T.; Ferrão, J.; Marques, A. C.; Oliveira, M. J.; Batra, N. M., Costa, P. M. et al. Based In-Situ Gold Nanoparticle Synthesis for Colorimetric, Non-Enzymatic Glucose Level Determination. *Nanomaterials* **2020,** *10* (10). DOI: 10.3390/nano10102027.

Pokropivny, V. V.; Skorokhod, V. V. Classification of Nanostructures by Dimensionality and Concept of Surface Forms Engineering in Nanomaterial Science. *Mater. Sci. Eng.: C* **2007,** *27*, 990–993.

Priyadarshini, E.; Pradhan, N. Gold Nanoparticles as Efficient Sensors in Colorimetric Detection of Toxic Metal Ions: A Review. *Sens. Actuators B: Chem.* **2017,** *238*, 888–902.

Redding, B.; Choma, M. A.; Cao, H. Speckle-Free Laser Imaging Using Random Laser Illumination. *Nature Photon.* **2012,** *6*, 355–359.

Saha, K.; Agasti, S. S.; Kim, C.; Li, X.; Rotello, V. M. Gold Nanoparticles in Chemical and Biological Sensing. *Chem. Rev.* **2012,** *112*, 2739–2779.

Santra, B.; Shneider, M. N.; Car, R. In Situ Characterization of Nanoparticles Using Rayleigh Scattering, *Sci. Rep.* **2017,** *7*. DOI: 10.1038/srep40230.

Sharma, F. M.; Singh, F. S.; Pandey, O. P. Excitation Induced Tunable Emission in Biocompatible Chitosan Capped ZnS Nanophosphors. *J. Appl. Phys.* **2010,** *107*, 104319–104328.

Shi, X.; Ge, K.; Tong, J.; Zhai, T. Low-Cost Biosensors Based on a Plasmonic Random Laser on Fiber Facet. *Opt. Exp.* **2020,** *28*, 12233–12242.

Shiohara, A.; Langer, J.; Polavarapu, L.; Liz-Marzán, L. M. Solution Processed Polydimethylsiloxane/Gold Nanostar Flexible Substrates for Plasmonic Sensing. *Nanoscale* **2014,** *6*, 9817–9823.

Sivakumar, P.; Kumar, G. K. G.; Sivakumar, P.; Renganathan, S. Studies on Sono-Chemical Biodiesel Production Using Smoke Deposited Nano MgO Catalyst. *J. Nanostruct. Chem.* **2014,** *4*, 107–115.

Smith, A. M.; Nie, S. Semiconductor Nanocrystals: Structure, Properties, and Band Gap Engineering. *Acc. Chem. Res.* **2010,** *43* (2), 190–200.

Soares, L.; Csáki, A.; Jatschka, J.; Fritzsche, W.; Flores, O.; Franco, R. et al. Localized Surface Plasmon Resonance (LSPR) Biosensing Using Gold Nanotriangles: Detection of DNA Hybridization Events at Room Temperature. *Analyst* **2014,** *139*, 4964–4973.

Sobakinskaya, E.; Busch, M. S.; Renger, T. Theory of FRET "Spectroscopic Ruler" for Short Distances: Application to Polyproline. *J. Phys. Chem. B* **2018,** *122* (1), 54–67.

Son, Y.; Park, M.; Son, Y.; Lee, J. S.; Jang, J. H.; Kim, Y. et al. Quantum Confinement and Its Related Effects on the Critical Size of GeO_2 Nanoparticles Anodes for Lithium Batteries. **2014,** *14* (2), 1005–1010.

Song, Q.; Xu, Z.; Choi, S. H.; Sun, X.; Xiao, S.; Akkus, O.; Kim, Y. L. Detection of Nanoscale Structural Changes in Bone Using Random Lasers. *Biomedical. Opt. Exp.* **2010,** *1*, 1401–1407.

Sreeprasad, T. S., Pradeep, T. Noble Metal Nanoparticles. In *Handbook of Nanomaterials*; Vajtai, R., Eds.; Springer: Berlin, Heidelberg, 2013.

Stetefeld, J.; McKenna, S. A.; Patel, T. R. Dynamic Light Scattering: A Practical Guide and Applications in Biomedical Sciences. *Biophys. Rev.* **2016,** *8* (4), 409–427.

Strianese, M.; Staiano, M.; Ruggiero, G.; Labella, T.; Pellecchia, C.; D'Auria, S. Fluorescence-Based Biosensors. *Methods Mol. Biol.* **2012,** *875*, 193–216.

Subash, B.; Krishnakumar, B.; Swaminathan, M.; Shanthi, M. ZnS–Ag–ZnO as an Excellent UV-Light-Active Photocatalyst for the Degradation of AV 7, AB 1, RR 120, and RY 84 Dyes: Synthesis, Characterization, and Catalytic Applications. *Ind. Eng. Chem. Res.* **2014,** *53*, 12953–12963.

Suber, L.; Imperatori, P.; Pilloni, L.; Caschera, D.; Angelini, N.; Mezzi, A. et al. Nanocluster Superstructures or Nanoparticles? The Self-Consuming Scaffold Decides. *Nanoscale* **2018,** *10*, 7472–7483.

Szunerits, S.; Boukherroub, R. Sensing Using Localised Surface Plasmon Resonance Sensors. *Chem. Commun.* **2012,** *48*, 8999–9010.

Tauc, J.; Menth, A. States in the Gap. *J. Non-Cryst. Solids* **1972,** *569*, 8–10.

Teengam, P.; Siangproh, W.; Tuantranont, A.; Vilaivan, T.; Chailapakul, O; Henry, C. S. Multiplex Paper-Based Colorimetric DNA Sensor Using Pyrrolidinyl Peptide Nucleic Acid-Induced AgNPs Aggregation for Detecting MERS-CoV, MTB, and HPV Oligonucleotides. *Anal. Chem.* **2017,** *89* (10), 5428–5435.

Tommasi, F.; Ignesti, E.; Fini, L.; Martelli, F.; Cavalieri, S. Random Laser Based Method for Direct Measurement of Scattering Properties. *Opt. Exp.* **2018,** *26*, 27616–27627.

Uddin, Md. T.; Hoquea, Md. E.; Bhoumick, M. C. Facile One-Pot Synthesis of Heterostructure SnO_2/ZnO Photocatalyst for Enhanced Photocatalytic Degradation of Organic Dye. *RSC Adv.* **2020,** *10*, 23554–23565.

Van Dijk, M. A.; Tchebotareva, A. L.; Orrit, M.; Lippitz, M.; Berciaud, S.; Lasne, D. et al. Absorption and Scattering Microscopy of Single Metal Nanoparticles. *Phys. Chem. Chem. Phys.* **2006,** *8*, 3486–3495.

Wang, G.; Huang, B.; Li, Z.; Lou, Z.; Wang, Z.; Dai, Y. et al. Synthesis and Characterization of ZnS with Controlled Amount of S Vacancies for Photocatalytic H_2 Production under Visible Light. *Sci. Rep.* **2015,** *5*, 8544–8550.

Wang, H. Plasmonic Refractive Index Sensing Using Strongly Coupled Metal Nanoantennas: Nonlocal Limitations. *Sci. Rep.* **2018,** *8*, 9589–9598.

Wang, X.; Shi, J.; Feng, Z.; Lia, M.; Li, C. Visible Emission Characteristics from Different Defects of ZnS Nanocrystals. *Phys. Chem. Chem. Phys.* **2011,** *13*, 4715–4723.

Wang, Z.; Hu, T.; Liang, R.; Wei, M. Application of Zero-Dimensional Nanomaterials in Biosensing. *Front. Chem.* **2020,** *8*. DOI: 10.3389/fchem.2020.00320.

Wang, Z.; Meng, X.; Kildishev, A. V.; Boltasseva, A.; Shalaev, V. M. Nanolasers Enabled by Metallic Nanoparticles: From Spasers to Random Lasers. *Laser Photonics Rev.* **2017.** DOI: 10.1002/lpor.201700212.

Wei, H.; Reyes-Coronado, A.; Nordlander, P.; Aizpurua, J.; Xu, H. Multipolar Plasmon Resonances in Individual Ag Nanorice. *ACS Nano* **2010,** *4*, 2649–2654.

Wiersma, D. S. The Physics and Applications of Random Lasers. *Nat. Phys.* **2008,** *4*, 359–367.

Wu, Q.; Miao, W.; Zhang, Y.; Gao, H.; Hui, D. Mechanical Properties of Nanomaterials: A Review. *Nanotechnol. Rev.* **2020,** *9*, 259–273.

Xie, J.; Wu, Q. Mater. One-Pot Synthesis of ZnO/Ag Nanospheres with Enhanced Photocatalytic Activity. *Letters* **2010,** *64*, 389–392.

Yang, C. C.; Jiang, Q. Size Effect on the Bandgap of II–VI Semiconductor Nanocrystals. *Mater. Sci. Eng. B* **2006,** *131* (1–3), 191–194.

Yang, F.; Song, P.; Ruan, M.; Xu, W. Recent Progress in Two-Dimensional Nanomaterials: Synthesis, Engineering, and Applications. *Flat Chem.* **2019,** *18*, 100133.

Yao, H.; Fukuia, C. Size and Morphology Effects on the Fluorescence Properties of π-Conjugated Poly(p-phenylene) Polyelectrolyte Nanoparticles Synthesized via Polyion Association. *J. Mater. Chem. C* **2016,** *4*, 2945–2953.

Yeshchenko, O. A.; Bondarchuk, I. S.; Losytskyy, Y. M.; Alexeenko, A. A. Temperature Dependence of Photoluminescence from Silver Nanoparticles. *Plasmonics* **2014,** *9*, 93–101.

Yetisen, A. K.; Montelongo, Y.; Qasim, M. M.; Butt, H.; Wilkinson, T. D.; Monteiro, M. J. et al. Photonic Nanosensor for Colorimetric Detection of Metal Ions. *Anal. Chem.* **2015,** *87*, 10, 5101–5108.

Yu, S. H.; Yoshimura, M. Shape and Phase Control of ZnS Nanocrystals: Template Fabrication of Wurtzite ZnS Single-Crystal Nanosheets and ZnO Flake-Like Dendrites from a Lamellar Molecular Precursor $ZnS{\cdot}(NH_2CH_2CH_2NH_2)_{0.5}$. *Adv. Mater*. **2002,** *14*, 296–300.

Yurkov, G. Y.; Fionov, A. S.; Koksharov, Y. A.; Koleso, V. V.; Gubin, S. P. Electrical and Magnetic Properties of Nanomaterials Containing Iron or Cobalt Nanoparticles *Inorg. Mater.* **2007,** *43*, 834–844.

Zak, A. K.; Majid, W. H. A.; Abrishami, M. E.; Yousef, R. X-ray Analysis of ZnO Nanoparticles by Williamson–Hall and Size–Strain Plot Methods. *Solid State. Sci.* **2011,** *13*, 251–256.

Zamiri, R.; Ahangar, A. H.; Tobaldi, D. M.; Rebelo, A.; Seabra, M. P.; Shabania, M. et al. *Phys. Chem. Chem. Phys*. **2014,** *16*, 22418–22425.

Zayats, A. V.; Smolyaninov, I. I.; Maradudin, A. A. Nano-Optics of Surface Plasmon Polaritons. *Phys. Rep*. **2005,** *408*, 131–314.

Zhang, H.; Ruochuan, L.; Tan, Y.; Xie, W. H.; Lei, H.; Cheung, H. Y. et al. A FRET-based ratiometric fluorescent probe for nitroxyl detection in living cells. *ACS Appl. Mater. Interfaces*, **2015a,** 7, 9, 5438–5443.

Zhang, D.; Pu, X.; Li, H.; Yu, Y. M.; Shim, J. J.; Cai, P. et al. Microwave-Assisted Combustion Synthesis of Ag/ZnO Nanocomposites and Their Photocatalytic Activities Under Ultraviolet and Visible-Light Irradiation. *Mater. Res. Bull*. **2015b,** *61*, 321–325.

Zhang, S.; Assaf, K. I.; Huang, C.; Hennig, A.; Nau, W. M. FRET-Based Ratiometric Fluorescent Probe for Nitroxyl Detection in Living Cells. *Chem. Commun*. **2019,** *55*, 671–674.

Zhang, Y.; Wang, Y. Nonlinear Optical Properties of Metal Nanoparticles: A Review. *RSC Adv.* **2017,** *7*, 45129–45144.

Ziashahabi, A.; Prato, M.; Dang, Z.; Poursalehi, R.; Naseri, N. The Effect of Silver Oxidation on the Photocatalytic Activity of Ag/ZnO Hybrid Plasmonic/Metal-Oxide Nanostructures under Visible Light and in the Dark. *Sci. Rep*. **2019,** *9*. DOI: 10.1038/s41598-019-48075-7.

CHAPTER 3

Nanotechnology for Selective Targeting and Cancer Drug Delivery

C. O. ADETUNJI[1*], O. T. OLANIYAN[2], O. O. AJAYI[3], O. A. ANANI[4], R. E. BODUNRINDE[5], A. INOBEME[6], J. B. ADETUNJI[7], and D. I. HEFFT[8]

[1]*Applied Microbiology, Biotechnology and Nanotechnology Laboratory, Department of Microbiology, Edo University Iyamho, PMB 04, Auchi, Edo State, Nigeria*

[2]*Laboratory for Reproductive Biology and Developmental Programming, Department of Physiology, Edo University Iyamho, Nigeria*

[3]*Department of Biochemistry, Edo University Iyamho, Edo State, Nigeria*

[4]*Laboratory of Ecotoxicology and Forensic Biology, Department of Biological Science, Faculty of Science, Edo University, Iyamho, Edo State, Nigeria*

[5]*Department of Microbiology, Federal University of Technology Akure, Nigeria*

[6]*Department of Chemistry, Edo University Iyamho, Nigeria*

[7]*Nutritional and toxicological Research Laboratory, Department of Biochemistry Sciences, Osun State University, Osogbo, Nigeria*

[8]*Edgbaston Campus, University of Birmingham, School of Chemical Engineering, Birmingham B15 2TT, UK*

[*]*Corresponding author. E-mail: adetunji.charles@edouniversity.edu.ng*

ABSTRACT

Several scientific breakthrough and inventions that could mitigate against uncountable barriers normally experienced in traditional drugs delivery system are utilized as therapeutic candidate for the management of cancer diseases.

The Science of Nanomaterials: Basics and Applications. Suresh C. Ameta & Rakshit Ameta (Eds.)

It has been observed that conventional chemotherapeutics are associated with several side effects and high level of toxicity, entailing impairment of the immune system as well as having numerous detrimental effects on organ with quick penetrating cells that might be linked to the absence of solubility, nonspecific targeting, and incapability of the drug to enter the core of the tumors, which eventually lead to compromised treatment with a decrease dose and with little subsistence rate. Nanotechnology has a very great prospect with direct impact on cancerous cells as well as higher selectively with enhanced cellular uptake and drug localization. Moreover, nanoparticles could be automated for accurate identification of cancerous cells permitting selective and more precise drug delivery, which could prevent interface with the healthy cells. This chapter provides detailed information on the cell identification capability of nanomaterials with the aids of special strategies that could differentiate them from the anticancer therapies. Special highlights are provided on specific nanodrugs that are highly effective for target delivery of the active constituents against numerous cancerous cells and how they could ameliorate several side effects of unadventurous therapies.

3.1 INTRODUCTION

Cancer is an age-long health challenge with an increasing global mortality rate. By 2030, about 30 million annual cancer deaths have been predicted (Li et al., 2020). Chemotherapy is one of the potent therapies in the management of cancer. Its minimal tumor-targeting ability is a limitation to its clinical efficacy (Li et al., 2020). Furthermore, multidrug resistance and severe side effects are peculiar to current cancer therapies (Shao et al., 2020). This necessitates the fabrication of drug delivery systems with the characteristics such as elevated concentration in the tumor tissues, absence of leakage while in circulation, appropriate release, and bioavailability upon the activation of the release stimuli.

Nanotechnology offers hope of better cancer diagnosis, monitoring, and therapy. Nanoparticles easily penetrate tumor tissues consequently on the improved permeability and retention influence of tumor blood vessels (Shao et al., 2020). This enhances drug accumulation at the tumor sites and conversely reduces drug distribution in noncancerous tissues (Chari, 1998).

Furthermore, response to pH, redox, enzymes, and reactive oxygen species (ROS) are also important features of nanocarriers, which improves drug release at the target sites (Hirata and Sahai, 2017; Guan et al., 2017).

There are evidence that sensitivity to pH, ROS, and enzymes could ensure adequate drug delivery at tumor sites (Li et al., 2020). A comprehensive review has been made on the application of several types of nanoparticles used in cancer diagnosis and treatment.

3.2 DIFFERENT TYPES OF NANOMATERIALS FOR TREATMENT OF CANCER DISEASES

3.2.1 VIRAL NANOPARTICLES

Viral nanoparticles (VNPs) are bionanomaterials with myriads of benefits over synthetic nanomaterials. They are obtained from the human papillomavirus, poliovirus, influenza, hepatitis E, and other viruses (Krause et al., 2018). They are low cost, highly biodegradable, stable, and biocompatible. They are suitable for drug delivery because of their versatility, absence of endosomal sequestering, and efficient cell entry (Krause et al., 2018; Esfandiari et al., 2018). In spite of these benefits, it has been reported that VNPs could induce innate immunity because of its viral protein particle (Grasso and Santi, 2010). A lot of VNPs show tropism to heparin sulfates and sialic acids, which reduce its use as a targeted nanocarrier (Krause et al, 2018).

The application of VNPs in cancer therapeutics is evolving. Bacteriophage MS2 VLP, altered with SP94 peptide, and encapsulated with doxorubicin/cisplatin/5 fluorouracil to inhibit human hepatocarcinoma cell is a typical example of VLP as a targeted therapeutic carrier (Ashley et al., 2011). Potato virus X (PVX) conjugated with HER (PVX-HER) was more effective in the arrest of HER2+ breast cancer cells in comparison with free Herceptin (Esfandiari et al., 2016). The conjugation of PVX with HER averted pathogenesis of the virus in plants, which raises hope of a similar effect in animals (Esfandiari et al., 2018). Its efficiency in Herceptin delivery to target cells after attaching to the virus produces PVX-HER nanoparticle.

3.2.2 HER-2 POSITIVE BREAST CANCER AND NANOTECHNOLOGY

Breast cancer is the most common female neoplasia known for its heterogeneity. HER-2 positive breast cancer overexpresses HER2. It has been reported to have a high rate of metastasis (White et al., 2020). Nanotechnology appears to be the panacea for the adverse effects of traditional therapies. Several reports showed the suitability of nanoparticles; however, the response of tumor cells to nanoparticles varies. Furthermore, the mechanisms of nanoparticles depend on their size, shape, and surface chemistry (Panariti et al., 2012). Hence, it is imperative to comprehend, how nanoparticles correlate with individual cell?

There are reports that cancer cells and noncancerous cells interact with surface modifications via dissimilar mechanisms of endocytosis. Polystyrene microparticles enhanced with fluorescent antibody using two types of polyethylenimine of different molecular weights and structures. Cancer cells preferred positively charged microparticles via micropinocytosis, while noncancerous cells preferred negatively charged microparticle (Patiño et al., 2015). Shape and surface modification could be major factors affecting the cellular uptake of gold nanoparticles. Cellular uptake of magnetic dextran-spermine nanoparticles conjugated with anti-HER2 antibody by HER 2 positive cells was assessed. More antibody-conjugated nanomaterials were taken up by the HER2 positive breast cancer cells, while none was taken up by the noncancerous cells (Avazzadeh et al., 2017).

There are reports of the use of dendrimeric nanoparticles to target HER-2 positive breast cancer (Miyano, 2020; Chan et al., 2013). The cellular uptake of docetaxel-loaded dendrimers and trastuzumab grafted joined to fluorescein isothiocyanate was studied in HER2 positive and negative breast cancer cells. More uptake of the nanoparticles occurred in the HER2 positive breast cancer cells relative to HER2 negative breast cancer cells (Kulhari et al., 2016).

3.2.3 IRON OXIDE NANOPARTICLES

Iron oxide nanoparticles were loaded in chitosan-conjugated pluronic-based nanocarriers and further conjugated with doxorubicin. A better targeted therapeutic effect as well as better imaging was observed in the cancer cells (Choi et al., 2014). Peptides may also be conjugated with nanoparticles for targeted chemotherapies. Report shows the use of Abraxane, which is an albumin-based nanoparticle laden with paclitaxel. Its efficacy, as well as low toxicity, were observed. However, its large size and absence of targeting ligand could be a demerit to its use (White et al., 2020). Iron oxide nanoparticle conjugated with anti-HER2 peptide (AHNP) laden with paclitaxel (AHNP-IONPPTX) appeared to resolve some of the drawbacks mentioned earlier (Mu et al., 2015).

3.2.4 HER2 SPECIFIC NANOBUBBLES

This is a new method that has application in cancer therapeutics. It uses nanobubbles with PEGylated phospholipid shell enhanced using Herceptin specifically for HER2 positive breast cancer (Jiang et al., 2016). It specifically targets the HER2 receptor. In addition to targeted drug delivery, it also improves contrast for diagnostic imaging.

3.2.5 MINI-NANODRUGS SPECIFIC FOR HER2

Enhanced effectiveness of trastuzumab conjugated with poly(beta-1-malic acid) and antisense oligonucleotide, that is specific for HER2-mRNA, has been reported (Ding et al., 2017). HER 2-mRNA-annealing morpholino antisense oligonucleotide produced is specific for HER 2 via anti-HER2/neu protein as a substitute for trastuzumab. Mininanodrugs offer a lot of benefits including enhanced drug delivery to target sites, safety, nonimmunogenicity, cost effectiveness, biodegradability, and stability (White et al., 2020).

3.2.6 LIPOSOMAL NANOPARTICLES

Liposomes are sphere-shaped structures with hydrophilic core enclosed by lipophilic bilayers. They are made of cholesterol and phospholipids. Their dual "philicities" make them multipurpose nanocarriers (Thakur and Kutty, 2019). Minimal toxicity, high loading ability, regulated drug release, stability, and targeted drug delivery are some qualities that make them ideal as nanocarriers (Montaseri et al., 2020; Zamboni, 2005). Chitosan-modified ursolic acid (UA) liposomes (CS-UA-L) was reported to have low adverse effects, high tumor targeting/directing ability, and regulated drug release (Wang et al., 2017a).

Moreover, tissue accumulation of UA in mice administered CS-UA-L was higher than mice administered free UA (Wang et al., 2017b). Ligand-targeted UA liposomes have been reported to have low side effects as well as high drug efficiency (Sapra et al., 2005). UA-loaded FA receptor-targeted liposomes aimed at KB cell lines via the FA receptor-mediated pathway (Yang et al., 2014). Polyethylene glycol (PEG)-modified–UA liposomes were described to have a higher drug-releasing potential as well as a higher inhibitory cell rates than free UA at the same concentration (Zhao et al., 2015).

The overexpression of integrin-α3 in triple-negative breast cancer (TNBC) was targeted with cyclic octapeptide LXY (Cys-Asp-Gly-Phe

(3,5-DiF)-Gly-Hyp-Asn-Cys) linked liposomes conveying doxorubicin and rapamycin. Better effectiveness was achieved with the dual drugs than the free drug (Dai et al., 2014). Furthermore, doxorubicin- and sorafenib-loaded liposomes were reported to exhibit more antineoplastic effects on TNBC cells (Lee et al., 2010a). Doxorubicin liposomal preparations are associated with cardiotoxicity. Hence, a need for safer preparations, better outcomes were observed in targeted liposomes regarding micrometastasis and tumor vasculature (Allen and Cullis, 2013).

3.2.7 TRIPLE-NEGATIVE BREAST CANCER AND NANOTECHNOLOGY

Triple-negative breast cancer is a destructive breast cancer subtype characterized by the absence of ER, PR, and HER 2. This hampers its treatment because current therapies are nonspecific (Zhou et al., 2020). Nanotechnology offers hope for targeted therapies.

3.2.8 QUERCETIN-LOADED PLGA/TPGS NANOPARTICLES

Quercetin (QT) is a flavonoid with anticancer properties. Its little bioavailability and meager solubility are its healing pitfalls. A report showed that polymeric nanoparticles for QT, PLGA-TPGS (Qu-NPs) exerted inhibitory effects on TNBC by inhibiting urokinase-type plasminogen activator (uPA) (Zhou et al., 2020). QT fused into solid nanoparticles (SLN) via microemulsification was reported to reduce drug release into cancer cells (Niazvand et al., 2019). QT-SLN suppressed breast cancer cell growth (Niazvand et al., 2019).

3.2.9 MICELLES

Micelles are potential drug delivery structures due to their amphipathic nature. A micelle-based nanosystem with styrene-co-maleic acid was developed for the delivery of RL71, a derivative of curcumin in TNBC cells. A higher cellular uptake and toxicity to the cancerous cells as well as minimal drug release were observed (Taurin et al., 2013). The major pitfall is a lack of specificity. Irinotecan (SN-38) poly(ethylene-glycol)-poly (glutamic acid) PEG-PGlu, that is, NK012 micelle appears to be the only micellar system for phase II clinical trial in TNBC patients because of its biofriendly properties (Matsumura, 2011; Krause et al., 2018).

3.2.10 POLYMERIC NANOPARTICLES

Polymeric nanoparticles can be made from both, regular and artificial polymers. They are suitable as nanocarriers because of their biodegradability, biocompatibility, and low toxicity (Elsabahy and Wooley, 2012). They are proven in the delivery of therapeutic drugs as well as miRNA and siRNA (Krause et al., 2018). Although several polymeric nanoparticles are known, it appears that Arg-Gly-Asp ligand is more efficacious in drug delivery and tumor metastasis in TNBC. Inhibition of connection and invasion of alpha–beta three integrin receptor, which was overexpressed in metastatic TNBC, was effected by cyclic RGD-functionalized solid–lipid nanoparticles (Shan et al., 2015).

3.2.11 METAL NANOPARTICLES

Titanium dioxide, platinum, gold, and zinc oxide are common metal and metal oxide nanoparticles. They are being explored for their therapeutic and diagnostic potentials because of their physicochemical features such as heat-producing, electrical, magnetic, and optical properties. Their ease of modification via conjugation with various functional groups makes them suitable for clinical cases. Metal nanoparticles use different mechanisms in exerting their therapeutic effects including apoptosis, production of heat, and oxidative stress (Su et al., 2014).

Extensive investigations have been done on gold nanoparticles. It was found to inhibit TNBC, when conjugated with cisplatin (Andey et al., 2015). Furthermore, gold nanoparticles increased oxidative stress via ROS production, which ultimately results in cell death (Andey et al., 2015). Zinc oxide NP acts by interfering with the cancer cell cycle, leading to apoptosis (Wahab et al., 2014). ZnO conveying asparaginase, a potent anticancer enzyme enhanced the specificity and stability, when administered along with paclitaxel and daunorubicin (Baskar et al., 2015). Conjugation of ZnO NPs with paclitaxel and cisplatin showed decreased toxicity in breast cancer cells (Hackenberg et al., 2012).

3.3 URSOLIC ACID AND ITS MECHANISM

UA is a pentacyclic triterpene carboxylic acid that has been recounted to retain a number of health benefits comprising antineoplastic properties.

Reports show that UA-induced cancer apoptosis and halted angiogenesis by mitigating the phosphorylation of p38 MAPK, EGFR, and Jun N-terminal kinase (Liu et al., 2019). The manifestation of proinflammatory factors including IL-2, IL-18, IL-6, and TNF α and γ were inhibited by UA (38). Furthermore, UA inhibited the expression of matrix metalloproteinases (MMP): MMP-9 and MMP-2. It also inhibited cell surface adhesive molecules: VCAM-1, ICAM-1, P, and E selectins, hence, hindering metastasis (Shao et al., 2020).

In spite of its ability to inhibit the growth of cancerous cells, poor hydrophilicity and permeability are its known limitations (Shao et al., 2020). When taken orally, UA is poorly absorbed and rapidly metabolized in the liver, hence, resulting in its low bioavailability (Liao et al., 2005; Chen et al., 2011). Moreover, its intravenous administration also results in organ nonspecific distribution (Zhang et al., 2013). These are limitations in its clinical use. The introduction of nanodrug delivery systems is promising in improving its bioavailability (Shao et al., 2020).

Chitosan, liposomes, dendrimer, and others have been used in constructing UA-based nanoformulation (Shao et al., 2020). UA coupled to polymeric compounds have also been reported for their anticancer efficacy. Gold-UA-loaded PLGA showed better cellular uptake as well as prevention of cervical cancer cell proliferation, invasion, and migration (Wang et al., 2017b). The antineoplastic effects of UA-loaded polyurethane nanoparticles on breast cancer cells were observed by Oprean et al (2016). The infusion of UA into polymeric nanoparticles improved drug retention at the tumor sites. Furthermore, the biodegradability of polymeric nanoparticles ensures their low toxicity (Shao et al., 2020).

The challenges of UA nanosystems include potential toxic effects of UA composite preparation to healthy tissues, unclear metabolism, and biodegradation difficulties of the nanopreparations, and possible restriction of UA bioavailability in tumor tissues by physiological factors. Further improvement is required for stability and storage of UA

3.4 NANOENGINEERED MATERIALS DESIGNED AS ANTICANCER THERAPY

Specific drug delivery systems utilizing nanoparticles such as lipid- or polymer-based nanoparticles have been known to reduce cost and side effects. Drug encapsulation plus nanoelectromechanical systems are utilized to transport and release molecules into the target site in cancer treatment to enhance drug bioavailability. Nanoengineered materials can be designed and in vivo imaging can be utilized to monitor nanoparticles in cancer biology to detect, diagnose, and treat. Newer approaches have emerged in drug delivery systems that can take nanoparticles directly through the cell membranes, cell cytoplasm, straight into the genetic component of cell, and regulate gene expression. Nanoparticles can activate signals, prevent tissue damage, alter pharmacokinetics, and improve drug solubility (Patra et al., 2018a).

Nie et al. (2007) reported the application of nanoparticles in drug delivery systems, most especially in breast and lung cancer treatment. Different metallic nanoparticles have been utilized to deliver drugs to cancer cell by them and subsequently inside the cell, tumor magnetic nanoparticles were vibrated to generate appropriate radiofrequency fields, causing a destruction of the liposome to deliver the drug in free form. Also, it has been shown that polyethylene glycol nanoparticles can be utilized to load antibiotics into a target cell in a pH-sensitive milieu for multidrug-resistant bacteria.

Nikalje (2015) showed a rat model biomimetic method, which is a design that utilizes multifunctional biomolecules like polymer *model* containing proteins that imitate the chemical composition, biologically functional, and physical structure moieties of the natural systems for the development of novel biomimetic biomaterials for in situ tissue treatment was utilized to drug coat nanoparticles against blood clots and infections. Minicell drug delivery systems are built to engulf pathogens and cancer cells with nanoparticle loaded with antibodies and membrane of mutant bacteria to destroy the tumor cell once inside the minicell.

Studies have revealed that nanoparticles have been utilized in the delivery systems of myelin antigens during immunological reactions in autoimmune diseases. Also, nanomaterials such as quantum dots, nanoshells, and nanowires are useful at tumor sites to emit light for magnetic resonance imaging to generate higher contrast image, DNA mutation, and biomarker detection for tracking and monitoring of intracellular process in real-time for a long duration of time (Jain, 2017).

Nikalje (2015) reported that certain functional groups can get attached to nanoparticle to enhance entry into tumor cells and accumulate due to the absence of lymphatic drainage in any tumor cell. It was revealed that the drug delivery system is established on the following effective transport to target cell, providing excellent encapsulation of the drugs and release.

Ochekpe et al. (2009) discussed the usage of nanoparticles for drug delivery in biomedical science with tremendous beneficial effects such as treatment of neurodegenerative disorders, tuberculosis treatment, clinical operative dentistry, surgery, visualization, tissue biomedical engineering, ophthalmology, immune response, antibiotic resistance, and vaccine design.

Patra et al. (2018a, 2018b) reported that the fast-growing area in the usage of nanotechnology is in the drug delivery system, where nanoscale materials are utilized as a diagnostic tool for the therapeutic treatment of several human diseases such as multiple sclerosis, acute ischemic stroke, cardiovascular diseases, inflammatory or infectious diseases (e.g., HIV encephalopathy) at a specific target site like delivery of dopamine to the brain, brain capillary endothelial cells, ocular drug delivery, and many more, utilizing optimization and biometric simulation of an intracranial nanoenabled scaffold device (Nikalje, 2015). It was noted that nanoparticles can be utilized as chemotherapeutic, immunotherapeutic, and biological agents using natural and synthetic products.

Evanoff and Chumanov (2005) reported that silver nanoparticles can be applied for the biodiagnosis and detection of cancer cell by array biosensor based on the shape, dielectric medium, and size. Using nanoshells, dual images can be produced to separate cancer areas from the noncancer zones so as to apply photothermal therapy on the cancer cells.

Loo et al. (2005) discovered that silver nanoparticles are currently being utilized for the treatment of cancer and a drug carrier mechanism to deliver drugs, such as erosion degradation, swelling, and diffusion plus implantation of nanodevices referred to as artificially stimulated cells or tissues in tissue engineering, which is believed to revolutionize artificial implants.

Tarafdar et al. (2013) suggested that many aspects of nanoscience have received massive attention for their beneficial role in current years centered on the physiochemical properties of many biolabeled nanoparticles treatments and diagnosis of cancer. Molecular change detection nanoparticles using different nanostructures and nanopores are the current area of research of cancer biology for the early diagnosis of cancer cells.

Kalangutkar (2015) suggested that in the fight against cancer, early detection is key and revealed that nanotechnology can help to early detect, screen, monitor, and effective treat various forms of cancer. It was demonstrated that complete entry of drug in a specific site could be achieved in nanobiotechnology tools coated with antibodies for cancer cells and then subsequently releasing anticancer drugs. When the nanomaterials are engulfed by cancer cells, electromagnetic field is generated to weaken them in chemotherapy. Recent studies have confirmed that nanoparticles have been altered to effectively distribute drugs or biomolecules to target areas in the human body, also in the tracking of drug buildup and metabolism in cells/tissues for better knowledge of the pharmacokinetics and pharmacodynamics of specific biomolecules or drugs (Azhar et al., 2020).

Khan et al. (2014) revealed that gold nanoparticles application in drug delivery system, which is one of the contemporary fast-advancing field in biomedical sciences, could be linked to their physiochemical characteristics such as stability, biocompatibility, inert nature, small size, noncytotoxicity, high surface-area-to-volume ratio, and high dispersity. Murawala et al. (2014) established that in clinical medicine, cancer treatment approaches are restricted such as chemotherapy, surgical restriction plus irradiation, which are very toxic, traumatic, less specific, and enormous adverse side effects. It was revealed that chemotherapy is a conventional approach and it is a form of restorative method, which possibly destroys health tissue alongside the cancer cells.

Han et al. (2017) showed that a specific drug distribution system is much improved than the orthodox drug treatment as it offers minimal side effects and effective distribution system, particularly in cancer treatment. It was revealed that a lot of work is ongoing in this area as scientists have developed in nanotechnology like nanocarriers targeted drug delivery covalent conjugation system owing to the beneficial physiochemical characteristics. Lee et al. (2010b) revealed that gold nanoparticles in drug delivery systems, diagnosis, imaging, and treatment have received massive attention in tumor therapies and cancer antigen due to unique optical, shape, high surface area, size, noncytotoxicity, biocompatible, and easy conjugation for effective dense drug loading in cancer therapy.

Maiti et al. (2019) described extensively on carbon nanotubes, nanocomposite in cancer treatment as they are utilized in delivering anticancer drugs, proteins, and genes during chemotherapy plus photothermal therapy, which might be attributed to their potential to absorb light and suppress tumor volume. It was revealed that cytosolic release of nanodrug caused cell death in lung cancer via abruption of the mitochondrial membrane.

Deng et al. (2016) demonstrated that in nanotechnology treatment of cancer, graphene oxide has been widely considered to promote a unique physiochemical modification and properties against cancer cells in photothermal and chemotherapy treatments. It was reported that in a design using graphene oxide and αvβ6-targeting peptide, this compound was able to stimulate dendritic cells, significantly inhibiting tumor development plus metastasis via infiltration of T lymphocytes cytotoxic CD8+ in the tumor microenvironment.

Chen et al. (2008) combined graphene oxide and quantum dots in an experiment to treat cancer cells due to the unique physiochemical characteristics of biomolecules both, as chemotherapeutics and radioisotopes. Their findings resulted into significant anticancer in vivo therapy by photodynamic approach with ROS formation and heat generation. This magnetic hyperthermia can target cancer cells and deliver nanodrugs specifically, thereby, suppressing tumor growth.

Valavanidis and Vlachogianni (2016) reported on the utilization of nanotechnology in biomedical science for the treatment of cancer and other malignant neoplasms through anticancer drug delivery systems. They noted that molecular imaging and early detection utilizing nanotechniques can only help in the treatment of cancerous tissues. It was revealed that photoluminescent nanoparticles can be utilized to visualize cancerous lesions, improve delivery time, and deliver anticancer drugs for effective therapy plus reduced toxicity. They showed that recently, there has been rigorous research output on the development of multifunctional nanomaterials not only for cancer but for multimodal imaging plus therapy of several diseases such as multimodal imaging, image-guided therapies, and therapeutics plus diagnostics.

3.5 NANOMATERIALS WITH SELECTIVE TARGETING AND CANCER DRUG DELIVERY

Meng et al. (2016) developed a minute smart nanocomposite labeled as G-CuS-DOX. This was made up of a nanogel that is sensitive to heat with a temperature for the critical solution of 42°C, using copper sulfide as a photothermal component while the anticancer agent was DOX. On introduction of the solution of nanocapsules into the cancerous cells of mice, DOX is discharged and the elevation of temperature within the tumor is controlled by ex vivo NIR laser device. High efficiency against tumor cells has been confirmed by using this technique.

A chitosan-derived nanomaterial was prepared by Gupta et al. (2017) by loading with *Paclitaxel* obtained from *Taxu brevifolia*, and this was used in the management of diverse types of tumors. It was established that the use of nanoparticles in the delivery of the drugs aid better release of the drugs and enhanced intake into the cells.

Lee et al. (2011) used chitosan conjugated with oleic acid and encapsulated oleic acid with iron oxide nanoparticles. They used this for the evaluation of the influence of nanoparticles in the cancerous cells and this was monitored using magnetic resonance sequencing. The study showed improvement in the usage of the nanoparticles in the handling of tumor cells.

Malam et al. (2009) emphasized that the use of nanotechnology in the area of drug delivery enhances the efficient and reliable discharge of drugs as well as lowering of overall systemic lethality. The new emphasis on this emerging area has been reflected in the availability of albumin nanoparticles and liposomal Doxil in commercial quantity. Recent development in the aspect of liposome techniques has provided more reliable therapy for cancerous cells that show resistance to different drugs. Nanotechnology in this regards aids in focusing for the treatment of breast and prostate cancer.

Elseoud et al. (2018) assessed the utilization of nanocrystals and nanoparticles from cellulose and chitosan, respectively, in oral discharging of repaglinide. They reported the average size distribution of the chitosan nanoparticles to be within 197 nm while for the cellulose nanocrystal, the average diameter was within 251–310 nm. A sustained discharge of the drug was observed, which was an outcome of the existence of hydrogen bonds between the drugs and the nanocrystals; hence, the nanoparticles formulated with cellulose showed a relatively lower discharge in comparison to nanoparticles from the native nanocrystal of cellulose.

Fabrication of a novel kind of nanoparticle from silicon was made by Park et al. (2009) so as to assist in the lessening of the rate of metabolism and improve the speed of circulation of the drugs when loaded into the body. In order to aid the buildup of the nanoparticles from place to place around the area of the tumor growth, silicon dioxide and silicon core of the nanomaterials were used. Nanoconjugates were used by injecting the mice through the vein in the lateral tail twice in a week. It was revealed that the conjugated methotrexate in the dendrimers brought about a lowering in the toxicity and much better efficiency when compared to the pure methotrexate in equal amount. The survival of the mice was better enhanced.

Moon et al. (2015) developed gold nanorods of reduced graphene oxide coating (r-GO-AuNRs) with a photoacoustic component that is highly amplified photoacoustic. The efficient performance was induced by the unique NIR light absorption alongside the layer of the decreased graphene oxide, which supplied the photothermal stability.

A nanosystem made up of albumin of human serum (HSA) was manufactured by Chen et al. (2009). The system is capable of self-assembling nanoparticles. It has a component that acts as photothermal agent and imaging probe fluorescence. The nanoparticles were observed to show long circulation alongside good stability. The heat from the system enhances the intake of the nanoparticles and the death of the tumor cell.

Praetorius and Mandal (2007) carried out a study in which they conjugated folate with nanoparticles through covalent bonding and the use of the carboxyl groups on the surface of the nanoparticles. They prepared nanocapsules of isobutyl-cyanoacrylate (IBCA), which were then coated using folate and targeted at the tumors. It was revealed that efficient targeting of folate receptors for selective drug delivery was observed through the conjugation of the nanoparticles to the folic acid.

Fabrication of characterization of a nanosystem comprising ligand–receptor using gold nanomaterials was made by Bhattacharyya (2011). Folate receptors shown in ovarian cancer were employed as the focused surface molecules while monoclonal antibodies were used against the receptor. They reported a higher efficiency in drug delivery, when the dual targeting system was employed in the releasing of the gold nanoparticles into the tumor cells, when each of the single ligand is employed.

Bellocq et al. (2003) investigated the effectiveness of drug delivery using transferring modified nanomaterials for handling of metastatic cancer. They observed that when the modification of the transferring was low, the stability of the nanoparticles was constant in transfect cells of leukemia.

A magnetoretic encapsulated magnetic nanoparticle was used by Lee et al. (2013) as a special therapeutic technique in improving drug delivery and reducing systemic toxicity to the cancerous cells. The self-assembling of the nanocomposites was from four major building components, which include Ad-PEG, iron oxide super magnetic nanomaterial, dendrimers of polymido-amine, and branched polyethylenimine.

Gao et al. (2012) did an attachment of α-cyclodextrin into a mesoporous nanoparticle of silica, using bonds of benzoic imine. The nanoparticles of mesoporous silica were shown to be hydrolyzed completely in the endosomes at a pH of 5. The released loaded of the doxorubicin drug was assessed at varying concentrations to confirm the functionality.

Synthesis of growth factor that is insulin-like in nature labeled as IGF1 was carried out by Zhou (2015). This was conjugated to nanoparticles of iron oxide with a payload of anthracycline doxoxubicin. The nanoparticle was then introduced intravenously into the xenograft from the patient resulting in an efficient penetration and focusing of tumor. It was revealed that the use of iron oxide nanoparticle brought about the inhibition of the pancreatic tumors.

A complex nanocomposite system comprising Au nanocomposite with Gd hybridized plasmonic was developed by Wang et al. (2016). Citrate Gd complexes were coloaded on the mesoporous coating of silica. The system of nanocomposite brought about improvement in the drug delivery.

3.6 NANODRUGS FOR THE MANAGEMENT OF CANCER CELLS

The use of conventional chemotherapeutic agent has been found to eradicate cancerous cells but have posed varied side effects over the years. Nanoparticles from metals have been joined together in pairs, for example, gold and iron oxide nanoparticles (Kogan et al., 2007). Several amino groups of amino acids with the carboxyl group of another and nanoparticles joined to display an ability of the nanomaterial affinity to combine with living tissue and having little/no cytotoxic effect (Nie et al., 2007; Veetil and Ye 2007). Cadmium selenide NPs are known to have an appearance of light reflection in the presence of UV light, and when administered, it passes gradually through small openings to the cancerous cells. This attribute makes it easier for recognition and removal of the affected cells (Sengupta and Sasisekharan, 2007).

The effectiveness of nanodrugs have long being judged by the possibility to manipulate the reaction of the body to the drugs, in terms of movement of the drugs through, into and out of the body during the period of administration, bioavailability, dissemination, breakdown, buildup, and excretion of the drugs. Nanoparticles were found to possess some attributes, which could help to increase the rate of delivery of drugs and its effectiveness. In cases that larger materials might have been eliminated from the body, the cells find it easier to carry out absorption of the nanodrugs due to their sizes. The treatment and elimination of cancer cells have been possible owing to the capability of the drugs to enter into the cell membrane and cytoplasm; thereby, impeding the in-cell processes (Banerjee and Verma, 2008). Tumor cells were found to gather together in areas tagged iron microplate with fluorescent, and this method makes diagnosis and treatment of cancer cells to be possible and easy. Ag NPs were found to selectively locate sites that are affected with cancer cells and administer the necessary therapeutic response at that site (Sengupta and Sasisekharan, 2007). The use of nanodrugs in combating cancer cells have been looked into by way of the relationship that exists between the effectiveness of the administered nanoagent and other response of the organism/human immune system (Salvioni et al., 2019; Riechel et al., 2019).

During the examination of the tissues and liquid from the human body, there was a relation between the amount of dose given and the quantity of intracellular confirmed nanoparticles (Zhang et al., 2011; Sega and Low, 2008). Alarifi et al. (2013) found nanoparticles to have an affinity to destroy cancer cells using different ways, which could be a change in the tide, causing disparity between antioxidants and free radicals in the cell and damaging the cell wall. Cerium oxide NPs are widely used in nanomedicine due to their capacity to persuade production of reactive oxygen and reducing the likelihood of premature mitosis (Ali et al., 2015). Cerium oxide NPs have been used against lungs cancer, liver cancer, and pancreatic cancer (Lin et al., 2006; Cheng et al., 2013; Wason et al., 2013) by infusing the death of the cancer cells. It was revealed that zinc oxide nanoparticles can use lethal effect on melanoma cells through imbalance between free radicals, antioxidants, cell destruction, and could also tackle liver and breast cancers concurrently (Wahab et al., 2014; Kc et al., 2016).

3.7 MODE OF ACTIONS OF NANOMATERIALS IN THE MANAGEMENT OF CANCER AND DRUG DELIVERY

Cancer has been reported by many health researchers as a serious malignant disease that can spread rapidly from one infected cell to the other. For the last 5 decades, the search for a permanent cure to this disease is still hanging (Dadwal et al., 2018). However, recently there has been serious employment of novel technology such as nanotechnology in the effective drug transfer for the management and possible cure of cancer. It is interesting to note how nanodrugs react. The mechanisms of action of nanodrugs use for the treatment of cancer vary, based on the nanoencapsulation used, and the target region sent to (Dadwal et al., 2018).

In an ideal situation, the efficacy of antitumor drugs for cancer treatment works in two prongs:

- First, the drug penetrates the body barriers and targets the desired tumor cells with slight damage done to the circulation and volume of the blood.
- Second, at the point of getting to the target sites, the nanodrugs should selectively terminate the cancer tissues without impacting the normal blood cells.

These two techniques are also linked with the enhancements of the quality of the living and survival of the patient by elevating the internal level of the nanodrugs and decreasing the dosage to avoid toxicities concurrently (Dadwal et al., 2018).

There are several concepts used in the evaluation of the mechanisms of drug transfer by nanomaterials in the management of cancer cells. Specific examples are the enhanced permeability and retention and mononuclear phagocyte system. It has been documented that cancer vasculatures are known for their abnormalities with leaky fortifications (Jain, 1994). The leakiness is a result of the reduction of the total pericytes and the endothelial tissues that are proliferated rapidly. These aberrant cells ranged between 100 and 1000 nm, but the normal cells are between 5 and 10 nm (de Menezes et al., 1998). The aberrant tissues have very large apertures that permit hydraulic and permeability conductivity in the cancer cells; thus, permitting the movement of nanomaterials' effect on them (Allen, 2002; Park et al., 2002). The spreading of impaired tumors from the lymphatic elevates the tumor permeability, thus, resulting in enhanced permeability and retention effects. However, macromolecules such as nanomaterials have been used to prolong the tumor retention times, which have shown to be elevated, more than the plasma concentration (de Menezes et al., 1998).

Nanomaterials also have the potential to clear the phagocytic mononuclear system. To talk about the benefit of the improved permeability and retention properties, the nanomaterials must continue to move for a particular time in the cells to gather the tumor. More so, the nanomaterials are susceptible to removal by the phagocytic mononuclear system known as reticuloendothelial structure. The phagocytic mononuclear system is made up of Kupffer tissues of the spleen, liver, monocytes of the blood, macrophages tissues, and from the bone marrow origin (Pastan et al., 2006). Nanomaterials can relate with the phagocytic mononuclear cells and elicit the process of opsonization since untimely removal from the flow will stop the nanomaterials from amassing in the cancer cells. More advanced stealth nanotechnology (hydrophilic polymer/PEG grafting) will be an alternative in resolving these issues (Peer et al., 2007; Dadwal et al., 2018). The mechanism of the hydrophilic polymer/PEG grafting, a stealth nanotechnology, enables cells immune interactions, avoiding the adsorption of protein, and steric balance.

The mode of actions of nanomaterials via avoidance of lysosomal degradation, the reliance of pH on drug transport, and cellular absorption of the nanodrugs are also paramount mechanisms in the management of cancerous cells (Sutradhar and Amin, 2014). Passive and active targeted nanomaterials encounter a chief issue in transporting drugs to neoplastic tissues because the activities of the lysosomes swiftly break both the drugs and the nanomaterials in the internal cells. After internalization, the transportation of the colloidal reaches the compartment of the lysosomal, therein, the hydrolytic enzymic of the lysosomes break both: the internal contents and the transporter carrier (Dadwal et al., 2018). Thus, the internal cell transportation of the tumor is changed, when the compressed drug is a base of the nucleic acid chain. Since the pH at the cancerous cells is highly acidic, the carriers can alter the solubility at reduced pH, hence, they can be utilized to transport and target drugs. Moreover, the external environment of the cancerous cells is also acidic. In this case, an alteration in the pH rise across the cell compartment is needed. Nanomaterials subtle to the pH rise are potential tumor drug transporter (Praetorius and Mandal, 2007; Sutradhar and Amin, 2014). A pH reactive nanomaterial is made up of a core and shell, which responds to different variations of pH rise and alter its solubility design. The shell–core nanomaterials polymer is planned with their lower critical solution temperature of pH 7.4, which causes the shell–core nanoparticles to precipitate and deform in the acid medium; thus, activating the discharge of chemotherapeutics as a conjugate nanomaterial molecule (Sutradhar and Amin, 2014). The nanoparticles can discharge the drugs over an estimated time frame by thawing and/or swelling via the reduction of the pH of the microspheres in the tumor cells (Barratt, 2003; Gordon and Hall, 2005; Praetorius and Mandal, 2007).

Nanomaterials have their own morphological, physical, and chemical characteristics that have a strong likeness for various drug charges in polarities via chemical relationships coordinated by hydrogen and covalent bonds or by weak forces, such as van der Waals and electrostatic interactions (Patra et al., 2018a, 2018b; Sethi et al., 2014; Kinnear et al., 2017; Mattos et al., 2017). When they evaluated the interaction of nanomaterials sourced from neem plants in conjunction with biogenic-silica, it was observed that the amalgam influenced the biological system as well as the discharge of biokinetic active substances in the tested organism both in in vitro and in vivo test settings.

Nanomaterials act under different specificity in the target cells of an organism to decrease the immunogenicity via chemical and coating functionalization with various materials, such as peptides, tunable surfactants, cell membrane, antibodies, normal polysaccharides, and polymers, which display affinity and binding properties with specific target cancerous cells that do not intercept specific barriers like fluid from the cerebrospinal and blood media (Gao et al., 2013, 2017; Kolhar et al., 2013; Jain and Jain, 2015; Martens et al., 2015; Pelaz et al., 2015; Almalik et al., 2017). These modified ligands of nanomaterials have been utilized to penetrate via cell membrane and permit drug transport in a programmed environment. The nanomaterials ligands also possess promising influence in the management of corona immunogenicity, retinal gene deformation, adenocarcinoma of the pulmonary cells, breast tumor, and melanoma (Shen et al., 2015; Gao et al., 2017; Wang et al., 2017c). Table 3.1 summarizes relevant nanomaterials that could be utilized for the treatment of cancer diseases.

3.8 LIMITATIONS OF NANOCARRIERS

The size and biodistribution of nanocarriers could affect their stability in blood circulation and bioaccumulation in nontarget tissues (Montaseri et al., 2020). Hypersensitivity reactions caused by the nanocarriers as well as osmotic pressure within cancer cells can reduce the concentration of accrued nanoparticles in cancer tissues. Immobilization of anticancer drugs on nanoparticles could exert toxic effects on noncancerous tissues and also cause unnecessary interactions with other physiological entities (Niemirowicz et al., 2015; Wang et al., 2015).

TABLE 3.1 Nanomaterials That Could Be Utilized for the Treatment of Cancer.

Nanomaterial/system	Unique feature/role in cancer treatment	References
CdSe	Reflects light in presence of UV	Sengupta and Sasisekharna (2007)
AgNPs	Selective location of tumor site	Salvioni et al. (2019)
Fe oxide NPs	Inhibition of pancreatic tumor	Zhou et al. (2015)
Au Nanocomposite	Enhancement of drug delivery	Wang et al. (2016)
Au Nanosystem with ligand–receptor	Better efficiency in drug delivery	Bhattacharyya (2011)
Nanocapsules of isobutyl-cyanoacrylate	Efficient targeting of folate receptors	Praetorius and Mandal (2007)
Nanosystem of albumin of human serum (HSA)	Better image probe, enhancement of intake of nanoparticles	Chen et al. (2009)
Au nanorods with graphene oxide	High photothermal stability	Moon et al. (2015)
Si NPs	Lowering of systemic toxicity and better circulation	Park et al. (2009)

3.9 CONCLUSION AND FUTURE RECOMMENDATION

The application of nanomaterials for adequate delivery of function and biologically active compounds to targeted cancer affected areas has been described. There is a need to provide more personalizing nanomedicine-based treatments as well as appropriate formulation techniques that help in mitigating against several challenges in achieving necessary pharmacokinetic features. There is a need to perform more experiments on toxicity, biodegradation, prevention strategy, and theranostic purposes. Also, there is a need to provide more information regarding the safety of different nanomaterials most especially regarding its modes of action involving by these therapeutic nanomaterials that could be utilized for effective delivery of (chemo-)therapeutic drug most especially to the pathological site. Furthermore, there is also a need to provide information on several techniques involved in preventing the buildup in healthy tissue and organ built on improved permeability and retention (EPR) influence.

KEYWORDS

- **nanomaterials**
- **cancer**
- **nanoparticles**
- **nanotechnology**
- **drug delivery**
- **drug localization**

REFERENCES

Alarifi, S.; Ali, D.; Alkahtani, S. Induction of Oxidative Stress, DNA Damage, and Apoptosis in a Malignant Human Skin Melanoma Cell Line after Exposure to Zinc Oxide Nanoparticles. *Int. J. Nanomed.* **2013,** *8*, 983–993.

Ali, D.; Alarifi, S.; Alkahtani, S. Cerium Oxide Nanoparticles Induce Oxidative Stress and Genotoxicity in Human Skin Melanoma Cells. *Cell Biochem. Biophys.* **2015,** *71* (3), 1643–1651.

Allen, T. M. Ligand-Targeted Therapeutics in Anticancer Therapy. *Nat. Rev. Cancer*. **2002,** *2*, 750–763.

Allen, T. M.; Cullis, P. R. Liposomal Drug Delivery Systems: From Concept to Clinical Applications. *Adv. Drug Deliv. Rev*. **2013,** *65*, 36–48.

Almalik, A.; Benabdelkamel, H.; Masood, A.; Alanazi, I. O.; Alradwan, I.; Majrashi, M. A. et al. Hyaluronic Acid Coated Chitosan Nanoparticles Reduced the Immunogenicity of the Formed Protein Corona. *Sci. Rep.* **2017,** *7*. DOI: 10.1038/s41598-017-10836-7.

Andey, T.; Sudhakar, G.; Marepally, S.; Patel, A.; Banerjee, R.; Singh, M. Lipid Nanocarriers of a Lipid-Conjugated Estrogenic Derivative Inhibit Tumor Growth and Enhance Cisplatin Activity Against Triple-Negative Breast Cancer: Pharmacokinetic and Efficacy Evaluation. *Mol. Pharm.* **2015,** *12* (4), 1105–1120.

Ashley, C. E.; Carnes, E. C.; Phillips, G. K.; Durfee, P. N.; Buley, M. D.; Lino, C. A. Cellspecific Delivery of Diverse Cargos by Bacteriophage MS2 Virus-Like Particles. *ACS Nano*. **2011,** *5*, 5729–5745.

Avazzadeh, R.; Vasheghani-farahani, E.; Soleimani, M, Amanpour, S, Sadeghi, M. Synthesis and Application of Magnetite Dextran-Spermine Nanoparticles in Breast Cancer Hyperthermia. *Prog. Biomater.* **2017,** *6* (3), 75–84.

Banerjere, N. H.; Verma, M. Application of Nanotechnology in Cancer. *Technol. Cancer Res. Treat.* **2008,** *7* (2), 149–157.

Barratt, G. Colloidal Drug Carriers: Achievements and Perspectives. *Cell. Mol. Life Sci.* **2003,** *60* (1), 21–37.

Baskar, G.; Chandhuru, J.; Fahad, K. S.; Praveen, A. S.; Chamundeeswari, M.; Muthukumar, T. Anticancer Activity of Fungal L-asparaginase Conjugated with Zinc Oxide Nanoparticles. *J. Mater. Sci. Mater. Med.* **2015,** *26* (1). DOI: 10.1007/s10856-015-5380-z.

Bellocq, N. C.; Pun, S. H.; Jensen, G. S.; Davis, M. E. Transferrin-Containing, Cyclodextrin Polymer-Based Particles for Tumor-Targeted Gene Delivery. *Bioconjugate Chem.* **2003,** *14*, 1122–1132.

Bhattacharyya, S. Efficient Delivery of Gold Nanoparticles by Dual Receptor Targeting. *Adv. Mater.* **2011,** *23* (43), 5034–5038

Chan, C.; Cai, Z.; Reilly, R.M. Trastuzumab Labeled to High Specific Activity with ^{111}In by Conjugation to G4 PAMAM Dendrimers Derivatized with Multiple DTPA Chelators Exhibits Increased Cytotoxic Potency on her2-Positive Breast Cancer Cells. *Pharm Res.* **2013,** *30*. DOI: 10.1007/s11095-013-1044-1.

Chari, R. V. Targeted Delivery of Chemotherapeutics: Tumor-Activated Prodrug Therapy. *Adv. Drug Deliv. Rev.* **1998,** *31*, 89–104.

Chen, K.; Xie, J.; Chen, X. RGD-Human Serum Albumin Conjugates as Efficient Tumor Targeting Probes. *Mol. Imag.* **2009,** *8* (2), 65–73.

Chen, Q.; Luo, S.; Zhang, Y.; Chen, Z. Development of a Liquid Chromatography-Mass Spectrometry Method for the Determination of Ursolic Acid in Rat Plasma and Tissue: Application to the Pharmacokinetic and Tissue Distribution Study. *Anal. Bioanal. Chem.* **2011,** *399*, 2877–2884.

Chen, J.; Chen, S.; Zhao, X.; Kuznetsova, L. V.; Wong, S. S.; Ojim, I. Functionalized Single-Walled Carbon Nanotubes as Rationally Designed Vehicles for Tumor-Targeted Drug Delivery. *J. Am. Chem. Soc.* **2008,** *130*, 16778–16785.

Cheng, G.; Guo, W.; Han, L.; Chen, E.; Kong, L.; Wang, L. et al. Cerium Oxide Nanoparticles Induce Cytotoxicity in Human Hepatoma SMMC-7721 Cells via Oxidative Stress and the Activation of MAPK Signaling Pathways. *Toxicol. In Vitro* **2013,** *27*, 1082 1088.

Choi, W. I.; Lee, J. H.; Kim, J. Y.; Heo, S. U.; Jeong, Y. Y.; Kim, Y. H. et al. Targeted Antitumor Efficacy and Imaging via Multifunctional Nano-Carrier Conjugated with Anti-HER2 Trastuzumab. *Nanomed. Nanotechnol. Biol. Med.* **2014,** *11* (2), 359–368.

Dadwal, A, Baldi, A.; Narang, R. K. Nanoparticles as Carriers for Drug Delivery in Cancer. *Artif. Cells Nanomed. Biotechnol.* **2018**. DOI: 10.1080/21691401.2018.1457039.

Dai, W.; Yang, F.; Ma, L.; Fan, Y.; He, B.; He, Q. Combined mTOR Inhibitor Rapamycin and Doxorubicin-Loaded Cyclic Octapeptide Modified Liposomes for Targeting Integrin α3 in Triple-Negative Breast Cancer. *Biomaterials* **2014,** *35* (20), 5347–5358.

De Menezes, D. E. L.; Pilarski, L. M.; Allen, T. M. In Vitro and in Vivo Targeting of Immuno-liposomal Doxorubicin to Human B-Cell Lymphoma. *Cancer Res.* **1998,** *58*, 3320–3330.

Deng, L.; Li, Q.; Rehili, S. A.; Omar, H.; Almalik, A.; Alshamsan, A. Hybrid Iron Oxide-Graphene Oxide-Polysaccharides Microcapsule: A Micromatryoshka for On-Demand Drug Release and Antitumor Therapy In Vivo. *ACS Appl. Mater. Interf.* **2016,** *8*, 6859–6868.

Ding, H.; Gangalum, P.R.; Galstyan, A.; Fox, I.; Patil, R.; Hubbard, P. HER2-Positive Breast Cancer Targeting and Treatment by a Peptide-Conjugated Mini Nanodrug. *Nanomed. Nanotechnol. Biol. Med.* **2017,** *13* (2), 631–639.

Elsabahy, M.; Wooley, K. L. Design of Polymeric Nanoparticles for Biomedical Delivery Applications. *Chem. Soc. Rev.* **2012,** *41* (7), 2545–2561.

Elseoud, W. S. A.; Hassan, M. L.; Sabaa, M. W.; Basha, M.; Hassan, E. A.; Fadel, S. M. Chitosan Nanoparticles/Cellulose Nanocrystals Nanocomposites as a Carrier System for the Controlled Release of Repaglinide. *Int. J. Biol. Macromol.* **2018,** *111*, 604–613.

Esfandiari, N.; Arzanani, M. K.; Soleimani, M.; Kohi-Habibi, M.; Svendsen, W. E. A New Application of Plant Virus Nanoparticles as Drug Delivery in Breast Cancer. *Tumor Biol.* **2016,** *37*, 1229–1236.

Esfandiari, N.; Arzanani, M. K.; Koohi-Habibi, M. The Study of Toxicity and Pathogenicity Risk of Potato Virus X/Herceptin Nanoparticles as Agents for Cancer Therapy. *Cancer Nano.* **2018,** *9*. DOI: 10.1186/s12645-018-0036-6.

Evanoff, D. D.; Chumanov, G. Synthesis and Optical Properties of Silver Nanoparticles and Arrays. *Chemphyschem*. **2005,** *6* (7), 1221–1231.

Gao, H.; Yang, Z.; Zhang, S.; Cao, S.; Shen, S.; Pang, Z.; et al. Ligand Modified Nanoparticles Increases Cell Uptake, Alters Endocytosis and Elevates Glioma Distribution and Internalization. *Sci. Rep.* **2013,** *3*, 25–34.

Gao, W. W, Zhang, L. F. Coating Nanoparticles with Cell Membranes for Targeted Drug Delivery. *J. Drug Target*. **2015,** *23*, 619–626.

Gao, X.; Zhang, J.; Xu, Q.; Huang, Z.; Wang, Y. Y.; Shen, Q. Hyaluronic Acidcoated Cationic Nanostructured Lipid Carriers for Oral Vincristine Sulfate Delivery. *Drug Dev. Ind. Pharm.* **2017,** *43*, 661–667.

Gao, Y.; Yang, C.; Liu, X.; Ma, R.; Kong, D.; Shi, L. A Multifunctional Nanocarrier Based on Nanogated Mesoporous Silica for Enhanced Tumor-Specific Uptake and Intracellular Delivery. *Macromol. Biosci*. **2012,** *12*(2), 251–259.

Gordon, E. M.; Hall, F. L. Nanotechnology Blooms, at Last. *Oncol. Rep.* **2005,** *13*, 1003–1007.

Grasso, S.; Santi, L. Viral Nanoparticles as Macromolecular Devices for New Therapeutic and Pharmaceutical Approaches. *Int. J. Physiol. Pathophysiol. Pharmacol.* **2010,** *2* (2), 161–178.

Guan, J.; Zhou, Z. Q.; Chen, M. H.; Li, H. Y.; Tong, D. N.; Yao, J. Folate-Conjugated and pH-Responsive Polymeric Micelles for Target-Cell-Specific Anticancer Drug Delivery. *Acta Biomater.* **2017,** *60*, 244–255.

Gupta, U.; Sharma, S.; Khan, I.; Gothwal, A.; Sharma, A. K.; Singh, Y. et al. Enhanced Apoptotic and Anticancer Potential of Paclitaxel Loaded Biodegradable Nanoparticles Based on Chitosan. *Int. J. Biol. Macromol.* **2017,** *98*, 810–819.

Hackenberg, S.; Scherzed, A.; Harnisch, W.; Froelich, K.; Ginzkey, C.; Koehler, C. Antitumor Activity of Photo-Stimulated Zinc Oxide Nanoparticles Combined with Paclitaxel or Cisplatin in HNSCC Cell Lines. *J. Photochem. Photobiol. B.* **2012,** *114*, 87–93.

Han, L.; Hao, Y. N.; Wei, X.; Chen, X. W.; Shu, Y.; Wang, J. H. Hollow Copper Sulfide Nanosphere–Doxorubicin/Graphene Oxide Core–Shell Nanocomposite for Photothermo-Chemotherapy. *ACS Biomater. Sci. Eng.* **2017,** *3*, 3230–3235.

Hirata, E.; Sahai, E. Tumor Microenvironment and Differential Responses to Therapy. *Cold Spring Harb. Perspect. Med.* **2017,** *7* (7), 1–14.

Jain, A.; Jain, S. K. Ligand-Appended BBB-Targeted Nanocarriers (LABTNs). *Crit. Rev. Ther. Drug. Carrier Syst.* **2015,** *32*, 149–180.

Jain, R. K. Barriers to Drug Delivery in Solid Tumors. *Sci. Am.* **1994,** *271*, 58–65.

Jain, K. K. Nanomolecular Diagnostics the Handbook of Nanomedicine. **2017,** *21*, 133–200.

Jiang, Q.; Hao, S.; Xiao, X.; Yao, J.; Ou, B.; Zhao Z, Production and Characterization of a Novel Long-Acting Herceptin-Targeted Nanobubble Contrast Agent Specific for Her-2-Positive Breast Cancers. *Breast Cancer*. **2016,** *23* (3), 445–455.

Kalangutkar, P. K. Nanotechnology–Advancing the Field of Biomedical Engineering. *Int. J. Cur. Res. Rev.* **2015,** *7*, 66–70.

Kc, B.; Paudel, S. N.; Rayamajhi, S.; Karna, D.; Adhikari, S.; Shrestha, B. G. et al. Enhanced Preferential Cytotoxicity through Surface Modification: Synthesis, Characterization and Comparative in Vitro Evaluation of TritonX-100 Modified and Unmodified Zinc Oxide Nanoparticles in Human Breast Cancer Cell (MDA-MB-231). *Chem. Cent. J.* **2016,** *10*, 16–32.

Khan, A. K.; Rashid, R.; Murtaza, G.; Zahra, A. Gold Nanoparticles: Synthesis and Applications in Drug Delivery. *Trop. J. Pharm. Res.* **2014,** *13* (7), 1169–1177.

Kinnear, C.; Moore, T. L.; Rodriguez-Lorenzo, L.; Rothen-Rutishauser, B.; Petri-Fink, A. Form Follows Function: Nanoparticle Shape and Its Implications for Nanomedicine. *Chem. Rev.* **2017,** *117*, 11476–11521.

Kogan, M. J.; Olmedo, I.; Hosta, L.; Guerrero, A. R.; Cruz, L. J.; Albericio, F. Peptides and Metallic Nanoparticles for Biomedical Applications. *Nanomedicine.* **2007,** *2*, 287–306.

Kolhar, P.; Anselmo, A. C, Gupta, V.; Pant, K.; Prabhakarpandian, B.; Ruoslahti, E.; Mitragotri, S. Using Shape Effects to Target Antibody-Coated Nanoparticles to Lung and Brain Endothelium. *Proc. Natl. Acad. Sci. USA* **2013,** *110*, 10753–10758.

Krause, P.; Boyaciyan, D.; Klitzing, R. V. Making Strong Polyelectrolyte Brushes pH-Sensitive by Incorporation of Gold Nanoparticles. *Soft. Matter.* 2018. DOI: 10.1039/c8sm00411k.

Kulhari, H.; Pooja, D.; Shrivastava, S.; Kuncha, M.; Naidu, V. G. M.; Bansal, V. Trastuzumab-Grafted PAMAM Dendrimers for the Selective Delivery of Anticancer Drugs to HER2-Positive Breast Cancer. *Sci. Rep.* **2016,** *6*. DOI: 10.1038/ srep2 3179.

Lee, C. M.; Jang, D.; Kim, J.; Cheong, S. J.; Kim, E. M.; Jeong, M. H. et al. Oleyl-Chitosan Nanoparticles Based on a Dual Probe for Optical/MR Imaging in Vivo. *Bioconjug. Chem.* **2011,** *22*,186–192.

Lee, J. H.; Chen, K. J.; Noh, S. H.; Garcia, M. A.; Wang, H.; Lin, W. Y. On-Demand Drug Release System for *in Vivo* Cancer Treatment through Self-Assembled Magnetic Nanoparticles. Angew. Chem. Int. Ed. Engl. **2013,** *52*, 4384–4348.

Lee, S. M.; Ahn, R. W.; Chen, F.; Fought, A. J.; O'Halloran, T. V.; Cryns, V. L. Biological Evaluation of pH-Responsive Polymer-Caged Nanobins for Breast Cancer Therapy. *ACS Nano*. **2010a,** *4* (9), 4971–4978.

Lee, K.; Lee, H.; Bae, K. H.; Park, T. G. Heparin Immobilized Gold Nanoparticles for Targeted Detection and Apoptotic Death of Metastatic Cancer Cells. *Biomaterial* **2010b,** *31*, 6530–6536.

Li, R.; Peng, F.; Cai, J.; Yang, D.; Zhang, P. Redox Dual-Stimuli Responsive Drug Delivery Systems for Improving Tumor-Targeting Ability and Reducing Adverse Side Effects. *Asian J. Pharm. Sci.* **2020,** *15*, 311–325.

Liao, Q.; Yang, W.; Jia, Y.; Chen, X.; Gao, Q.; Bi, K. LC-MS Determination and Pharmacokinetic Studies of Ursolic Acid in Rat Plasma after Administration of the Traditional Chinese Medicinal Preparation Lu-Ying Extract. *Yakugaku Zasshi.* **2005,** *125* (6), 509–515.

Lin, W.; Huang, Y. W.; Zhou, X. D.; Ma, Y. Toxicity of Cerium Oxide Nanoparticles in Human Lung Cancer Cells. *Int. J. Toxicol.* **2006,** *25*, 451–457.

Liu, P.; Du, R.; Yu, X. Ursolic Acid Exhibits Potent Anticancer Effects in Human Metastatic Melanoma Cancer Cells (SK-MEL-24) via Apoptosis Induction, Inhibition of Cell Migration and Invasion, Cell Cycle Arrest, and Inhibition of Mitogen-Activated Protein Kinase (MAPK)/ERK Signaling Pathway. *Med. Sci. Monit.* **2019,** *25*, 1283–1290.

Loo, C.; Lowery, A.; Halas, N.; West, J.; Drezek, R. Immuno Targeted Nanoshells for Integrated Cancer Imaging and Therapy. *Nano. Lett*. **2005,** *5* (4), 709–711.

Maiti, D.; Tong, X.; Mou, X.; Yang, K. Carbon-Based Nanomaterials for Biomedical Applications: A Recent Study. *Front. Pharmacol.* **2019,** *9*. DOI: 10.3389/fphar.2018.01401.

Malam, Y.; Loizidou, M.; Seifalian, A. Liposomes and Nanoparticles: Nanosized Vehicles for Drug Delivery in Cancer. *Trends Pharmacol. Sci.* **2009,** *30*, 592–599.

Martens, T. F.; Remaut, K.; Deschout, H.; Engbersen, J. F. J.; Hennink, W. E.; van Steenbergen, M. et al. Coating Nanocarriers with Hyaluronic Acid Facilitates Intravitreal Drug Delivery for Retinal Gene Therapy. *J. Control. Release* **2015,** *202*, 83–92.

Matsumura, Y. Preclinical and Clinical Studies of NK012, an SN-38- Incorporating Polymeric Micelles, Which Is Designed Based on EPR Effect. *Adv. Drug Deliv. Rev.* **2011,** *63* (3), 184–192.

Mattos, B. D.; Rojas, O. J.; Magalhaes, W. L. E. Biogenic Silica Nanoparticles Loaded with Neem Bark Extract as Green, Slow-Release Biocide. *J. Clean. Prod.* **2017,** *142*, 4206–4213.

Meng, Z. Q.; Wei, F.; Wang, R. H.; Xia, M. G.; Chen, Z. G.; Wang, H. P. NIR-Laser-Switched *in Vivo* Smart Nanocapsules for Synergic Photothermal and Chemotherapy of Tumors. Adv. Mater. **2016,** *28*, 245–253.

Miyano, T. Anionic Amino Acid Dendrimer–Trastuzumab Conjugates for Specific Internalization in HER2-Positive Cancer Cells. *ACS Pub.* **2020.** https://pubs.acs.org/doi/abs/10.1021/mp100 105c

Montaseri, H.; Kruger, C. A.; Abrahamse, H. H. Organic Nanoparticle Based Active Targeting for Photodynamic Therapy Treatment of Breast Cancer Cells. *Oncotarget*, **2020,** *11,* 2120–2136.

Moon, H.; Kumar, D.; Kim, H.; Sim, C.; Chang, J. H.; Kim, J. M. Amplified Photoacoustic Performance and Enhanced Photothermal Stability of Reduced Graphene Oxide Coated Gold Nanorods for Sensitive Photo Acoustic Imaging. ACS Nano. **2015,** *9*, 2711–2719.

Mu, Q.; Kievit, F. M.; Kant, R. J.; Lin, G.; Jeon, M.; Zhang, M. Anti-HER2/neu Peptide-Conjugated Iron Oxide Nanoparticles for Targeted Delivery of Paclitaxel to Breast Cancer Cells. *Nanoscale.* **2015,** *7* (43), 18010–18014.

Murawala, P.; Tirmale, A.; Shiras, A.; Prasad, B. L. V. In Situ Synthesized BSA Capped Gold Nanoparticles: Effective Carrier of Anticancer Drug Methotrexate to MCF-7 Breast Cancer Cells. *Mater. Sci. Eng. C.* **2014,** *34*, 158–167.

Niazvand, F.; Orazizadeh, M.; Khorsandi, L.; Abbaspour, M.; Mansouri, E.; Khodadadi, A. Effects of Quercetin-Loaded Nanoparticles on MCF-7 Human Breast Cancer Cells. *Medicina* **2019,** *55*. DOI: 10.3390/medicina55040114

Nie, S.; Xing, Y.; Kim, G. J.; Simons, J. W. Nanotechnology Applications in Cancer. *Ann. Rev. Biomed. Eng.* **2007,** *9*, 257–288.

Niemirowicz, K.; Surel, U.; Wilczewska, A. Z.; Mystkowska, J.; Piktel, E.; Gu, X. et al. Bactericidal Activity and Biocompatibility of Ceragenin-Coated Magnetic Nanoparticles. *J. Nanobiotechnol.* **2015,** *13*. DOI: 10.1186/s12951-015-0093-5.

Nikalje, A. P. Nanotechnology and Its Applications in Medicine. *Med. Chem.* **2015,** *5*, 81–89.

Ochekpe, N. A; Olorunfemi, P. O.; Ngwuluka, N. C. Nanotechnology and Drug Delivery Part 1: Background and Applications. *Trop. J. Pharm. Res.* **2009,** *8* (3), 265–274.

Oprean, C.; Zambori, C.; Borcan, F.; Soica, C.; Zupko, I.; Minorics, R. Anti-Proliferative and Antibacterial in Vitro Evaluation of the Polyurethane Nanostructures Incorporating Pentacyclic Triterpenes. *Pharm. Biol.* **2016,** *54* (11), 2714–2722.

Panariti, A.; Miserocchi, G.; Rivolta, I. The Effect of Nanoparticle Uptake on Cellular Behavior: Disrupting or Enabling Functions? *Nanotechnol. Sci. Appl.* **2012,** *5*, 87–100.

Park, J. H.; Gu, L.; Von Maltzahn, G.; Ruoslahti, E.; Bhatia, S. N.; Sailor, M. J. Biodegradable Luminescent Porous Silicon Nanoparticles for *in Vivo* Applications. Nat. Mat. **2009,** *8*, 331–336.

Park, J. W.; Hong, K.; Kirpotin, D. B. Anti-HER2 Immunoliposomes. *Clin. Cancer Res.* **2002,** *8*, 1172–1181.

Pastan, I.; Hassan, R.; Fitzgerald, D. J. Immunotoxin Therapy of Cancer. *Nat. Rev. Cancer* **2006,** *6*, 559–565.

Patiño, T.; Soriano, J.; Barrios, L.; Ibáñez, E.; Nogués, C. Surface Modification of Microparticles Causes Differential Uptakeresponses in Normal and Tumoral Human Breast Epithelial Cells. *Sci. Rep.* **2015,** *5*. DOI: 10.1038/srep11371.

Patra, J. K.; Das, G.; Fraceto, L. F.; Campos, E. V. R.; Rodriguez-Torres, M. D. P. et al. Nano Based Drug Delivery Systems: Recent Developments and Future Prospects. *J. Nanobiotechnol.* **2018a,** *16*, 1–33. https://doi.org/10.1186/s12951-018-0392-8.

Patra, J. K.; Gitishree, D.; Fernandes, F. L.; Estefania, V.; Maria, del P.; Laura, S. et al. Campos EVR. *J. Nanobiotechnol.* **2018b,** *16*. DOI: 10.1186/s12951-018-0392-8.

Peer, D.; Zhu, P.; Carman, C. V. Selective Gene Silencing in Activated Leukocytes by Targeting siRNAs to the Integrin Lymphocyte Function-Associated Antigen-1. *Proc. Natl. Acad. Sci. USA* **2007,** *104*, 4095–4100.

Pelaz, B.; del Pino, P.; Maffre, P.; Hartmann, R.; Gallego, M.; Rivera-Fernandez, S. et al. Surface Functionalization of Nanoparticles with Polyethylene Glycol: Effects on Protein Adsorption and Cellular Uptake. *ACS Nano.* **2015,** *9*, 6996–7008.

Praetorius, N. P.; Mandal, T. K, Engineered Nanoparticles in Cancer Therapy. *Recent Pat. Drug Deliv. Formul.* **2007,** *1*, 37–51.

Reichel, D.; Tripathi, M.; Perez, J. M. Biological Effects of Nanoparticles on Macrophage Polarization in the Tumor Microenvironment. *Nanotheranostics* **2019,** *3*, 66–88.

Salvioni, L; Rizzuto, M. A; Bertolini, J. A; Pandolfi, L; Colombo, M.; Prosperi, D. Thirty Years of Cancer Nanomedicine: Success, Frustration, and Hope. *Cancers*. **2019,** *11*. DOI: 10.3390/cancers11121855

Sapra, P.; Tyagi, P.; Allen, T. M. Ligand-Targeted Liposomes for Cancer Treatment. *Curr. Drug Deliv.* **2005,** *2*, 369–381.

Sega, E. I.; Low, P. S. Tumor Detection Using Folate Receptor- Targeted Imaging Agents. *Cancer Metastasis Rev.* **2008,** *27*, 655–664.

Sengupta, S.; Sasisekharan, R. Exploiting Nanotechnology to Target Cancer. *Br. J. Cancer.* **2007,** *96*, 1315–1319.

Sethi, M.; Sukumar, R.; Karve, S.; Werner, M. E.; Wang, E. C.; Moore, D. T. et al. Effect of Drug Release Kinetics on Nanoparticle Therapeutic Efficacy and Toxicity. *Nanoscale* **2014,** *6*, 2321–2327.

Shan, D.; Li, J.; Cai, P.; Prasad, P.; Liu, F.; Rauth, A. M. RGD-Conjugated Solid Lipid Nanoparticles Inhibit Adhesion and Invasion of αvβ3 Integrin Overexpressing Breast Cancer Cells. *Drug Deliv. Transl. Res.* **2015,** *5* (1), 15–26.

Shao, J.; Fang, Y.; Zhao, R.; Chen, F.; Yang, M.; Jiang, J. et al. Evolution from Small Molecule to Nano-Drug Delivery Systems: An Emerging Approach for Cancer Therapy of Ursolic Acid. *Asian J. Pharm. Sci.* **2020,** *1*, 52–75.

Shen, H. X.; Shi, S. J.; Zhang, Z. R.; Gong, T.; Sun, X. Coating Solid Lipid Nanoparticles with Hyaluronic Acid Enhances Antitumor Activity against Melanoma Stem-Like Cells. *Theranostics*. **2015,** *5*, 755–771.

Su, X. Y.; Liu, P. D.; Wu, H.; Gu, N. Enhancement of Radiosensitization by Metalbased Nanoparticles in Cancer Radiation Therapy. *Cancer Biol. Med.* **2014,** *11* (2), 86–91.

Sutradhar, K. B.; Amin, M. L. Nanotechnology in Cancer Drug Delivery and Selective Targeting. *ISRN Nanotechnol.* **2014,** *2014*. DOI: 10.1155/2014/939378.

Tarafdar, J. C.; Sharma, S.; Raliya, R. Nanotechnology: Interdisciplinary Science of Applications *Afr. J. Biotechnol.* **2013,** *12* (3), 219–226.

Taurin, S.; Nehoff, H.; Diong, J.; Larsen, L.; Rosengren, R. J.; Greish, K. Curcumin Derivative Nanomicelles for the Treatment of Triple Negative Breast Cancer. *J. Drug. Target.* **2013,** *21* (7), 675–683.

Thakur, V.; Kutty, R. V. Recent Advances in Nanotheranostics for Triple Negative Breast Cancer Treatment. *J. Exp. Clin. Cancer Res.* **2019,** *38*. DOI: 10.1186/s13046-019-1443-1

Valavanidis, A.; Vlachogianni, T. Engineered Nanomaterials for Pharmaceutical and Biomedical Products New Trends, Benefits and Opportunities. *J. Pharma. Rep.* **2016,** *1* (105) 1–8.

Veetil, J. V.; Ye, K. Development of Immunosensors Using Carbon Nanotubes. *Biotechnol. Prog.* **2007,** *23*, 517–531.

Wahab, R.; Siddiqui, M. A.; Saquib, Q.; Dwivedi, S.; Ahmad, J.; Musarrat, J. ZnO Nanoparticles Induced Oxidative Stress and Apoptosis in HepG2 and MCF-7 Cancer Cells and Their Antibacterial Activity. *Colloids Surf. B Biointerf.* **2014,** *117*, 267–276.

Wang, J.; Liu, J.; Liu, Y.; Wang, L. M.; Cao, M. J.; Ji, Y. L. Gd-hybridized Plasmonic Au-Nanocomposites Enhanced Tumor-Interior Drug Permeability in Multimodal Imaging-Guided Therapy. Adv. Mater. **2016,** *28*, 8950–8958.

Wang, M.; Zhao, T.; Liu, Y.; Wang, Q.; Xing, S.; Li, L. Ursolic Acid Liposomes with chitosan Modification: Promising Antitumor Drug Delivery and Efficacy. *Mater. Sci. Eng. C Mater. Biol. Appl.* **2017a,** *71*, 1231–1240.

Wang, S.; Meng, X.; Dong, Y. Ursolic Acid Nanoparticles Inhibit Cervical Cancer Growth in Vitro and in Vivo via Apoptosis Induction. *Int. J. Oncol.* **2017b,** *50* (4), 1330–1340.

Wang, T.; Hou, J. H.; Su, C.; Zhao, L.; Shi, Y. J. Hyaluronic Acid-Coated Chitosan Nanoparticles Induce ROS-Mediated Tumor Cell Apoptosis and Enhance Antitumor Efficiency by Targeted Drug Delivery via CD44. *J. Nanobiotechnol.* **2017c,** *15*. DOI: 10.1186/s12951-016-0245-2.

Wang, Y.; Santos, A.; Evdokiou, A.; Losic, D. An Overview of Nanotoxicity and Nanomedicine Research: Principles, Progress and Implications for Cancer Therapy. *J. Mater. Chem. B.* **2015,** *3*, 7153–7172.

Wason, M. S.; Colon, J.; Das, S.; Seal, S.; Turkson, J.; Zhao, J.; Baker, C. H. Sensitization of Pancreatic Cancer Cells to Radiation by Cerium Oxide Nanoparticle-Induced ROS Production. *Nanomedicine.* **2013,** *9*, 558–569.

White, B. E.; White, M. K.; Adhvaryu, H.; Makhoul, I.; Nima, Z. A.; Biris, A. S. et al. Nanotechnology Approaches to Addressing HER2 Positive Breast Cancer. *Cancer Nano.* **2020,** *11*. DOI: 10.1186/s12645-020-00068-2.

Xu X; Ho W; Zhang X; Bertrand, N. Farokhzad, O. Cancer Nanomedicine: From Targeted Delivery to Combination Therapy. *Trends Mol. Med.* **2015,** *21*, 223–232.

Yang, G.; Yang, T.; Zhang, W.; Lu, M.; Ma, X.; Xiang, G. In Vitro and in Vivo Antitumor Effects of Folate-Targeted Ursolic Acid Stealth Liposome. *J. Agric. Food Chem.* **2014,** *62* (10), 2207–2215.

Zamboni, W. C. Liposomal, Nanoparticle, and Conjugated Formulations of Anticancer Agents. *Clin. Cancer Res.* **2005,** *11* (23), 8230–8234.

Zhang, H.; Li, X.; Ding, J.; Xu, H.; Dai, X. Hou, Z. Delivery of Ursolic Acid (UA) in Polymeric Nanoparticles Effectively Promotes the Apoptosis of Gastric Cancer Cells through Enhanced Inhibition of Cyclooxygenase 2 (COX-2). *Int. J. Pharm.* **2013,** *441* (1–2), 261–268.

Zhang, X. Q. Multimodal Nanodiamond Drug Delivery Carriers for Selective Targeting, Imaging, and Enhanced Chemotherapeutic Efficacy. *Adv. Mater.* **2011,** *23*, 4770–4775.

Zhao, T.; Liu, Y.; Gao, Z.; Gao, D.; Li, N.; Bian, Y. et al. Self-Assembly and Cytotoxicity Study of PEG-Modified Ursolic Acid Liposomes. *Mater. Sci. Eng. C Mater. Biol. Appl.* **2015,** *53*,196–203.

Zhou, H. IGF1 Receptor Targeted Theranostic Nanoparticles for Targeted and Image-Guided Therapy of Pancreatic Cancer. *ACS Nano.* **2015,** *9* (8), 7976–7991.

Zhou, Y.; Chen, D.; Xue, G.; Yu, S.; Yuan, C.; Huang, M. et al. Improved Therapeutic Efficacy of Quercetin-Loaded Polymeric Nanoparticles on Triple-Negative Breast Cancer by Inhibiting uPA. *RSC Adv.* **2020,** *10*. DOI: 10.1039/D0RA04231E.

CHAPTER 4

Cogent and Strategic Approaches for Gene Delivery with Nanostructures

A. UMAMAHESWARI[1], M. VIJAYALAKSHMI[1], M. DHIVYA SREE[2], S. LAKSHMANA PRABU[1*], and A. PURATCHIKODY[1]

[1]*Department of Pharmaceutical Technology, University College of Engineering (BIT Campus), Anna University, Tiruchirappalli 620024, Tamil Nadu, India*

[2]*Department of Biomedical Engineering, PSG College of Technology, Coimbatore 641004, Tamil Nadu, India*

[*]*Corresponding author. E-mail: slaxmanvel@gmail.com*

ABSTRACT

Delivering drugs to the target site in the body with a desired quantity is an important task, and it has been considered as a major issue in providing effective treatment for a number of diseases through the drug-delivery system. On the other hand, conventional dosage have limitations which are incomplete effectiveness, reduced biodistribution, and deficiency in the selectivity. In recent times, controlled drug-delivery systems have had several advantages relative to conventional dosage forms. Recently, a considerable interest in nanotechnology products development with different class of biopolymer has seized the promising significant improvements in patient's health care. Due to their versatility and enormous nanotechnology development, now it is possible to deliver the drug at the target site also enable to receive the right dose at right time in right manner. Lately, gene delivery has been considered as one of the auspicious in treating different diseases like AIDS, cancer, and cardiovascular diseases in human, which acquired genetically.

The Science of Nanomaterials: Basics and Applications. Suresh C. Ameta & Rakshit Ameta (Eds.)

In gene delivery, nanobiomaterials surface properties are modified; then utilized along with the genetic components. These genetic materials are protected by this nanoparticle until it reaches the target site. Biopolymers have an important role in biocompatibility, biodegradability, and nanoparticle surface features like particle surface charge, size, and shape. This chapter summarizes the different preparation techniques of nanocarriers utilized as/in drug delivery device, biopolymers in gene delivery, its delivery mechanisms, and its applications.

4.1 INTRODUCTION

In health care, delivering drugs in anticipated concentration at the desired site to induce the anticipated therapeutic effect is a crucial one. Limitations of conventional dosage are incomplete effectiveness, reduced biodistribution, and deficiency in the selectivity, may damage normal cells and organs due to its toxic property (Nevozhay et al., 2007). But these limitations are overawed by the target drug-delivery system. Some of the outstanding achievements at the end of the last century are the nanotechnology, which deals about the study of biological and synthetic materials properties in the field of medicine and pharmaceuticals. The word "nano" is a word from Greek, which means "dwarf". The meaning for nano is very small or miniature size (Nikalje, 2015). National Nanotechnology Initiative(NNI) defined that nanoparticles are structures of size from 1 to 100 nm in at least one dimension. Several hundred nanometers in size of the particles also come under the term "nano." Nanotechnology is the study of extremely small structures by manipulation of matter on a near-atomic scale to produce new structures, materials, and devices, which can offer a promising tool in an unprecedented scientific advancement for many sectors, such as medicine, consumer products, energy production, and in engineering manufacturing (Approaches to safety nanotechnology).

Nanotechnology can improve the existing technologies and also enhance intensely the effectiveness of innovative applications in all the sectors. The nanomaterials are made from inorganic or organic materials, which has peculiar characteristic properties with respect to its physicochemical properties such as particle shape, size, solubility, surface charge, surface area, generating oxidant potential, agglomeration, magnetic, electronic properties, and biological properties (ability to penetrate tissue and cell barriers) which are different from the properties of both macro- and microscale particles (Chen et al., 2013; Suri et al., 2007). These peculiar properties of the nanostructure material made considerable interest in developing nanotechnological products in different biomedical applications and in health care. Surface modification of nanomaterials can improve the binding property to the receptors or tissue, which increase the cellular uptake, immune compatibility, and solubility. Many private and public research organizations are involved to advance the medical research by developing nanoproducts to improve health care (Nikalje, 2015). Suitably designed nanotechnology products can hold the potential role and significant improvements in the patient's health care system like in detection of disease, delivering the drug in controlled manner, and in tissue engineering as biosensors (Wilczewska et al., 2012).

Avoiding the body's defense mechanism is a main advantage of nanoparticle (Bertrand and Leroux, 2012). Due to nanoparticles unique property different nano moieties includes nanocapsules, nanotubes, nanogels, dendrimers, silicon or carbon materials, magnetic nanoparticles, micelles, emulsions, and lipid-based nanoparticle (solid lipid nanoparticle, liposomes) are utilized as specific targeted drug delivery/controlled drug delivery of different classes of molecules such as drugs, macromolecules like protein, peptides, and genetic materials for treating several diseases (Nikalje, 2015). Targeted therapy depends upon the design and the conjugation between the drug and nanocarrier. In targeted therapy, either the drug can be attached covalently or adsorbed to the surface or can be encapsulated inside of the nanomaterial. Covalent bonding has the advantage when compared with other bonding, as it allows controlling the number of drug molecules attaching to the nanomaterial. Once the drug-nanomaterial conjugates reach the target site, controlled release of drug can be attained by changing the physiological environment like acidity, basicity, temperature, activity of enzyme, and osmolality. Comparison of scale units with various nanomaterials in drug-delivery system is shown in Figure 4.1 (Wilczewska et al., 2012). Size distribution of the different nano products is shown in Table 4.1.

4.2 NANOPARTICLES IN DRUG DELIVERY

Transporting and delivering the drug to the target site is a fundamental task and it is a main problem in the delivery of drug. Researchers focused on the targeted drug delivery as this task provides a huge area of development. Nanotechnological products are widely used in nanomedicine as drug delivery. Avoiding the body's defense mechanism is one of the main strengths of nanoparticle. Nanoparticle can alter and improve the pharmacokinetics and biodistribution of the drug, by escaping from the bloodstream via intracellular or transcellular pathway and paracellular pathway and depositing the drugs in the morbid region. Different morphological nanomaterials such as nanoparticles, nanocapsules, nanotubes, nanogels, micelles, drug conjugates, metallic nanoparticles, and dendrimers are used in drug delivery.

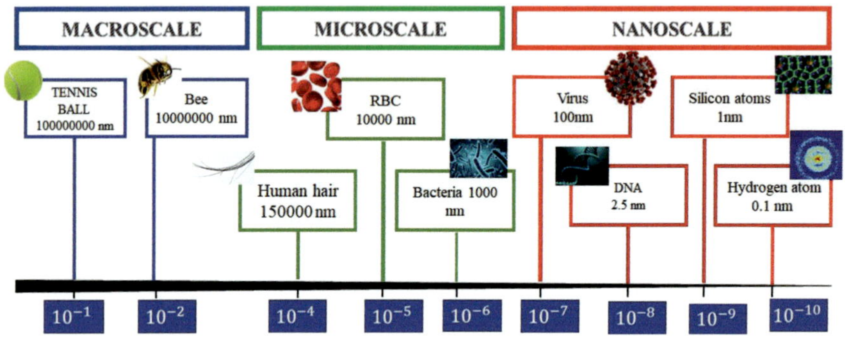

FIGURE 4.1 Comparison of scale units with various nanocarriers in drug-delivery system. *Source:* Adapted from Wilczewska et al. (2012). With permission. © 2012 by Institute of Pharmacology Polish Academy of Sciences.

Other than the size, the shape of the nanoparticle plays a critical role in drug delivery and transport to the target site. The nanoparticle shape can affects biocompatibility, cellular uptake, and retention of the drug in tissues and organs. Drugs can be efficiently delivered in the specific target site by appropriate designing of the preparation technique, suitable lipid, and polymer as carrier with desired size and shape of the nanoparticle.

Any particle having a size less than 1 μm diameter is called as nanoparticle. Various techniques are followed to develop particles <100 nm due to its unique properties like physical, useful biological, and potentially different properties.

Attaining particle size <100 nm is feasible from the hard materials like silica, diamonds, and metal oxides when compared with drug and polymer. Hard materials are having the melting points above 1000°C, whereas drugs as soft materials having the melting point below 300°C, hence preparation of nanoparticles in the range of 1–100 nm is more difficult due to its stickiness. Preparation of nanoparticle in the particle size between 50 and 300 nm for drug delivery requires 10^4–10^8 molecules. This particle size range can be prepared from either in solution phase or particles in millimeter size (Gupta, 2006).

4.2.1 NANOPARTICLE–DRUG DELIVERY PREPARATION TECHNIQUES

Nanoparticle for the drug delivery can be prepared by the following methods:

1. Dispersion of preformed polymers
   i. Solvent evaporation method
   ii. Solvent diffusion method
   iii. Nanoprecipitation method
   iv. Salting-out method
2. Polymerization of monomers
3. Ionic gelation method for hydrophilic polymers (Mahapatra and Singh, 2011; Soppimath et al., 2001; Mohanraj and Chen, 2006; Pinto Reis et al., 2006; Garg et al., 2011).

4.2.1.1 DISPERSION OF PREFORMED POLYMERS

Solvent Evaporation Method: In this technique, the polymer is solubilized in an organic solvent in which the polymer is completely soluble. Subsequently in the polymer solution, the drug is solubilized or dispersed to form an oil phase (internal phase). This oil phase is slowly added into a continuous phase containing a suitable emulsifying agent, further continuous mixing of the solution to induce emulsification and to form an oil/water (o/w) emulsion. After producing a stable emulsion, evaporated the organic solvent by raising the pressure or temperature with continuous stirring.

The particle size of the nanoparticles depends upon the various process parameters like concentration of surfactant/emulsifying agent, polymer, and rate of stirring speed to remove the solvent (Jaiswal et al., 2004; Hoa et al., 2012; Tice and Gilley, 1985). Systematic solvent evaporation method of nanoparticle preparation is shown in Figure 4.2.

Spontaneous Emulsification/Solvent Diffusion Method: Solvent diffusion method is a modified method of solvent evaporation method. In this method, the oil phase consists of acetone or methanol as the water-miscible solvent and small quantity of dichloromethane or chloroform as the water-immiscible organic solvent. An interfacial turbulence between the two solvents is created due to spontaneous diffusion, which leads to form small particles. Particle in small size are achieved by increasing the water miscible solvent concentration. This method can be used for both hydrophobic and hydrophilic drugs. A multiple emulsion like w/o/w needs to be produced, if the drug is hydrophilic in nature.

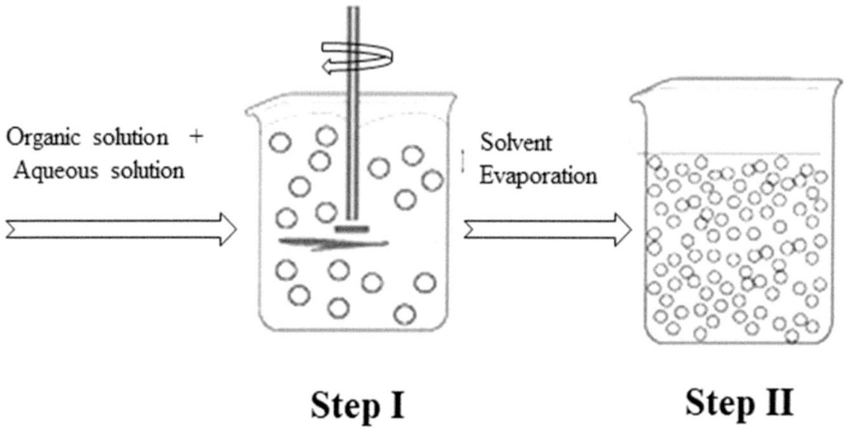

FIGURE 4.2 Preparation of nanoparticle by solvent evaporation method.

In solvent diffusion technique, nanoparticles are formed by the following steps:

- ➢ Emulsion droplets are produced when the polymeric solution is added in the aqueous phase.
- ➢ Solvents like acetone or methanol diffuses quickly from each emulsion droplet then reduces the particle size into nano size.
- ➢ Following the "solvent-evaporation" process, remaining water-immiscible solvent like dichloromethane or chloroform is removed from the system, which makes the droplets solidify to finally form polymeric nanoparticles (Mahapatra and Singh, 2011; Soppimath et al., 2001; Mohanraj and Chen, 2006; Jaiswal et al., 2004; Niwa et al., 1993; Murakami et al., 1999; Leroux et al., 1995; Vauthier and Bouchemal, 2011).

  Systematic spontaneous emulsification/solvent diffusion method of nanoparticle preparation is shown in Figure 4.3.

Nanoprecipitation Method: Fessi first introduced the nanoprecipitation technique for the preparation of polymeric nanoparticles. Generally, hydrophobic drug is entrapped in this method, but hydrophilic drug is also adapted in this method. In this method, initially, polymer is precipitated from an organic solution followed by solicitation due to interfacial deposition by displacement of semipolar solvents with water with or without surfactant.

In general, both drug and polymer are solubilized in a polar, water-miscible solvent of intermediate polarity like acetone, acetonitrile, ethanol, or methanol. This phase is injected (i.e., drop-by-drop addition) into an aqueous solution with continuous stirring consisting a stabilizer as a surfactant. Deposition of polymer is occurred by rapid diffusion of solvent between water and organic solvents, leads to produce a colloidal suspension. Subsequently, under the reduced pressure, the solvent is removed. In the first step, the development of colloidal polymer particle and a phase separation are achieved by using water-miscible solvent. The developed nanosuspension is converted into a fine nanoparticle powder by mixing with a cryoprotectant like 5% mannitol and then freeze dried (Nagavarma et al., 2012; Govender et al., 1999; Mehrotra and Pandit, 2012; Barichello et al., 1999; Fessi et al., 1989; Hornig et al., 2009; Moorthi and Kathiresan, 2013; Galindo-Rodriguez et al., 2004; Karmi et al., 2011). Preparation of nanoparticle by nanoprecipitation technique is shown in Figure 4.4.

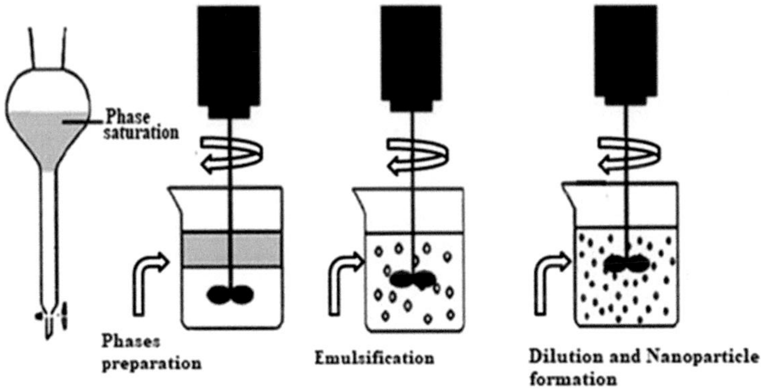

FIGURE 4.3 Preparation of nanoparticle by spontaneous emulsification/solvent diffusion method.

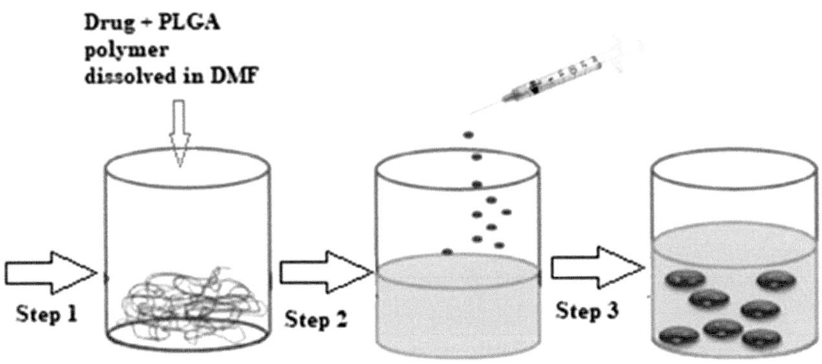

FIGURE 4.4 Nanoparticle preparation by nanoprecipitation method.

Salting-Out Method: Salting-out technique is an alternative method of the emulsion process, which can be overwhelmed the usage of hazardous organic solvents to the environment as well as to physiological system. Through the salting-out process, water-miscible solvent is removed from the aqueous solution.

In a water-miscible organic solvent, both drug and polymer are solubilized. This resulting solution is added into an aqueous solution consists of salting-out agent and a colloidal stabilizer. Magnesium chloride hexahydrate 60% w/w or 1:3 polymer to salt ratio of magnesium acetate tetrahydate is used as salting-out agent whereas hydroxyethylcellulose or polyvinylpyrrolidone are used as colloidal stabilizer.

In this method, due to the presence of salts, no solvent diffusion takes place. This salting-out agents inhibits the miscibility between organic solvent and aqueous phase leads to form an emulsion.

This formed emulsion is diluted with excess water in stirring condition to produce a reverse salting-out effect, which induces polymer precipitation by migration of water miscible organic solvent into the aqueous phase. Further the drug in the solution is encapsulated by this precipitate polymer leads to produce nanoparticles. Salting-out agents and residual solvents in the formulation are removed by cross-flow filtration technique.

The selection of the salting-out agent is important, because it plays a crucial role important role in drug encapsulation efficiency also this process can minimizes the encapsulation of protein. In addition, it can be used for encapsulation of heat-sensitive substances. The main disadvantages of this method are that it cannot be useful for lipophilic drugs and extensive washing steps (Allemann et al., 1992, 1993; Galindo-Rodriguez et al., 2004; Garg et al., 2011; Pinto Reis et al., 2006; Nagavarma et al., 2012). Salting-out method of nanoparticle preparation is shown in Figure 4.5.

4.2.1.2 POLYMERIZATION METHOD

NPs can be prepared from monomers by polymerization to form NPs in an aqueous solution. The techniques are

a) *Emulsion polymerization*: Involves monomer emulsification in nonsolvent phase.
   It has been further classified into two types, based on the phases.
b) *Dispersion polymerization*: Involves monomer dispersion in nonsolvent phase.

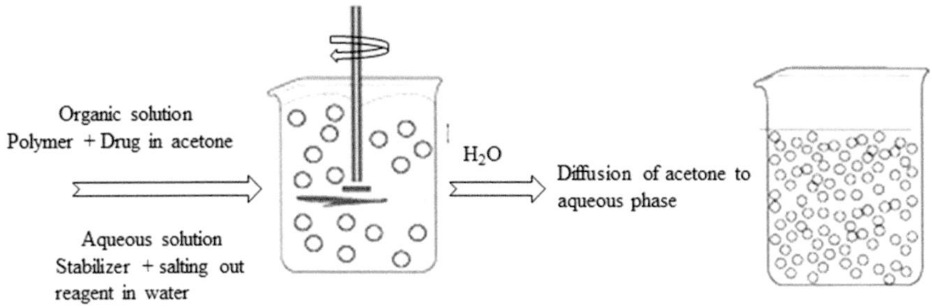

FIGURE 4.5 Nanoparticle preparation by salting-out method.

Emulsion Polymerization: It is a quick and readily scalable method for the preparation of nanoparticle. This method involves monomer dispersion into an emulsion or inversion of microemulsion or nonsolvent in which monomer is not soluble. Various surfactants and stabilizers are used to prevent aggregation in the early stages of polymerization.

Monomer is solubilized in the aqueous continuous phase containing an initiator (ion or free radical). Based on the anionic polymerization mechanism, monomer is ionized or collide the monomer radicals with other monomers leads to start the chain growth. Drug is incorporated into the nanoparticle either by solubilizing the drug in polymerization medium or adsorption of drug in the nanoparticle. Both surfactants and stabilizers are used, which leads to produce nanoparticle in suspension form.

Incorporation of drugs in the NPs is by either of the following:

- Drug is dissolved in the polymerization medium.
- Drug is adsorbed onto the fully formed nanoparticle surfaces.

Various stabilizers such as dextran 10, 40, and 70, poloxamer 188, 184, and 237; surfactants such as polysorbate 20, 40, or 80 are used during the preparation of nanoparticle by polymerization method.

The concentration of stabilizer and surfactant can disturb the particle size of the nanoparticle. Ultracentrifugation or by suspending the nanoparticle in isotonic medium can remove the stabilizers and surfactants leads to produce surfactant-free nanoparticles. The size and molecular mass of NPs depend upon the pH of the polymerization medium but NP production is not possible above a pH of 3.0 (Soppimath et al., 2001; Garg et al., 2011; Nagavarma et al., 2012; Wang et al., 2016; Ekman and Sjfholm, 1978; Kreuter, 1982, 1991; Krause et al., 1986). Preparation of nanoparticle by polymerization method is shown in Figure 4.6.

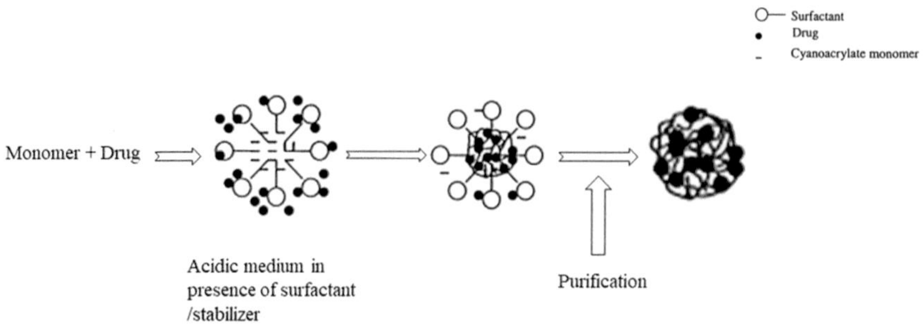

FIGURE 4.6 Preparation of nanoparticle by polymerization method.

4.2.1.3 IONIC GELATION METHOD FOR HYDROPHILIC POLYMERS

In ionic gelation method, polymers like gelatin, chitosan, and sodium alginate are used, which are hydrophilic natural polymers. Due to ionic interaction at room temperature, materials undergo transition from liquid to gel. Based on the drug and polymer solubility, both are solubilized either in weak acidic medium or water. This mixture is added dropwise with continuous stirring into a solution containing both counter ions and stabilizer. Oppositely charged species together forms complexation leads to produce spherical shaped particles. Due to ionic force, an electrostatic interaction is formed between two aqueous phases leads to produce coacervates. Particle size can be reduced to the nanometric range by sonicating the resultant solution (Mahapatra and Singh, 2011; Garg et al., 2011; Nagavarma et al., 2012; Bagul et al., 2012; Fan et al., 2012; Kannan and Manikandan, 2015). Preparation of nanoparticle by ionic gelation method is shown in Figure 4.7.

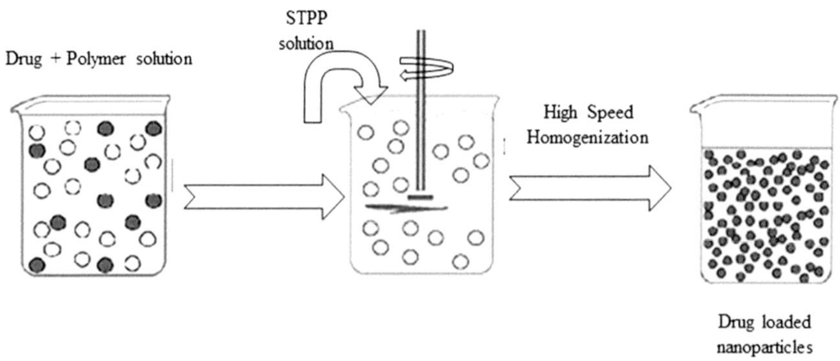

FIGURE 4.7 Preparation of nanoparticle by Ionic gelation method.

4.3 CLASSIFICATION OF NANOPARTICLE

Nanoparticle in drug delivery device is classified into three categories:

1. Polymeric nanoparticle
2. Solid lipid nanoparticle
3. Inorganic nanoparticle

4.3.1 POLYMERIC NANOPARTICLE

In polymeric nanoparticle, drugs are encapsulated in nanoparticle core, surrounded by a shell-like polymeric membrane, entrapped in the polymer matrix, or bound to the particles surface by adsorption (Hans and Lowman, 2006). Polymers used in nanoparticle drug delivery are polyethylene glycol, polyesters, polyanhydrides, polyamides, polyorthoesters, polyaprolctone, poly-vinylpyrrolidone, polyvinyl alcohol, poly(lactide-co-glycolide), poly(lactic acid), poly-ε-caprolactone, PGA(polyglycolic acid), poly-l-glutamic acid, polylactic acid, PNIPAAm [Poly(*N*-isopropylacrylamide)], pHEMA[poly 2-hydroxyethyl methacrylate], PPy [polypyrrole], PAMAM [poly (amido-amine)], dextran, chitosan, and so on (Liechty et al., 2010; Vilar et al., 2012; Srivastava et al., 2016).

4.3.2 LIPID-BASED NANOPARTICLES

Liposomes, oil dispersions (micelles), lipid nanoparticles, and solid lipid nanoparticles are the major classes of lipid-based nanocarriers used in drug delivery. Lipid particulate drug-delivery systems are phytosomes, liposomes, liposphers, niosomes, ethosomes, cubosomes, cochleates, tranfersomes, and nanoemulsions.

In general, lipid-based formulations are used to improve the solubility of Class II and Class IV drugs under Biopharmaceutics Classification System (BCS).

Commonly such as sterols, cholesterol, phospholipids, oils, waxes, vitamins (fat soluble), and glycerides are used as lipids in lipid-based nanoparticles. These lipids are readily soluble in nonpolar solvents and not in polar solvents.

Lipids are classified into the following:

- *Homolipids*—Homolipids are fatty acid esters with alcohols, contains only hydrogen, carbon, and oxygen. Cerides, glycerides, and sterides are examples of homolipids.

- *Heterolipids*—In addition to hydrogen, carbon and oxygen, heterolipids contains nitrogen (N), and phosphorus (P). Phospholipids, glycolipids, and sulfolipids are examples of heterolipids.
- *Complex lipids*—Complex lipids are occurs by closely linkage with cell membrane proteins and subcellular particles (Puri et al., 2009; Attama et al., 2012).

4.3.3 INORGANIC NANOPARTICLE

Inorganic nanoparticles are classified into quantum dots (QDs) and metal nanoparticle.

Quantum Dots: Quantum dots (QD) are luminescent nanocrystals with the size of 1–10 nm, which have exclusive optical property. Elements belonging to the group from II to VI such as Ag, Hg, Cd, Ln, Zn, P, Pb, Te, and Se are generally used in QDs preparation. These QDs have special characteristic properties, which includes improved singal brightness, size-tunable light emission, instantaneous excitation of multiple fluorescence colors, and photo bleaching. Depends upon the laser light source size these fluorescent QDs can show different colors like Maple red-orange, fort orange, orange, yellowish orange, yellowish green, yellow, green, adirondack green, greenish blue, and blue.

QD is a spherical semiconductor made up of three parts such as core, shell, and cap (Qi and Ga, 2008; Lakshmana Prabu and Suriya Prakash, 2017). Core is made on semiconductor material, that is, CdSe, shell is ZnS coat that surrounds the core, cap encapsulates as double layer (Figure 4.8).

The important application of QD in drug delivery is that it can be traceable; therefore it is applied in elucidating the absorption, distribution, metabolism, excretion, pharmacodynamics properties of drug candidate, in vivo transfection, and in disease diagnosis. These QDs are successfully utilized in gene delivery by conjugation of QD with genetic material either DNA or messenger RNA (mRNA), which can protect the genetic material from degradation (Qi and Ga, 2008; Lakshmana Prabu and Suriya Prakash, 2017; Zhang et al., 2011; Probst et al., 2013; Gulia and Kakkar, 2013; Dabbousi et al., 1997; Modani ct al., 2013; Dey and Rao, 2011; Rathore et al., 2006; Mishra et al., 2012; Jamiesona et al., 2007; Drbohlavova et al., 2009; Zrazhevskiyn and Gao, 2009; Wang and Chen, 2011; Vengala et al., 2012; Mukherjee and Das, 2011; Cheki et al., 2013; Ozkan, 2004; Dorfs et al., 2011; Mokari and Banin, 2003).

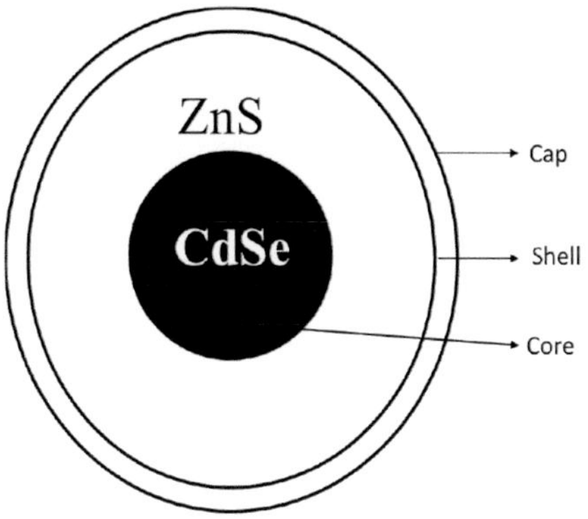

FIGURE 4.8 Structure of a quantum dot.

Metal Nanoparticle: The association of organic compounds with inorganic nanoparticles leads to form hybrids are called metal nanoparticle. Metal nanoparticles are made with either iron oxide or inorganic elements like gold, silver, zinc, cadmium, and silica. These metal nanoparticles have unique physicochemical, optical, and electrical properties, which have been utilized in different biomedical applications (Lakshmana Prabu and Suriya Prakash, 2017).

Extensively gold nanoparticles (GNPs) are utilized when compared with other metal nanoparticles due to its distinctive characteristic property like flexibility of synthesis, inert, less cytotoxic, biocompatibility, ability in fictionalization, high functionalization, high absorption coefficient, stability, surface modification, and potential capability of site-specific delivery/target delivery. Due to the above-said properties, GNPs are utilized in gene delivery. The following different methods can be adopted for the preparation of GNP:

1. Physical method—Microwave irradiation method and γ-irradiation method
2. Chemical method—Seeding growth method
3. Green synthesis method—Sunlight irradiation method and natural biomaterial eggshell membrane

GNPs display in different shapes like spherical, irregular shape, suboctahedral, octahedral, tetrahedral, decahedral, icosahedral multiple twined, multiple twined, nanorods, nanotriangles, and nanoprisms (Zhang et al., 2011; Arruebo et al., 2007; Tiwari and Lee, 2013; Alanazi et al., 2010; Kim and Jon, 2012; Stobiecka and Hepel, 2011; Giljohann et al., 2010;

Ghosh et al., 2013; Pissuwane et al., 2011; Khan and Vishakante, 2013; Tarnawski and Ulbricht, 2011).

4.4 GENE DELIVERY

Deletion of genes or mutation in human beings leads to create different genetic disorders like regulation of cell cycle, metabolic pathway disorder, receptor function, cell skeleton, protein function, and its structure (Erdal et al., 2012). Recently, genetic material such as plasmids, DNA, RNA, and siRNA have been used in gene delivery system to replace the undesired genes or damaged genes or production of required proteins. Gene factors such as length, sequence size, charge, stability, and half-life are needed to consider in the gene delivery (Fattal and Barratt, 2009).

In gene delivery, two barriers such as cellular and extracellular barriers obstruct the delivery of genetic material. Delivering the therapeutic genetic material into target site through plasma membrane is the major hurdle in this delivery system. Sequences in this delivery system are bypassing the host immune system, crossing the plasma membrane, endosomal membrane interruption followed by delivering the genetic material into the nucleus. The effectiveness depends on the delivery of genetic material into the specific target site. To carry the hydrophilic and negatively charged DNA and deliver into the cell membrane, a vector/carrier is essential. To any genetic material, a good vector/carrier must provide satisfactory protection and stability during the extracellular delivery phase, passing through cell membranes, cytoplasm, subsequently strip, and nuclear delivery (Putnam, 2006; Pack et al., 2005; Mansouri et al., 2004; Yang et al., 1995; Giannoukakis et al., 1999; Liu et al., 2001; Mahato, 1999). Among the various carriers, nanoparticles have been considered as a suitable carrier due to the following reasons:

- Nanoparticle can occur in the same size as proteins.
- Nanosize material provides greater surface area, which gives higher binding ability with number of functional groups in the surface.
- Also, it has modified drug release and surface characteristic properties (Nitta and Numata, 2013).
- Polymer source and its physicochemical property play a critical role in either encapsulate the genetic material or conjugate with genetic material in nanoparticle and to deliver the genetic material in treating the different genetic diseases/disorders (Luten et al., 2008; Merdan et al., 2002; Wong et al., 2007; Rana, 2007).

4.4.1 MECHANISMS OF GENE DELIVERY

Plasmid DNA:Plasmid DNA (pDNA) is most preferred genetic material in gene delivery. This pDNA, when enters into the nucleus, it transcribed and translate the coding gene to the protein, subsequently, it will be expressed from the cell.

RNA Interference: RNA interference (RNAi) is triggered by double-stranded RNA (dsRNA), stimulates the antiviral interferon, which can inhibit the synthesis of protein by mRNA degradation. This mRNA is a small noncoding nucleic acid; through posttranslation regulation protein expression is regulated.

Small Interfering RNA: Small interfering RNA (siRNA) consists of 21–23 nucleotides and more specific than long dsRNA, which can inhibit the gene expression (Rana, 2007; Guo et al., 2010; Van Rooij, 2011).

4.4.2 LIMITATION IN GENE DELIVERY

There are two major limitations in gene delivery system:

1. The first limitation is about the carrier. A good carrier to carry the nucleic acid to the target site without affecting its integrity and potential risks. In general, viruses due to its defense mechanism (reticuloendothelial system) they can have the ability to locate and identify the defined target cells. Whereas chemical vectors can identify and conjugate with defined target through different methods.
2. The second limitation is the nucleic acid penetration through plasma membrane into the cell. Viruses enter through its natural mechanisms, whereas chemical vector achieve through by disturbing the plasma membrane or internal vesicular membrane (Jin et al., 2014).

4.4.3 CATEGORIES OF GENE DELIVERY SYSTEMS

There are two types of gene delivery systems:

1. Viral transduction systems
2. Nonviral transfection systems

Viral Transduction Systems: During the initial stage of gene delivery, viruses were used. Mechanism in this delivery system is circulation of virus in the blood, cell surface receptor binding, passing into the cell; evade destruction of lysosomal, protection in the cytosol, and delivering the genetic material into the nucleus.

But viruses have certain limitations/disadvantages in this delivery system. The following are the limitations:

- Low loading capacity
- Size of the DNA
- Large scale manufacturing
- Cost toward the quality control testing
- Immunogenicity and potential oncogenicity as safety factors (Itaka and Kataoka, 2009)

Nonviral Transfection Systems: Considering the above limitations/ disadvantages, researchers have given more attention in developing nonviral vector in gene delivery (Jordan, 2003; Luten et al., 2008; Merdan et al., 2002; Wong et al., 2007; Mrsny, 2005).

Nonviral transfection systems consist of plasmid DNA condensed into nanoparticle by a suitable polymer (Dizaj et al., 2014).

Advantages of nonviral transfection systems are:

- Simple preparation technique
- Utilization of synthetic polymers and its structural modification
- Target delivery (cell/tissue)
- Less oncogenic and immunogenic effect
- Loading capacity
- Unhindered size of plasmid
- Production cost
- Ease of vector modification to carry the plasmid (Jin and Ye, 2007; Nitta and Numata, 2013; Lee et al., 2013)

Different stages of polymeric gene delivery are as follows:

1. DNA/polymer complexation. Preferably cationic polymers are used to neutralize the DNA phosphate backbone and prevent the charge repulsion between the cell membrane having negatively charge and DNA, subsequently condense the DNA bulkiness into nanosize complex.
2. DNA/polymer complex is also termed as polyplex passes through the cell membrane. The complex is transported into the cell, through the cell membrane, by either a nonspecific or a receptor-mediated endocytosis.

3. Then the DNA/polymer complex enters into the cytoplasm typically in an endosome, followed by the release of the DNA/polymer complex into the cytoplasm.
4. Cytosolic transport to the nucleus.
5. Delivering the genetic material into the nucleus. Once the genetic material is delivered into the nucleus, now it is free to translate into therapeutic protein or introduced into the genome (Zhang et al., 2011; Jin and Ye, 2007; Nitta and Numata, 2013).

Category of nonviral vector delivery:

1. Organic—Cationic polymers, conjugated polymers, and lipid complexes
2. Inorganic—QDs, magnetic nanoparticles, carbon nanotubes, and metal nanoparticles. (Lee et al., 2013).

4.4.4 POLYMERS IN GENE DELIVERY

Polymers are materials that have been either isolated from natural sources or chemically synthesized. Based on its different environmental conditions such as temperature, pH, ionic strength, magnetic field, electric field, chemical and biological stimuli, polymers can respond and deliver the drug. Biopolymer is a novel biomaterial that has been used very widely in drug-delivery system due to its unique properties like biodegradability, biocompatibility, and immunogenicity. Between natural and synthetic biopolymers, synthetic biopolymers are having good mechanical property, definite structure, degradation kinetic, and so on. Due to the above-said properties, synthetic biopolymers have been used extensively in preparing biodegradable nanoparticles in delivering drugs, genes, and vaccines (Ward and Georgiou, 2011; Twaites et al., 2005; Mahapatra and Singh, 2011; Eliyahu et al., 2005; Pack et al., 2005; Dang and Leong, 2006; Lakshmana Prabu and Ruckmani, 2017).

Polymeric Gene Delivery Vector: Polymers in polymeric gene delivery vector should be biocompatible, biodegradable, and condensate with the genetic material through electrostatic interactions and to release the same into the cytoplasm. Surface modification of NP is an approach; it can assist the electrostatic interaction and condensate the genetic material with cationic polymers to form DNA–nanoparticle complex or conjugate (Jin and Ye, 2007).

Polymer Properties: Polymer with strong positive charges on its surface are having strong condensation property with DNA and it will retard the release, therefore polymers having *pK* values from 5 to 7 are considered in polymeric vector delivery. Molecular structure, molecular weight, and its composition are also considered while selection of polymer in this delivery system (Jin et al., 2014; Lakshmana Prabu and Ruckmani, 2017; Godbey and Mikos, 2001; Xu et al., 2009). A schematic diagram for the gene delivery from nanoparticle is shown (Jin et al., 2014) in Figure 4.9.

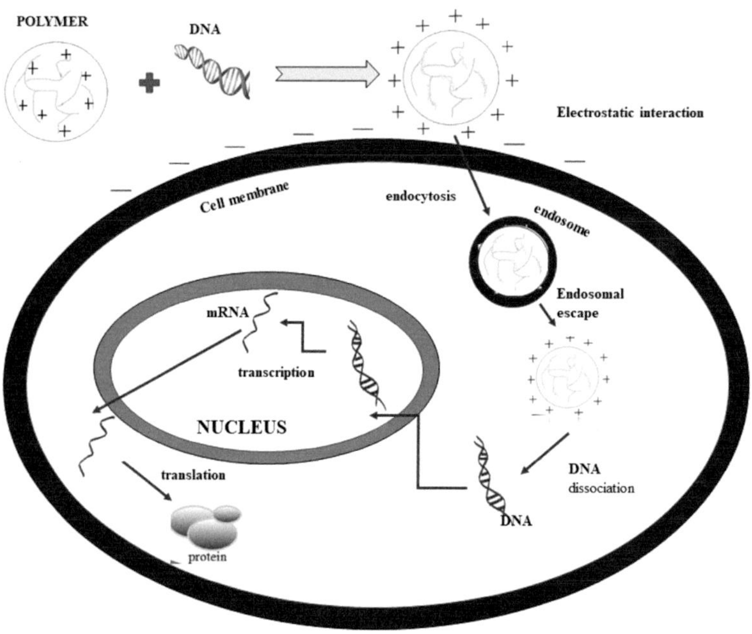

FIGURE 4.9 Gene delivery process of polymeric nanoparticle.

Source: Redrawn from Jin et al., 2014. http://creativecommons.org/licenses/by-nc-nd/3.0/ © 2014. Ivyspring International Publisher.

Preparation of Polymeric Gene Vector: Polymeric gene vector is prepared by mixing cationic polymer with plasmid DNA to form DNA-NP complex by condensation. During condensation between polycation and DNA, DNA undergoes a conformational alteration in its size from 200–300 to 100 nm.

Techniques for condensation of DNA with polymeric material:

1. Evaporation under vacuum
2. Freeze drying

The charge ratio between DNA and polymer is calculated based on the phosphates in DNA and amines in polymer (Oyewumi and Rice, 2006).

4.4.5 DENDRIMERS

Dendrimer is a macromolecule in monodisperse form having regular branched structure with minimum of one branched junction at each repeated unit of three. In dendrimers, DNA and dendrimers are encapsulated in water-soluble polymer and then deposited or sandwiched in functional polymer film.

Dendrimers are prepared by:

1. Divergent method
2. Convergent method

Divergent Method: In this method, dendrimer will grow from multifunctional core molecules to outwards. First-generation dendrimer is formed by reaction between the core and monomer molecules, and then this step is repetitive leads to build layer by layer. Formation of dendrimer by divergent method is shown in Figure 4.10.

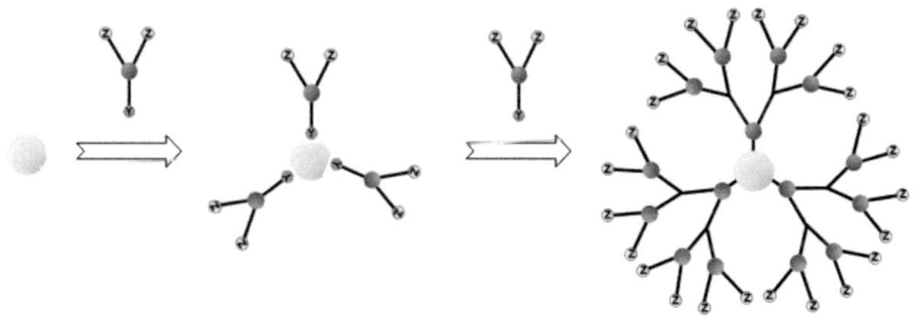

FIGURE 4.10 Formation of dendrimer by divergent methods.

Convergent Method: In this method, the dendrimer is constructed by stepwise process initially from the end groups and then inward progressing. Formation of dendrimer by convergent method is shown in Figure 4.11.

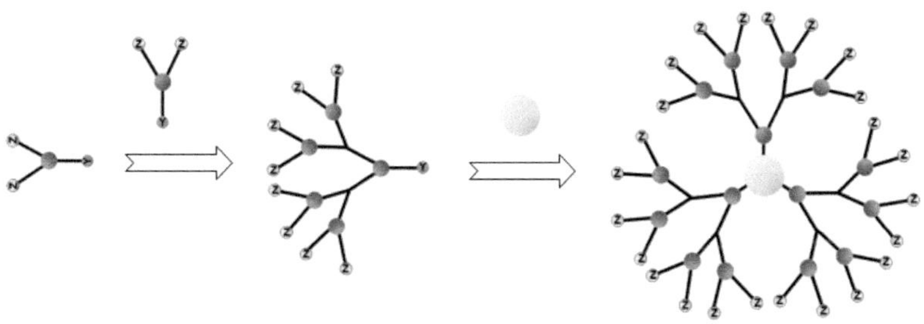

FIGURE 4.11 Formation of dendrimer by the convergent method. Here Y is the chemically active focal point and Z is the functional chemical group of another monomer.

4.4.5.1 OTHER TYPES OF DENDRIMERS

Amino-acid-based dendrimers: Amino acids like phenylalanine, tryptophase, aspartic acid, glutamic acid, glycine, valine, leucine, and alanine are used in the preparation of amino acid-based dendrimers.

Synthesis of amino-acid-based dendrimers are:

1. Grafting of amino acid or peptide and then show on the conventional dendrimer surface.
2. Amino acids or peptides attachment to an organic or peptide core.

Glycodendrimers: Glycodendrimers can be synthesized by coupling of aminoterminated dendrimers with isothiocyanate functionalized glycosyl and mannopyranoside ligands; also with *N*-hydroxysuccinimide (NHS) activated galactopyranosyl derivative.

Hydrophobic dendrimers: Dendrimers having hydrophilic surface and hydrophobic interior is called hydrophobic dentrimers, which can provide improved encapsulation and solubility of hydrophobic drug. These types of dendrimer-branching architectures also hold the hydrophobic drug and imitate the micells of amphiphilic polymer.

Asymmetric dendrimers: Asymmetric dendrimers are prepared by linking the different generations of dendrons to a linear core through coupling reaction leads to form a branched dendrimer (Boas and Heegaard, 2004; Cheng et al., 2008; Dufes et al., 2005; Gao et al., 2008; Gillies and Frechet, 2005; Klajnert and Bryszewska, 2001; Waite et al., 2009; Biswas et al., 2013; Medina and El-Sayed, 2009; Tomalia and Frechet, 2002).

4.5 POLYMERS CLASSIFICATION IN GENE DELIVERY

4.5.1 CATIONIC POLYMERS

Most commonly cationic polymers are used in gene drug delivery. When genetic materials like DNA combined with cationic polymers lead to condensate into distinct entity, they is called polyplexes. These polyplexes are compact nanoparticles, produced by electrostatic interaction between the DNA phosphate negative charge and amines positive charge; N:P ratio determines the strength and weakness of binding property of DNA and polymers (Zhang et al., 2011; Pack et al., 2005; Synowiecki and Al-Khateeb, 2003; Krikorian et al., 2002; Lungwitz et al., 2005; Nimesh, 2012). Cationic polymers and their structure are shown in Table 4.1.

TABLE 4.1 Structure of Cationic Polymers.

S. no	Name of the cationic olymer	Structure
1	Chitosan	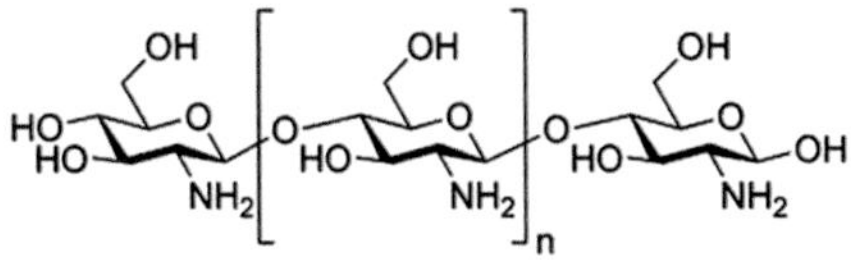
2	Poly L-lysine	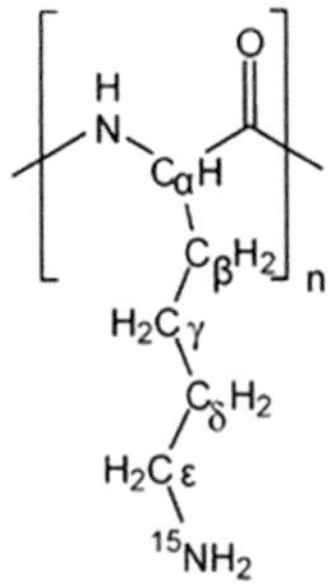
3	Polyethyleneimine	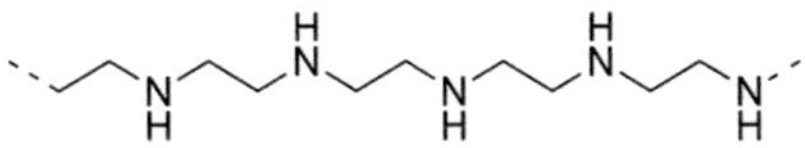
4	Branched polyethyleneimine	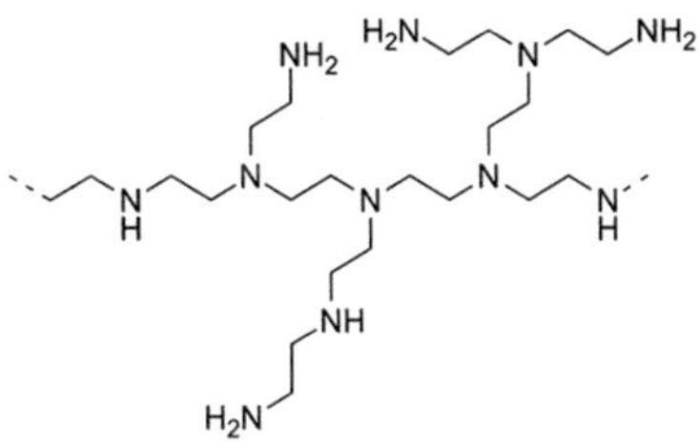
5	Dendrimer polyethyleneimine	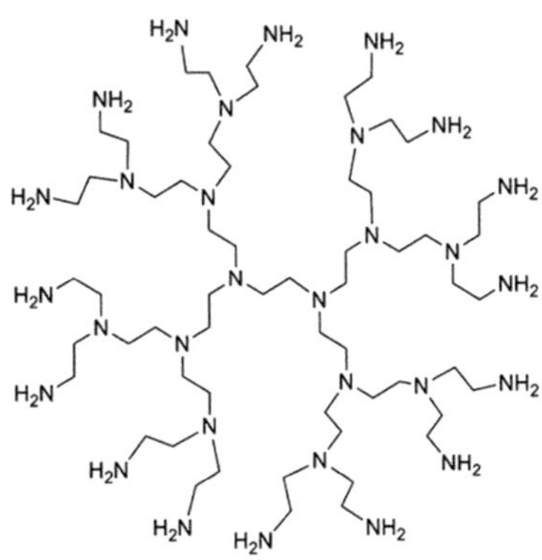

4.5.2 CATIONIC LIPIDS

Cationic lipids are positive amphiphilic in nature. These cationic lipids have four foremost components:

1. Cationic polar head—For self-assembly interaction with DNA.
2. A hydrophobic chain—It affects lipid bilayer and its physical properties.
3. A spacer—Present between cationic polymer head and hydrophobic chain and has an important role like stability, biodegradation, and transferring of genetic material.
4. A backbone is frequently made by glycerol-based domain as a scaffold (Volavanidis and Vlachogianni, 2016).

Cationic lipids and their structure (Zhang et al., 2011; Lakshmana Prabu and Ruckmani, 2017) are shown in Table 4.2.

TABLE 4.2 Structure of Cationic Lipids.

S. no	Cationic lipids	Structure
	Monovalent cationic lipids	
1	DOTMA (N-[1-(2,3-dioleyloxy) propyl]-N,N,N-trimethylammonium chloride)	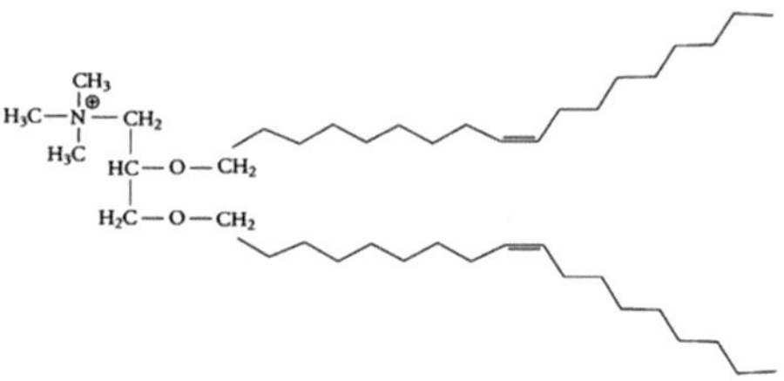
2	DOTAP, [1,2-bis(oleoyloxy)-3-(trimethylammonio) propane]	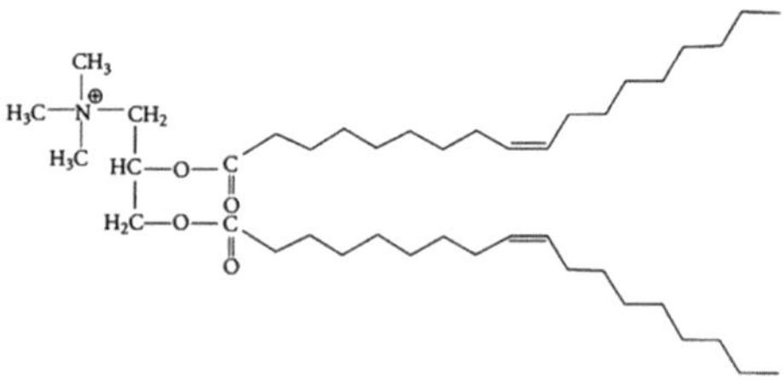
3	DC-Chol 3β[*N*-(*N*′,*N*′-dimethylaminoethane)-carbamoyl]	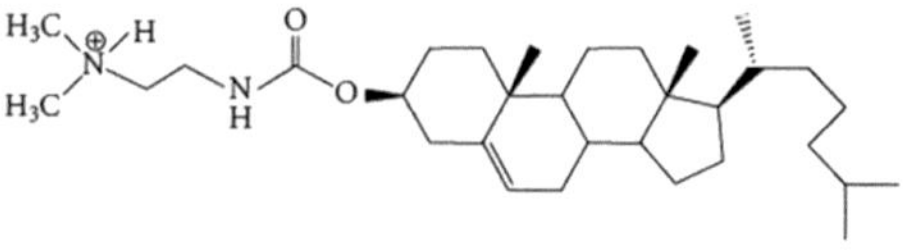

TABLE 4.2 *(Continued)*

S. no	Cationic lipids	Structure
	Multivalent cationic lipids	
1	DOSPA (2,3-dioleyloxy-N-[2(sperminecarboxamido) ethyl]-N,N-dimethyl-l-propanaminium trifluoroacetate)	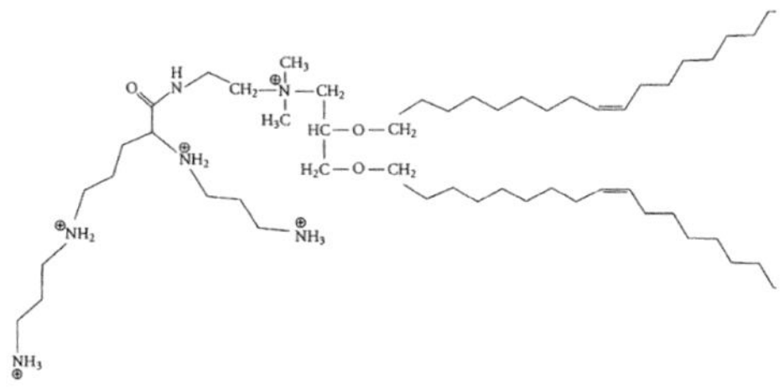
2	DOGS (di-octadecyl-amido-glycyl-spermine)	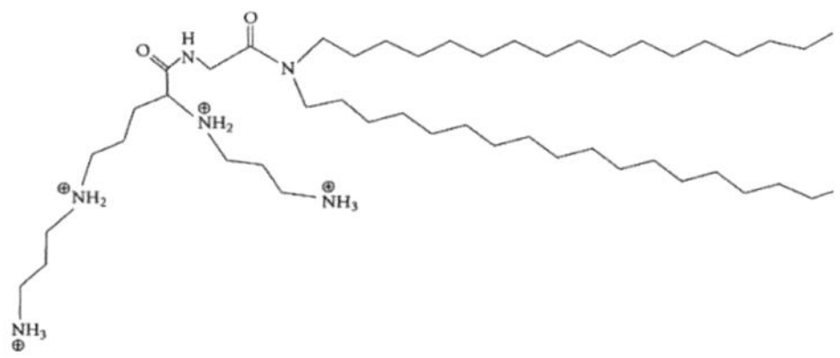

4.5.3 NEUTRAL LIPIDS

Neutral lipids are lipids used along with cationic polymer like DC-Chol, DOTAP, and other cholesterol derivatives; which increases the gene transfection efficiency. At lower or acidic pH, these neutral lipids undergo conformational change to an inverted hexagonal packing structure which looks like a honeycomb; this can undergoes electrostatic interaction and condense the DNA inside the honeycomb structure (Zhang et al., 2011; Lakshmana Prabu and Ruckmani, 2017). Structure of neutral lipids is shown in Table 4.3.

4.5.4 POLYETHYLENE GLYCOL

Polyethylene glycol is made by ethylene oxide and its physical property depends upon chain length. PEG is available in different forms like star, comb, and branched as well as nontoxic and eliminated through kidney. Surface modification can protect its degradation and it has been used as liposomes in target drug delivery (Zhang et al., 2011; Lakshmana Prabu and Ruckmani, 2017).

TABLE 4.3 Structure of Neutral Lipids.

S. no	Neutral lipids	Structure
1	DOPE (dioleoylphosphatidylethanolamine)	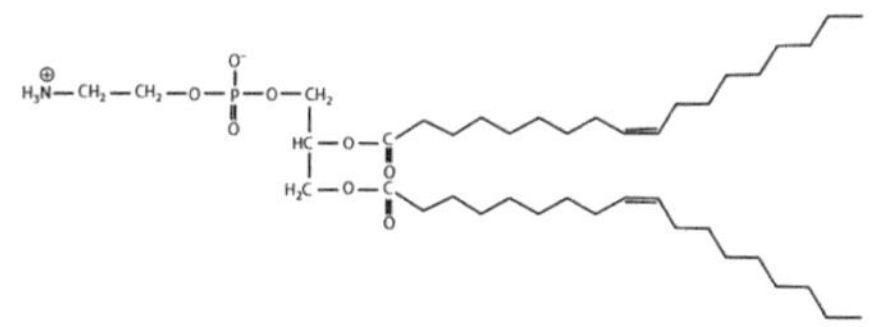
2	DOPC (dioleoylphosphatidylcholine)	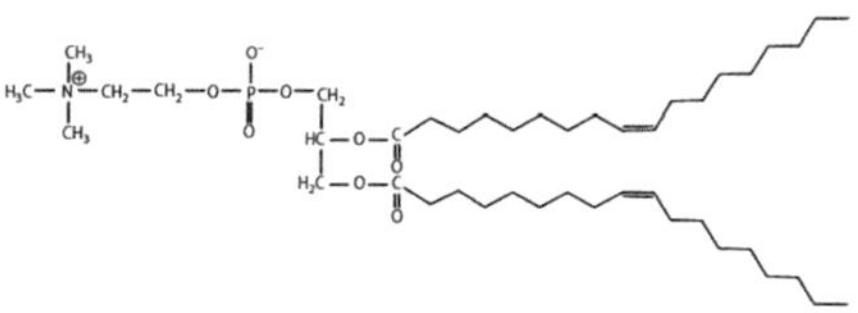

4.6 CONCLUSION

In the last few decades, nanotechnology as a science has gained much importance in all the sectors. Both inorganic and organic materials are utilized in synthesizing the nanomaterials. This nanomaterial has very peculiar characteristic properties, which has been adopted for improving the existing technologies also in health care profession especially in disease diagnosis as well as treating various diseases through drug-delivery system. Nanomaterials are prepared either "top-down" or "bottom-up" method; however, based on this technical concept different methods have been developed to synthesis these nanostructured products. In spite of new breakthroughs techniques in synthesis of nanostructure material as well as its characterization, still the global community faces many barriers in formulating high efficient with nontoxic nanomaterials in the delivery systems. In drug-delivery system, recently gene delivery system utilizing genetic material such as plasmids, DNA, RNA, and siRNA have gained much importance as well as an attractive therapeutic method in treating various genetic disorders. In gene delivery system, between the nonviral and viral transduction method, nonviral transduction method has gained more advantages. The efficiency in delivering the genetic material into the target site is the major hurdle but it can be resolved by selecting a suitable carrier/vector to carry the genetic material and to delivery in the target site. For an ideal carrier, nanoparticles must make with suitable biopolymers as carrier/vector, which can protect the genetic material within the polymer complex and delivery into the target site. An accurate knowledge of structural features of the nanoparticle, pharmacological interactions, method of preparation, stability, and its sterilization has to be given significant importance in addition to the qualifying criterion such as safety and bioavailability.

It is predicted that innovations in these pioneering technologies will aid in overcoming these challenges to enhance the transfer of the genetic material into the specific target site to normalize the damaged gene and to create more therapeutic applicability of nanodrug-delivery systems.

KEYWORDS

- **gene delivery**
- **cogent**
- **nanostructure**
- **nanobiomaterial**
- **biopolymer**

REFERENCES

Alanazi, A. K.; Radwan, A. A.; Alsarra, I. A. Biopharmaceutical Applications of Nanogold. *Saudi Pharm. J.* **2010,** *18*, 179–193.

Allemann, E.; Gurnay, R.; Doelker, E. Preparation of Aqueous Polymeric Nanodispersions by a Reversible Salting-Out Process: Influence of Process Parameters on Particle Size. *Int. J. Pharm.* **1992,** *87*, 247–253.

Allemann, E.; Leroux, J. C.; Gurnay, R.; Doelker, E. In Vitro Extended-Release Properties of Drug-Loaded Poly(D, L-lactic) Acid Nanoparticles Produced by a Salting-Out Procedure. Pharm. Res. **1993,** 10, 1732–1737.

Approaches to Safety Nanotechnology. https://www.cdc.gov/niosh/topics/nanotech/pdfs/Approaches_to_Safe_Nanotechnology.pdf (accessed June 13, 2020).

Arruebo, M.; Pacheco, R. F.; Ibarra, M. R.; Santamaria, J. Magnetic Nanoparticle for Drug Delivery. *Nano Today* **2007,** *2* (3), 22–32.

Attama, A. A.; Momoh, M. A.; Builders, P. F. Lipid Nanoparticulate Drug Delivery Systems: A Revolution in Dosage Form Design and Development. In *Recent Advances in Novel Drug Carrier Systems*; Sezer, A. D., Ed.; InTech: Croatia, **2012,** 107–140.

Bagul, R.; Mahajan, V.; Dhake, A. New Approaches in Nanoparticlate Drug Delivery System—A Review. *Int. J. Curr. Pharm. Res*. **2012,** *4* (2), 29–38.

Barichello, J. M.; Morishita, M.; Takayama, K.; Nagai, T. Encapsulation of Hydrophilic and Lipophilic Drugs in PLGA Nanoparticles by the Nanoprecipitation Method. *Drug Dev. Ind. Pharm*. **1999,** *25*, 471–476.

Bertrand, N.; Leroux, J. C. The Journey of a Drug-Carrier in the Body: An Anatomo-Physiological Perspective. *J. Control. Release* **2012,** *161*, 152–163.

Biswas, S.; Vladimir, P. Torchilin. Dendrimers for siRNA Delivery. *Pharmaceuticals* **2013,** *6*, 161–183.

Boas, U.; Heegaard, P. M. Dendrimers in Drug Research. *Chem. Soc. Rev.* **2004,** *33*, 43–63.

Cheki, M.; Moslehi, M, Assadi, M. Marvelous Applications of Quantum Dots. *Eur. Rev. Med. Pharmacol. Sci.* **2013,** *17*, 1141–1148.

Chen, Y. C.; Huang, X. C.; Luo, Y. L.; Chang, Y. C.; Hsieh, Y. Z.; Hsu, H. Y. Non-Metallic Nanomaterials in Cancer Theranostics: A Review of Silica- and Carbon-Based Drug Delivery Systems. *Sci. Technol. Adv. Mater.* **2013,** *14*, 404–407.

Cheng, Y.; Wang, J.; Rao, T.; He, X.; Xu, T. Pharmaceutical Applications of Dendrimers: Promising Nanocarriers for Drug Delivery. *Front. Biosci.* **2008,** *13*, 1447–1471.

Dabbousi, B. O.; Viejo, R. J.; Mikulec, F. V.; Heine, J. R.; Mattoussi, H.; Ober, R. (CdSe) ZnS Core-Shell Quantum Dots: Synthesis and Characterization of a Size Series of Highly Luminescent Nanocrystallites. *J. Phys. Chem. B.* **1997,** *101*, 9463–9475.

Dang, J. M.; Leong, K. W. Natural Polymers for Gene Delivery and Tissue Engineering. *Adv. Drug. Del. Rev.* **2006,** *58* (4), 487–499.

Dey, N. S.; Rao, M. E. B. Quantum Dot: Novel Carrier for Drug Delivery. *Int. J Res. Pharm. Biomed. Sci.* **2011,** *2* (2), 448–458.

Dizaj, S. M.; Jafari, S.; Khosroushahi, A. Y. A Sight on the Current Nanoparticle-Based Gene Delivery Vectors. *Nanosc. Res. Lett.* **2014,** *9*, 252. DOI: 10.1186/1556-276X-9-252.

Dorfs, D.; Krahne, R.; Falqui, A.; Manna, L.; Giannini, C.; Zanchet, D. Quantum Dots: Synthesis and Characterization. In *Comprehensive Nanosciences and Technology*; David, L. A., Gregory, D. S., Gary, P. W., Eds.; Academic Press: Amsterdam, 2011; pp 219–270.

Drbohlavova, J.; Adam, V.; Kizek, R.; Hubalek, J. Quantum Dots-Characterization, Preparation and Usage in Biological Systems. *Int. J. Mol. Sci.* **2009,** *10*, 656–673.

Dufes, C.; Uchegbu, I. F.; Schatzlein, A. G. Dendrimers in Gene Delivery. *Adv. Drug Deliv. Rev.* **2005,** *57*, 2177–2202.

Ekman, B.; Sjfholm, I. Improved Stability of Proteins Immobilized in Microparticles Prepared by Modified Emulsion Polymerization Technique. *J. Pharm. Sci.* **1978,** *67*, 693–696.

Eliyahu, H.; Barenholz, Y.; Domb, A. J. Polymers for DNA Delivery. *Molecules* **2005,** *10*, 34–64.

Erdal, C.; Ali, D. S.; Emre, S. C. Gene Delivery Systems: Recent Progress in Viral and Non-Viral Therapy. In *Recent Advances in Novel Drug Carrier Systems*; Ali, D. S., Ed.; InTech: Croatia, 2012; pp 437–470. DOI: 10.5772/53392. http://www.intechopen.com/books/recent-advances-in-novel-drug-carrier-systems/gene-delivery-systems-recent-progress-in-viral-and-non-viral-therapy

Fan, W.; Yan, W.; Xu, Z.; Ni, H. Formation Mechanism of Monodisperse, Low Molecular Weight Chitosan Nanoparticles by Ionic Gelation Technique. *Colloids Surf. B Biointerf.* **2012,** *90*, 21–27.

Fattal, E.; Barratt, G. Nanotechnologies and Controlled Release Systems for Delivery and Antisense Oligonucleotides and Small Interfering RNA. *Br. J. Pharmacol.* **2009,** *157* (2), 179–194.

Fessi, H.; Puisieux, F.; Devissaguet, J. P.; Ammoury, N.; Benita, S. Nanocapsule Formation by Interfacial Polymer Deposition Following Solvent Displacement. *Int. J. Pharm.* **1989,** *55*, R1–R4.

Galindo-Rodriguez, S.; Allemann, E.; Fessi, H.; Doelker, E. Physicochemical Parameters Associated with Nanoparticle Formation in the Salting-Out, Emulsification-Diffusion, and Nanoprecipitation Methods. *Pharm. Res.* **2004,** *21*, 1428–1439.

Gao, Y.; Gao, G.; He, Y.; Liu, T.; Qi, R. Recent Advances of Dendrimers in Delivery of Genes and Drugs. *Mini Rev. Med. Chem.* **2008,** *8*, 889–900.

Garg, A.; Visht, S.; Sharma, P. K.; Kumar, N. Formulation, Characterization and Application on Nanoparticle: A Review. *Der Pharmacia Sinica* **2011,** *2* (2), 17–26.

Ghosh, R.; Singh, L. C.; Shohet, J. M.; Gunaratne, P. H. A Gold Nanoparticle Platform for the Delivery of Functional MicroRNAs into Cancer Cells. *Biomaterials* **2013,** *34*, 807–816.

Giannoukakis, N.; Thomson, A.; Robbins, P. Gene Therapy in Transplantation. *Gene Ther.* **1999,** *6* (9), 1499–1511.

Giljohann, D. A.; Seferos, D. S.; Daniel, W. L.; Massich, M. D.; Patel, P. C.; Mirkin, C. A. Gold Nanoparticles for Biology and Medicine. *Angew. Chem.* **2010,** *49*, 3280–3294.

Gillies, E. R.; Frechet, J. M. Dendrimers and Dendritic Polymers in Drug Delivery. *Drug Discov. Today* **2005,** *10*, 35–43.

Godbey, W. T.; Mikos, A. G. Recent Progress in Gene Delivery Using Non-Viral Transfer Complexes. *J. Cont. Rel.* **2001,** *72* (1–3), 115–125.

Govender, T.; Stolnik, S.; Garnett, M. C.; Illum, L.; Davis, S. S. PLGA Nanoparticles Prepared by Nanoprecipitation: Drug Loading and Release Studies of a Water Soluble Drug. *J. Cont. Rel.* **1999,** *57* (2), 171–185.

Gulia, S.; Kakkar, R. ZnO Quantum Dots for Biomedical Applications. *Adv. Mat. Lett.* **2013,** *4* (12), 876–887.

Guo, P.; Coban, O.; Snead, N. M.; Trebley, J.; Hoeprich, S.; Guo, S. Engineering RNA for Targeted siRNA Delivery and Medical Application. *Adv. Drug Deliv. Rev.* **2010,** *62*, 650–666.

Gupta, R. B. Fundamentals of Drug Nanoparticles. In *Nanoparticle Technology for Drug Delivery*; Gupta, R. B., Kompella, U. B., Eds.; Taylor & Francis Group, LLC: New York, 2006; pp 1–19.

Hans, M. L.; Lowman, A. M. *Nanoparticles for Drug Delivery*; Taylor & Francis Group, LLC: New York, 2006.

Hoa, L. T. M.; Chi, N. T.; Nguyen, L. H.; Chien, D. M. Preparation and Characterization of Nanoparticles Containing Ketoprofen and Acrylic Polymers Prepared by Emulsion Solvent Evaporation Method. *J. Exp. Nanosci.* **2012,** *7* (2), 189–197.

Hornig, S.; Heinze, T.; Remzi Becer, C.; Schubert, U. S. Synthetic Polymeric Nanoparticles by Nanoprecipitation. *J. Mater. Chem.* **2009,** *19*, 3838–3840.

Itaka, K.; Kataoka, K. Recent Development of Nonviral Gene Delivery Systems with Virus-Like Structures and Mechanisms. *Eur. J. Pharm. Biopharm.* **2009,** *71*, 475–483.

Jaiswal, J.; Gupta, S. K.; Kreuter, J. Preparation of Biodegradable Cyclosporine Nanoparticles by High-Pressure Emulsification-Solvent Evaporation Process. *J. Cont. Rel.* **2004,** *96* (1), 169–178.

Jamiesona, T.; Bakhshi, R.; Petrovaa, D.; Pococka, R.; Imanib, M.; Seifaliana, A. M. Biological Applications of Quantum Dots. *Biomaterials* **2007,** *28*, 4717–4732.

Jin, L.; Zeng, X.; Liu, M.; Deng, Y.; He, N. Current Progress in Gene Delivery Technology Based on Chemical Methods and Nano-Carriers. *Theranostics* **2014,** *4* (3), 240–255.

Jin, S.; Ye, K. Mediated Drug Delivery and Gene Therapy. *Biotechnol. Prog.* **2007,** *23*, 32–41.

Jordan, M. Synthetic and Semisynthetic Polymers as Vehicles for in Vitro Gene Delivery Into Cultured Mammalian Cells. In *Synthetic Polymers for Biotechnology and Medicine*; Freitag, R., Ed.; Eurekah.com/Landes Bioscience: Georgetown, DC, 2003; pp 19–39.

Kannan, K.; Manikandan, M. Selection of a Suitable Method for the Preparation of Polymeric Nanoparticles: Multi-Criteria Decision Making Approach. *Adv. Pharm. Bull.* **2015,** *5* (1), 57–67.

Karmi, A.; Husseini, GA.; Faroun, M.; Sowwan, M. Multifunctional Nano Vehicles for Combined 5-fluorouracil and Gold Nanoparticles Based on the Nanoprecipitation Method. *J. Nanosci. Nanotechnol.* **2011,** *11* (6), 4675–4683.

Khan, M. S.; Vishakante, G. D.; Siddaramaiah, H. Gold Nanoparticles: A Paradigm Shift in Biomedical Applications. *Adv. Colloid Interf. Sci.* **2013,** *199–200*, 44–58.

Kim, D.; Jon, S. Gold Nanoparticles in Image-Guided Cancer Therapy. *Inorgan. Chim. Acta.* **2012,** *393*, 154–164.

Klajnert, B.; Bryszewska, M. Dendrimers: Properties and Applications. *Acta Biochemica Polonica* **2001,** *48*, 199–208.

Krause, H. J.; Schwarz, A.; Rohdewald, P. Interfacial Polymerization, an Useful Method for the Preparation of Polymethylcyanoacrylate Nanoparticles. *Drug Dev. Ind. Pharm.* **1986,** *12*, 527–552.

Kreuter, J. Nanoparticle-Based Drug Delivery Systems. *J. Cont. Rel.* **1991,** *16*, 169–176.

Kreuter, J. The Mechanism of Termination in Heterogeneous Polymerization. *J. Poly. Sci.* **1982,** *20*, 543–545.

Krikorian, V.; Kurian, M.; Galvin, M. E.; Nowak, A. P.; Deming, T. J.; Pochan, D. J. Polypeptide-Based Nanocomposite: Structure and Properties of Poly(L-lysine)/Na-Montmorillonite. *J. Poly. Science: Part B: Poly. Phys.* **2002,** *40*, 2579–2586.

Lakshmana Prabu, S.; Suriya Prakash, T.N.K.; Thirumurugan, R. Medicated Nanoparticle for Gene Delivery. In *Advanced Technology for Delivering Therapeutics*; Maiti, S., Sen, K. K., Eds.; InTech Publication: Croatia, 2017; pp 13–29. ISBN 9789535131229.

Lakshmana Prabu, S.; Ruckmani, R. Biopolymer in Gene Delivery. In *Advanced Technology for Delivering Therapeutics*. Maiti, S., Sen, K. K., Eds.; InTech Publication: Croatia, 2017; pp 137–158.

Lee, J. M.; Yoon, T. J.; Cho, Y. S. Recent Developments in Nanoparticle-Based siRNA Delivery for Cancer Therapy. *Biomed Res. Int.* **2013,** Article ID 782041. DOI: 10.1155/2013/782041.

Leroux, J. C.; Allemann, E.; Doelker, E.; Gurnay, R. New Approach for the Preparation of Nanoparticles by an Emulsification–Diffusion Method. *Eur. J. Pharm. Biopharm.* **1995,** *41*, 14–18.

Liechty, W. B.; Kryscio, D. R.; Slaughter, B. V.; Peppas, N. A. Polymers for Drug Delivery Systems. *Ann. Rev. Che. Biomol. Eng.* **2010,** 1, 149–173.

Liu, F.; Liang, K. W.; Huang, L. Systemic Administration of Naked DNA: Gene Transfer to Skeletal Muscle. *Mol. Interv.* **2001,** *1* (3), 168–172.

Lungwitz, U.; Breunig, M.; Blunk, T.; Gopferich, A. Polyethylenimine-Based Non-Viral Gene Delivery Systems. *Eur. J. Pharm. Biopharm.* **2005,** *60*, 247–266.

Luten, J.; Van Nostruin, C. F.; De Smedt, S. C.; Hennink, W. E. Biodegradable Polymers as Nonviral Carriers for Plasmid DNA Delivery. *J. Cont. Rel.* **2008,** *126* (2), 97–110.

Mahapatra, A.; Singh, D. K. Biodegradable Nanoparticle Are Excellent Vehicle for Site Directed in-Vivo Delivery of Drugs and Vaccines. *J. Nanobiotech.* **2011,** *9*, 55.

Mahato, R. I. Non-Viral Peptide-Based Approaches to Gene Delivery. *J. Drug. Target.* **1999,** *7* (4), 249–268.

Mansouri, S.; Lavigne, P.; Corsi, K.; Benderdour, M.; Beaumont, E.; Fernandes, J. C. Chitosan-DNA Nanoparticles as Nonviral Vectors in Gene Therapy: Strategies to Improve Transfection Efficacy. *Eur. J. Pharm. Biopharm.* **2004,** *57* (1), 1–8.

Medina, S. H.; El-Sayed MEH. Dendrimers as Carriers for Delivery of Chemotherapeutic Agents. *Chem. Rev.* **2009,** *109*, 3141–3157.

Mehrotra, A.; Pandit, J. K. Critical Process Parameters Evaluation of Modified Nanoprecipitation Method on Lomustine Nanoparticles and Cytostatic Activity Study on L132 Human Cancer Cell Line. *J. Nanomed. Nanotech.* **2012,** *3* (8), 1000–1149.

Merdan, T.; Kopecek, J.; Kissel, T. Prospects for Cationic Polymers in Gene and Oligonucleotide Therapy against Cancer. *Adv. Drug Deliv. Rev.* **2002,** *54* (5), 715–758.

Mishra, S.; Tripathy, P.; Sinha, S. P. Advancements in the Field of Quantum Dots. *Int. J. Adv. Res. Tech.* **2012,** *1* (3), 1–5.

Modani, S.; Kharwade, M.; Nijhawan, M. Quantum Dots: A Novelty of Medical Field with Multiple Applications. *Int. J. Curr. Pharm. Res.* **2013,** *5* (4), 55–59.

Mohanraj, V. J.; Chen, Y. Nanoparticles—A Review. *Trop. J. Pharm. Res.* **2006,** *5* (1), 561–573.

Mokari, T.; Banin, U. Synthesis and Properties of CdSe/ZnS Core/Shell Nanorods. Chem. Mater. **2003,** 15, 3955–3960.

Moorthi, C.; Kathiresan, K. Fabrication of Highly Stable Sonication Assisted Curcumin Nanocrystals by Nanoprecipitation Method. *Drug Invention Today.* **2013,** *5* (1), 66–69.

Mrsny, R. J. Tissue- and Cell-Specific Targeting for the Delivery of Genetic Information. In *Polymeric Gene Delivery: Principles and Applications*; Amiji, M. M., Ed.; Taylor & Francis: Boca Raton, FL, 2005; pp 4–30.

Mukherjee, S.; Das, U. Quantum Dots: An Optimistic Approach to Novel Therapeutics. *Int. J. Pharm. Sci. Rev. Res.* **2011,** *7*, 59–64.

Murakami, H.; Kobayashi, M.; Takeuchi, H.; Kawashima, Y. Preparation of Poly (DL-lactide-co-glycolide) Nanoparticles by Modified Spontaneous Emulsification Solvent Diffusion Method. *Int. J. Pharm.* **1999,** *87*, 143–152.

Nagavarma, B. V. N.; Yadav, H. K. S.; Ayaz, A.; Vasudha, L. S.; Shivakumar, H. G. Different Techniques for Preparation of Polymeric Nanoparticles- a Review. *Asian J. Pharm. Clin. Res.* **2012,** *5* (3), 16–23.

Nevozhay, D.; Kanska, U.; Budzynska, R.; Boratynski, J. Current Status of Research on Conjugates and Related Drug Delivery Systems in the Treatment of Cancer and Other Diseases (Polish). *Postepy HigMed. Dosw.* **2007,** *61*, 350–360.

Nikalje, A. P. Nanotechnology and Its Applications in Medicine. *Med. Chem.* **2015,** *5* (2), 81–89.

Nimesh, S. Polyethylenimine as a Promising Vector for Targeted siRNA Delivery. *Curr. Clin. Pharmacol.* **2012,** *7*, 121–130.

Nitta, S. K.; Numata, K. Biopolymer-Based Nanoparticles for Drug/Gene Delivery and Tissue Engineering. *Int. J Mol. Sci.* **2013,** *14* (1), 1629–1654.

Nitta, S. K.; Numata, K. Biopolymer-Based Nanoparticles for Drug/Gene Delivery and Tissue Engineering. *Int. J. Mol. Sci.* **2013,** *14*, 1629–1654.

Niwa, T.; Takeuchi, H.; Hino, T.; Kunou, N.; Kawashima, Y. Preparations of Biodegradable Nanospheres of Water-Soluble and Insoluble Drugs with D, L-lactide/Glycolide Copolymer by a Novel Spontaneous Emulsification Solvent Diffusion Method, and the Drug Release Behavior. *J. Cont. Rel.* **1993,** *25* (1–2), 89–98.

Oyewumi, M. O.; Rice, K. G. DNA Nanoparticle Gene Delivery Systems. In *Nanoparticle Technology for Drug Delivery*; Gupta, R. B., Kompella, U. B., Eds.; Marcel Dekker, **2006,** *159*, 361–379.

Ozkan, M. Quantum Dots and Other Nanoparticles: What Can They Offer to Drug Discovery? *Drug Discov. Today.* **2004,** *9*, 1065–1071.

Pack, D. W.; Hoffman, A. S.; Pun, S.; Stayton, P. S. Design and Development of Polymers for Gene Delivery. *Nat. Rev. Drug. Discov.* **2005,** *4*, 581–593.

Papasani, M. R.; Wang, G.; Hill, R. A. Gold Nanoparticles: The Importance of Physiological Principles to Devise Strategies for Targeted Drug Delivery. *Nanomed. Nanotechnol. Biol. Med.* **2012,** *8*, 804–814.

Pinto Reis, C.; Neufeld, R. J.; Ribeiro, A. J.; Veiga, F. Nanoencapsulation I. Methods for Preparation of Drug-Loaded Polymeric Nanoparticles. *Nanomed.: Nanotech. Bio. Med.* **2006,** *2* (1), 8–21.

Pissuwan, D.; Niidome, T.; Cortie, M. B. The Forthcoming Applications of Gold Nanoparticles in Drug and Gene Delivery Systems. *J. Cont. Rel.* **2011,** *149*, 65–71.

Probst, C. E.; Zrazhevskiy, P.; Bagalkot, V.; Gao, X. Quantum Dots as a Platform for Nanoparticle Drug Delivery Vehicle Design. *Adv. Drug Deliv. Rev.* **2013,** *65* (5), 703–718.

Puri, A.; Loomis, K.; Smith, B.; Lee, J. H.; Yavlovich, A.; Heldman, E.; Blumenthal, R. Lipid-Based Nanoparticles as Pharmaceutical Drug Carriers: From Concepts to Clinic. *Crit. Rev. Ther. Drug Carrier Syst.* **2009,** *26* (6), 523–580.

Putnam, D. Polymers for Gene Delivery across Length Scales. *Nat. Mater.* **2006,** *5*, 439–451.

Qi, L.; Ga, X. Emerging Application of Quantum Dots for Drug Delivery and Therapy. *Expert Opin. Drug Deliv.* **2008,** *5* (3), 263–267.

Rana, T. M. Illuminating the Silence: Understanding the Structure and Function of Small RNAs. *Nat. Rev. Mol. Cell Biol.* **2007,** *8*, 23–36.

Rathore, K. S.; Lowalekar, R.; Nema, R. K. Quantum Dots: A Future Drug Delivery System. *Pharma Rev.* **2006,** *4*, 30–32.

Soppimath, K. S.; Aminabhavi, T. M.; Kulkarni, A. R.; Rudzinski, W. E. Biodegradable Polymeric Nanoparticles as Drug Delivery Devices. *J. Cont. Rel.* **2001,** *70* (1–2), 1–20.

Srivastava, A.; Yadav, T.; Sharma, S.; Nayak, A.; Kumari, A.; Mishra, N. Polymers in Drug Delivery. *J. Biosci. Med.* **2016,** *4*, 69–84.

Stobiecka, M.; Hepel, M. Double-Shell Gold Nanoparticle Based DNA-Carriers with Poly-L-Lysine Binding Surface. *Biomaterials* **2011,** *32*, 3312–3321.

Suri, S. S.; Fenniri, H.; Singh, B. Nanotechnology-Based Drug Delivery Systems. *J. Occup. Med. Toxicol.* **2007,** *2*, 16.

Synowiecki, J.; Al-Khateeb, N. A. Production, Properties, and Some New Applications of Chitin and Its Derivatives. *Crit. Rev. Food Sci. Nutr.* **2003,** *43* (2), 145–171.

Tarnawski, R.; Ulbricht, M. Amphiphilic Gold Nanoparticles: Synthesis, Characterization and Adsorption to PEGylated Polymer Surfaces. *Colloid Surf. A* **2011,** *374*, 13–21.

Tice, T. R.; Gilley, R. M. Preparation of Injectable Controlled-Release Microcapsules by Solvent-Evaporation Process. *J. Cont. Rel.* **1985,** *2*, 343–352.

Tiwari, P. K.; Lee, Y. S. Gene Delivery in Conjunction with Gold Nanoparticle and Tumor Treating Electric Field. *J Appl. Phys.* **2013,** *114* (5), 054902–054905. DOI: 10.1063/1.4817090.

Tomalia, D. A.; Frechet, J. M. J. Discovery of Dendrimers and Dendritic Polymers: A Brief Historical Perspective. *J. Polym. Sci. Part A: Polym Chem.* **2002,** *40* (16), 2719–2728.

Twaites, B. R.; Alarcon, C. D. H.; Lavigne, M.; Saulnier, A.; Pennadam, S. S.; Cunliffe, D. Thermoresponsive Polymers as Gene Delivery Vectors: Cell Viability, DNA Transport and Transfection Studies. *J. Cont. Rel.* **2005,** *108*, 472–473.

Van Rooij, E. The Art of MicroRNA Research. *Circ. Res.* **2011,** *108*, 219–234.

Vauthier, C.; Bouchemal, K. Processing and Scale-up of Polymeric Nanoparticle. In *Intracellular Delivery—Fundamental Biomedical Technologies*; Prokop, A., Eds., Vol 5; Springer: Dordrecht, 2011; pp 433–456.

Vengala, P.; Dasari, A.; Yeruva, N. Quantum Dots for Drug Delivery and Therapy. *Int. J. Pharm. Tech.* **2012,** *4* (2), 2055–2074.

Vilar, G.; Tulla-Puche, J.; Albericio, F. Polymers and Drug Delivery Systems. *Cur. Drug Del.* **2012,** *9*, 367–394.

Volavanidis, A.; Vlachogianni, T. Engineered Nanomaterials for Pharmaceutical and Biomedical Products New Trends, Benefits and Opportunities. *Pharm. Bioprocess* **2016,** *4* (1), 13–24.

Waite, C. L.; Sparks, S. M.; Uhrich, K. E.; Roth, C. M. Acetylation of Pamam Dendrimers for Cellular Delivery of Sirna. *BMC Biotechnol.* **2009,** *9*, 38. DOI: 10.1186/1472-6750-9-38

Wang, Y.; Chen, L. Quantum Dots, Lighting Up the Research and Development of Nanomedicine. *Nanomed: Nanotech. Biol. Med.* **2011,** *7*, 385–402.

Wang, Y.; Li, P.; Tran, T. T.; Zhang, J.; Kong, L. Manufacturing Techniques and Surface Engineering of Polymer Based Nanoparticles for Targeted Drug Delivery to Cancer. *Nanomaterials* **2016,** *6* (2), 26.

Ward, M. A.; Georgiou, T. K. Thermoresponsive Polymers for Biomedical Applications. *Polymers* **2011,** *3*, 1215–1242.

Wilczewska, A. Z.; Niemirowicz, K.; Markiewicz, K. H.; Car, H. Nanoparticles as Drug Delivery Systems. *Pharmacol. Rep.* **2012,** *64*, 1020–1037.

Wong, S. Y.; Pelet, J. M.; Putnam, D. Polymer Systems for Gene Delivery-Past, Present, and Future. *Progress Poly. Sci.* **2007,** *32* (8–9), 799–837.

Xu, F. J.; Zhang, Z. X.; Ping, Y.; Li, J.; Kang, E. T.; Neoh, K. G. Star-Shaped Cationic Polymers by Atom Transfer Radical Polymerization from Beta-Cyclodextrin Cores for Nonviral Gene Delivery. *Biomacromolecules* **2009,** *10* (2), 285–293.

Yang, Y.; Li, Q.; Erte, H. C. J.; Wilson, J. M. Cellular and Humoral Immune Responses to Viral Antigens Create Barriers to Lung Directed Gene Therapy with Recombinant Adenoviruses. *J. Virol.* **1995,** *69*, 2004–2015.

Zhang, X.; Balazs, D. A.; Godbey, W. T. Nanobiomaterials for Nonviral Gene Delivery. In *Nanobiomaterials Hand Book*; Sitharaman, B., Ed.; CRC Press, Taylor & Francis, 2011; pp 13-1 to 13-25.

Zrazhevskiyn, P.; Gao, X. Multifunctional Quantum Dots for Personalized Medicine. *Nano Today*. **2009,** *4* (5), 414–428.

CHAPTER 5

Nanomaterials for Nanomedical and Nanobiotechnological Studies: Current Status and Applications of Carbon-Based Nanomaterials and Quantum Dots

ANCA ARMĂȘELU*

Department of Electrical Engineering and Applied Physics, Faculty of Electrical Engineering and Computer Science, Brașov, Romania

**E-mail: anca_armaselu@yahoo.com*

ABSTRACT

Nanomaterials (NMs) possess uncommon physical and chemical properties, such as extremely small size, high surface energy, large surface-to-volume ratio, and high reactivity which are dissimilar from their bulk counterparts with the identical structure. The smart union of these properties will mark a novel epoch of the NMs with numerous possible uses in the area of nanomedicine and nanobiotechnology, such as imaging, diagnosis, prevention, disease therapy, drug delivery, tissue engineering, catalysis, photochemistry sensors, tagging, optical, electronic, and magnetic devices, and so forth.

The goal of this chapter is to provide an innovative review about the process of the development of some novel types of NMs with a higher efficiency and with a lower toxicity and their applications in various areas of nanomedicine and of the nanobiotechnology.

Among NMs, carbon-based nanomaterials (CBNs) have received appreciable interest in nanomedical sciences and nanobiotechnological studies in connection with their numerous applications owing to their uncommon combinations of excellent properties.

The Science of Nanomaterials: Basics and Applications. Suresh C. Ameta & Rakshit Ameta (Eds.)

This chapter summarizes the most recent research advances in developing of CBNs (including of carbon nanotubes, graphene oxide, and CBNs combined with diverse categories of NMs) and aims to highlight their emerging applications in the fields of cell labeling, imaging methods, and drug delivery.

5.1 INTRODUCTION

There are numerous ideal works like the ones, which are quoted here, having to do with natural, incidental, and projected nanomaterials (NMs) (Boverhof et al., 2015; Hardy et al., 2018; Hochella Jr. et al., 2019). Many recent studies recall that in accordance with the direction of the European Commission, the term "nanomaterials" (NM) describes a natural, incidental, or manufactured material incorporating particles, either in an unbound state or as an aggregate in which one or more external dimensions is in the size scale of 1–100 nm for ≥ 50% of the particles, depending on the number size distribution and if this indicator is not satisfied, the material is categorized like non-NM (Bleeker et al., 2012; Schwirn and Völker, 2016; Soares et al., 2018; Schwirn et al., 2020).

Engineered nanomaterials (ENMs) are a category of intentionally designed and prepared materials with nanoscale dimensions produced for commercial scopes, to have certain properties or structures, and which show a quickly growth domain of applications in daily life (Gubala et al., 2018). Definitions of nanoscale materials, which typically include the same limitations on ENM dimension as well as the usage in regulatory assessments have been, and continue to be, a field of dynamic scientific and policy dispute (Yokel and MacPhail, 2011; Boverhof et al., 2015).

NMs present the constitutional features between those of atoms and bulk materials. This fact appears as a consequence of the nanometric size of the materials that determine a large fraction of surface atoms, a high surface energy, a spatial confinement of electron movements and reduced imperfections which do not occur in the appropriate bulk materials (Razali et al., 2016; Pandey, 2017).

Since these nanoscale materials typically have very different properties from their bulk counterparts, due to the importance of quantum and surface effects, some unusual, physical, chemical, and biological properties may occur. NMs exhibit unique and extraordinary physicochemical properties, being increasingly used for commercial scopes in different fields due to their controlled dimension, high surface-to-volume ratio, physical appearance (distinctive shape), composition, existence of biochemical moieties on surface, special hydrophilic or hydrophobic character and aggregation (Navya and Daima, 2016).

This review furnishes information on the recent advances which have been carried out in studies on ENM—containing products, on the methodology for the manufacture of such devices and on their diverse applications. The work also highlights that it is extremely essential to better appreciate biological reactions to intentionally produced NMs in order to construct prognosticative forms for a more reliable prototype of NMs.

Cellular compartments of ENMs comprising attachment, signaling, spreading, proliferation differentiations, apoptosis, all are influenced by the interactions between them and cells (Liu and Tang, 2017). Due to nanoscale results and increased surface area, ENMs, which are able to travel effortlessly in the inner parts of the human body, have been examined as promising instruments for the progress of diagnostic biosensors, drug and gene delivery systems, biodetection and biomedical imaging, as well as in other various innovative applications used in nanomedicine and nanobiotechnology (Dong et al., 2012; Valavanidis and Vlachogianni, 2016).

The uncommon properties of different categories of ENMs ("intentionally produced" or "manufactured" could be utilized identically) offer them some new electrical, catalytical, magnetic, mechanical, thermal, or imaging characteristics, which are advantageous for applications in diverse fields of biology, nanomedicine, and nanobiotechnology.

Jain (2008) emphasized that taking into account the intrinsic nanoscale functional constituents of living cells, it was unavoidable that nanotechnology would be applied in biotechnology, thus, creating the term called nanobiotechnology; Jain presented nanomedicine as the application of nanobiotechnologies to medicine and explained the relationship of nanobiotechnology to nanomedicine and other biotechnologies.

There are many norms of classification of ENMs, such as depending on their composition, crystallinity, sizes, and forms. One of the most extensively exploited categorizations of the ENMs is based on their composition (Ion et al., 2014). Most actual ENMs can be organized into four categories: carbon-based, metal-based, dendrimers, and composites (Bratovcic, 2019).

The aim of this chapter is to review the types, the characterization techniques, and the applications in biomedical field related to carbon-based nanomaterials (CBNMs).

Further, in this review, the reports will be made in the name of "NMs" instead of "ENMs."

5.2 CARBON-BASED NANOMATERIALS

This section discusses a comprehensive analysis of the recent progress in the field of CBNMs for biomedical applications, the future perspectives, and opportunities in this domain have been indicated.

Functional CBNMs prove uncommon physical and chemical properties, including thermal, mechanical, electrical, optical, and structural variety. Due to these advantageous characteristics, CBNMs are utilized in a large gamut of domains comprising biology, medicine, energy transfer, and energy storage (Urooj et al., 2016; Blais-Oullete et al., 2020; Zhou et al., 2020; Gazzi et al., 2020). Notably, in recent years, as a result of the presence of various allotropes of carbon, from famous allotropic phases, such as amorphous carbon, graphite, and diamonds to lately discovered promising carbon nanotubes (CNTs), graphene oxide (GO), and fullerene, the carbon-based materials have been developed and widely used in several sectors of biomedical engineering including applications for targeted drug/gene delivery, fluorescence biosensing, bioimaging, therapeutics, and other newer applications (Lacerda et al., 2007; Maiti et al., 2018; Zhang et al., 2018; Rout et al., 2018; Mishra et al., 2018; Mousavi et al., 2020).

CBNMs are a developing category of biomaterials utilized to expand noninvasive accurate and manageable light-responsive medical devices in a broad area of biomedical applications by incorporating photothermal therapy, photodynamic therapy, drug delivery, and regenerative medicine. Some remarkable scientific results show that the field of nanomedicine requires the use of CBNMs, which should be surface modified by chemical, physical, and biological procedures to enhance the solubility of the CBNMs in aqueous media, and to become more compatible with biological systems (Yamashita et al., 2012; Wan et al., 2016; Ajitha et al., 2018; Mohajeri et al., 2018; Casais-Molina et al., 2018; Rajakumar et al., 2020; Şimşek et al., 2020; Anzar et al., 2020).

The unusual features of CBNMs permit them to interact with molecules via noncovalent forces, such as hydrogen bonding, electrostatic forces, van der Waals forces, and hydrophobic interactions (Chatzimitakos and Stalikas, 2017). These NMs are mainly made of carbon, the geometric construction of the particles in NMs being fundamental indicator for their categorization. The particles can have forms of tubes, horns, spheres, ellipsoids (Chatzimitakos and Stalikas, 2017; Talreja and Kumar, 2018). According to their shape, nanoparticles possessing tube-like shape are named as CNTs, horn-shaped particles are nanohorns, and spheres or ellipsoids are included in the group of fullerenes (Beg et al., 2010; Gonzales-Garcia et al., 2016; Salazar-Rios et al., 2018; Chaturvedi et al., 2020; Bhalla et al., 2021).

5.2.1 CARBON NANOTUBES

Ever since their invention by Iijima (1991), CNTs have become the most extensively used CBNMs and their unusual physical, chemical, and electronic properties have contributed to a diversity of technological uses in functional nanodevices, particularly as field-effect transistors and chemical and biological sensors, in specialty electronics, molecular memories, optics to functionalized DNA, high efficiency adsorbent electrode material for energy storage tool, and protein functionalization (Aydin and Akins, 2013; Cardano et al., 2018; Manzetti and Gabriel, 2019; Dudina et al., 2019; Park et al., 2020).

The CNTs have big surface area, optical property and small sizes, which possibly encourage the binding of the great number of biomolecules on it. The CNTs can permeate cell membranes, thus paving the way for the utilization as a delivery factor of drugs, nucleic acids, and proteins in the cytoplasm (Costa et al, 2016; Perez-Luna et al., 2018; Rafiq et al., 2020).

Saifuddin et al. (2012) furnished a comprehensive review about the chemistry of CNTs, which has made tremendous progress lately, showing that the functionalization of CNTs, and especially CNTs of defined length, diameter, and chirality will conduct to a higher supervision of CNT-based materials, and instruments at the molecular level in a problem that has to do with the applications of CNTs in biomedical imaging drug delivery, biosensing, and the projection of functional nanocomposites. Rai et al. (2020) exhibited a new remarkable and very fruitful review work on the field of CNTs that is used in the biomedical sector.

The CNTs present a cylindrical carbon structure made of one layer of graphene (i.e., only one cylinder, then they are named single-walled CNTs, SWCNTs) or many layers of graphene (i.e., more concentric cylinders—then they are called multiwalled CNTs, MWCNTs) (Camilli and Passacantando, 2018). Double-walled CNTs (DWCNTs) are located at the interface between SWCNTs and MWCNTs and show that their properties are rather close to those of SWCNTs but can as MWCNT be covalently functionalized without degrading much of their electrical conductivity due to the existence of a second external wall (Leon et al., 2012; Neves et al., 2012; Fujisawa et al., 2016; Simon et al., 2019; Lyu et al., 2020).

Ameta et al. (2019a) examined some exceptional novel research, approaches, and applications of CNTs, demonstrating that CNTs can be used as a photocatalytic material in pure form or in the form of a composite with some photocatalyst. The same authors reviewed the use of CNTs like a photocatalytic material in the case of environmental cleaning.

5.2.1.1 CELL LABELING AND IMAGING METHODS

CNTs can operate like labeling and imaging agents due to their unique optical features. Hong et al. (2009) summarized the important progress that has been made in the development of SWCNTs technology in molecular imaging applications, highlighting that SWCNTs display unique properties which recommend them for utilization in a diversity of imaging methods, such as magnetic resonance imaging, near infrared fluorescence, Raman spectroscopy, photoacoustic tomography, and radionuclide-based imaging.

Several scientific works, based on the characteristic of CNTs, to have optical transition in near infrared region (NIR), which has been manifested to be helpful in biological tissue, for the reason that NIR has greater penetration depth and lower excitation scattering (Harrison and Atala, 2007), have reviewed various new successful applications of CNTs as multifunctional nanoprobes for biomedical imaging (Cherukuri et al., 2004; Cha et al., 2013; Dinarvand et al., 2020).

In this regard, Dinarvand et al. (2020) outlined some of the newest developing techniques used in the case of the optical detection of neurotransmitter release and analyzed the case of the NIR fluorescent CNT-based sensors and engineered fluorescent proteins for monoamine imaging, presenting in detail the exemplification of SWCNT-based fluorescent sensors for dopamine and serotonin.

Mann et al. (2020) reported that SWCNTs have attracted much attention not only because of their uncommon chemical pattern, but also because of their notable photophysical features including nonbleaching/blinking NIR fluorescence. They have also presented a novel method for covalent decoration of SWCNTs with biomolecules using light-driven diazonium chemistry procedure that acts as anchor points for the attachment of biomolecules.

Katz and Willner (2004) presented recent developments in the manufacture of hybrid systems containing CNTs and biomolecules (proteins and DNA) and discussed about the research that has been done in connection with the combination of biomaterials with CNTs, this fact permitting the utilization of hybrid systems as active field-effect transistors or biosensors devices (enzyme electrodes, DNA sensors). They reviewed some of the current scientific achievements that have been obtained in this interdisciplinary domain CNT-based nanobioelectronics and nanobiotechnology showing that the integration of biomaterials into nanoelectronic and nanomechanic systems based on CNTs will lead to spectacular progress.

Some important studies have revealed that many of the examples which have been successfully used in the field of the nanomedical exploration refer to the use of CNTs as adapted multifunctional "nanodevices" that can serve as diagnostic tools, tissue-specific drug delivery structure or vehicles for gene therapy, perfectly blending this fact with multimodal imaging abilities (Liu et al., 2007; Hong et al., 2009; Kostarelos et al., 2009; Lu et al., 2009; Liu et al., 2011; Vittorio et al., 2011; Komane et al., 2016).

Because CNTs are normally fabricated using the catalytic chemical vapor deposition method and include metals, especially Fe at their closed ends, CNTs are paramagnetic (Kim et al., 2005; Vittorio et al., 2011). Vittorio et al. (2009) have shown that the magnetic properties of MWCNTs perform a significant role for numerous biomedical uses, an attractive use being the probability of labeling the cells with the magnetic nanoparticles (MNPs) and to lead their location through an exterior magnetic field. The same type of the procedure was used by these authors to quide and enhance nerve regeneration after the cure of the axon with MNPs.

Ameta et al. (2019b) reported some outstanding results of advanced scientific research, in which they presented the applications of CNTs in the field of the chemical sensors and biosensors, illustrating that CNT sensors are utilized to accurately detect very small concentrations of alcohol, ammonia, aldehyde, as well as to detect the presence of pathogenic bacteria. In addition, CNTs are indicated as exceptional biomarkers for the early and rapid detection of different cancers.

The confluence between biotechnology and nanotechnology has led to the progress of some new fast and sensitive analytical methodologies that are based on the system of nanostructured biosensors, this system being usually applied in many food industries for packaging and detection of pathogens. Nanobiotechnology provides more cost-effective, rapid in situ/on-site and more trusty detection techniques to supervise biological risks due to the small dimension of biosensor instrument and their reaction features, compared with the traditional strategies that are used to detect microbial contaminants, these traditional strategies being sensitive and cheap, but requiring a longer time to obtain the final results (Yasmin et al, 2016; Pan et al., 2019).

Ameta et al. (2019c) explained that food undergoes a diversity of changes in postharvest and processing operations that could influence their biological and biochemical characteristics and that the evolution of nanotechnology could ultimately influence the food industry field as well. They also presented various noteworthy cases in the domain of biosensors research, which is related to the fast and sensitive detection of foodborne infections, emphasizing that the biosensors can quickly transmit the results based on a progressive organic reaction and showing the principal role of CNTs-based sensors in food packaging and processing (Ameta et al., 2020).

5.2.1.2 DRUG DELIVERY PLATFORMS

Many fascinating studies have demonstrated that CNTs are secure and more powerful options to former drug delivery techniques as they can cross membranes, bearing therapeutic drugs, vaccines, biomolecules, genes deep into the target cells which are previously unreachable (He et al., 2013; Basu and Mehta, 2014). Diverse investigations have highlighted modern advancements in the design of drug administration based on functionalized carbon nanotubes (f–CNTs), presenting the description methods that can be utilized to obtain a significant therapeutic efficiency of targeted drug delivery and explaining different analytical strategies that can be used to describe the basic construction, chemical pattern, and functional groups related to CNTs after the change of the surface (Marque et al., 2010; Tan et al., 2014).

Eatemadi et al. (2014) have discovered a new SWCNT-based tumor-targeted drug delivery system (DDS) which involves tumor-targeting ligands, anticancer activity drugs and functionalized SWCNTs and revealing that if this system interacts with cancer cells, then it can determine receptor-mediated endocytosis with the identification of the cancer-specific receptors on the surface of cancer cells and distribute chemotherapeutic agents (2014).

Sudhakar et al. (2019) exhibited the actual situation of information about the biomedical applications of CNTs, providing some details in connection with their use in the case of various diseases. They described that CNTs operate like a drug load, where it can capture the potent to toxic drug by small change in the composition of CNTs and it can transport to diverse body tissues without affecting the normal cells and tissue.

5.2.2 GRAPHENE OXIDE

Graphene and the graphene family of materials (GFMs) present some exceptional features, which produce an extraordinary interest in different areas like energy storage sensors, photocatalysis, lithium batteries, and biomedicine (Chaudhary et al., 2019a; 2019b). Several groups of researchers used this fact that the most frequently utilized structure of the GFMs in biomedical applications is GO, which is the oxidized form of graphene and which has more oxygen-containing surface groups, mainly COOH moieties on the layer edges as well as C–O–C and OH on the basal planes (Guerrero-Contreras and Caballero-Briones, 2015; Li et al., 2018; Diez-Pascual et al., 2020). In addition to some unique physicochemical attributes, such as small size, large surface area, special strength in 2D structure, intriguing optical and electronic features, and high purity, GO shows superior aqueous processability, amphiphilicity, surface functionalization capacity, fluorescence quenching ability, power of easy cell membranes permeation for encouraging nanobiotechnological uses, which comprise imaging, biosensors, drug delivery, and cancer therapy (Santhosh et al., 2017; Banerjee, 2018).

5.2.2.1 CELL LABELING AND IMAGING METHODS

Senapati et al. (2019) reported the model of a novel fluorophore using GO functionalization, which shows aggregation-induced emission along with pH-responsive nanoprobe, the biocompatible property of the recently developed material being an appropriate one for its use in biomedical imaging for cancer diagnosis with better correctness at a lower price.

Dinda et al. (2016) discovered an over simple technique, which can be used for drug delivery and cell labeling. This refers to the synthesis of highly fluorescent manganese-doped zinc sulfide (ZnS:Mn) nanocrystals covalently functionalized with polyethylene glycol conjugated graphene oxide (GO–PEG), then the anticancer drug doxorubicin (DOX) was loaded onto these GO–PEG/ZnS:Mn composite particles. They obtained some novel composite particles [GO–PEG/ZnS:Mn] that have special properties related to the great efficiency of drug uptake and cell labeling ability.

5.2.2.2 DRUG DELIVERY PLATFORMS

Extensive researches carried out in the case of graphene and GFNMs, in the field of nanomedicine showed that starting from simple drug delivery model, the application of graphene and GFNMs was expanded to the field of "smart" platforms of multiple therapeutic modalities involving photothermal therapy, photodynamic therapy, and magnetic hyperthermia therapy (Gu et al., 2019).

Numerous investigations have presented GO and its functional derivatives as an extremely attractive structure to be used in new applications in the field of drug administration, emphasizing that the positive result of a GO-based drug delivery mechanism is determined by the following three aspects:

- Manufacture of a carrier with an ideal loading volume.
- Confirmation of the level of toxicity.
- Biocompatibility prior to clinical examination and the projection of a device able to deliver the drugs in a directed way to a chosen site for an efficient treatment (Liu et al., 2013; Makharza et al., 2013; Wu et al., 2015,).

Chen et al. (2014) described the structure of a triple-functional stimuli-responsive nanosystem based on the cointegration of superparamagnetic Fe_3O_4 and paramagnetic MnO_x NPs onto exfoliated GO nanosheets using a productive double-redox strategy (DRS). They also explained that the biocompatible GO-based nanoplatform stop the cancer cell metastasis process by downregulating the expression of metastasis-related proteins.

5.2.3 CNMS COMBINED WITH DIFFERENT TYPES OF NMS

Chen and Lu (2010) suggested that CNTs decorated with NPs constitute a new category of hybrid NMs that are considered good building blocks for different uses and that could exhibit not only the special features of NPs and nanotubes (NTs) but also supplemental physical and chemical features due to the interaction between CNTs and attached NPs, explaining that the assembly of the diverse metallic and semiconducting NPs onto the external areas of CNTs was accomplished using the electrostatic force-directed assembly process.

Modugno et al. (2015) revealed the fact that the metal nanoparticles (MNPs) are successfully used in drug delivery, noninvasive tumor imaging and rapid detection and because (carbon nanomaterials) CNMs represent excellent platforms for the binding of MNPs, MNP/CNM hybrids not only blend the uncommon features of MNPs and CNMs but also present new features, which appear from the interactions between the two structures.

Semiconductor quantum dots (QDs) constitute a new category of inorganic fluorophore with higher photophysical properties: size-adjustable light emission, strong light absorption, very high signal brightness, resistance against photobleaching and simultaneous excitation of multiple fluorescence colors (Zhao and Zeng, 2015). The crucial ability to tune the emission color of a QD has to do with the quantum confinement effect, where the nanocrystal dimensions determine the energies of excited-state charge carriers (Lim et al., 2015).

Several elegant and interesting approaches have reported the successful recent advances in the field of applications of QDs technology (Ameta et al., 2019c; Bhatt et al., 2019, 2020, 2021; Jat et al., 2019). Some of the important biomedical uses of QDs have to do with the next domains: tumor and microRNA detection, drug administration, photodynamic therapy, microbial labeling, multiplexed imaging, dual imaging, therapeutic platform, real-time in vivo and cellular process imaging (Wagner et al., 2019).

Vasudevan et al. (2015) showed that core-shell QDs were developed to enhance the photoluminescence efficiency of single QDs, emphasizing not only the significance and the utilizations of core-shell QDs, but also their changes and surface utilizations in medicine and sensor utilizations for biochemical analyzes. CdSe/ZnS core-shell QDs present a remarkable interest due to their uncommon strong luminance and raised photostability (Hoshino et al., 2004; Beachley and Wen, 2009).

Some significant works have sought to furnish various modern perspectives on the applications of QDs in the fields of biomedical detection and optical imaging, showing that CdSe/ZnS core-shell QDs is an ideal substitute to traditional fluorophores utilized to label the microbial cells (Damian et al., 2010; Armăşelu et al., 2011a, 2011b; Ardelean et al., 2011).

Tan et al. (2011) presented that QDs can be used to image cancer cell and the CNTs can be used as feasible possibility for a modern chemotherapeutic drug, indicating that the process of conjugating QDs with CNTs is possible. This process allowed the utilization of new qualities in the field of cancer theranostics. They also explained that the conjugation of QDs to CNTs allow them to operate like novel hybrid nanocomposite material and discussed the involvements of nanomedicine in a clinical setting.

Kim et al. (2011) communicated the process of the attachment of the CdSe/ZnS QDs on to the surface of shortened MWCNTs (sMWCNTs) in order to increase luminescence intensities of CNTs. The resultant QDs-sMWCNTs nanohybrid complex was described by Fourier transform infrared (FTIR) spectroscopy, ultraviolet (UV) light, scanning electron microscopy (SEM), and optical microscopy (OM).

In order to propose some new materials with applicability in the biomedical field, including biological imaging and IR detectors, Das et al. (2014) discussed the case of the noncovalent attachment of PbS QD to SWCNTs and MWCNTs, the resultant composites being analyzed with the help of TEM and FTIR spectroscopy.

5.3 CONCLUSION

So far, there is a broad gamut of NMs that have been invented and analyzed in the situation of the applications in the biomedical domain. Carbon nanotechnology has advanced quickly during the past few years and carbon allotropes, particularly graphene, GO, and CNTs have been successfully applied in several areas of biology and medicine, such as biosensors, cell labeling, cellular imaging, antimicrobial agents, drug and DNA delivery, and anticancer therapy. This present chapter provided a comprehensive review, which is focused on some principal categories of CBNMs and their emerging biomedical applications. It also explored the main achievements in hybrids based on CNTs and GO. The knowledge that has been exposed in this study should conduct to a higher perception of the key elements that may control future directions of exploration.

KEYWORDS

- **nanomaterials**
- **nanomedicine**
- **nanobiotechnology**
- **carbon-based nanomaterials**
- **carbon nanotubes**
- **graphene oxide**
- **quantum dots**
- **current status**
- **applications**

REFERENCES

Ajitha, A. R.; Akhina, H.; Aswathi, M. K.; Mathew, L.; Thomas, S. Carbon Nanotubes: An Ideal Candidate for Biomedical Applications. *JSM Nanotechnol. Nanomed.* **2018,** *6* (2), 1065–1075.

Ameta, R.; Jat, K. K.; Bhatt, J.; Ameta, S. C. Quantum Dots and Their Applications. In *Advances in Nanotechnology and the Environmental Sciences: Applications, Innovations, and Visions for the Future Series: AAP Research Notes on Nanosciences and Nanotechnology.*

Vakhrushev, A. V., Ameta, S. C., Susanto, H., Haghi, A. K., Eds.; CRC Press/Taylor & Francis; Apple Academic Press: New York, 2019c; pp 119–150.

Ameta, R.; Jat, K. K.; Bhatt, J.; Rai, A.; Nargawe, T.; Soni, D.; Ameta, S. C. Carbon Nanotubes as Chemical Sensors and Biosensors: A Review. In *Advances in Nanotechnology and the Environmental Sciences: Applications, Innovations, and Visions for the FutureSeries: AAP Research Notes on Nanosciences and Nanotechnology*; Vakhrushev, A. V., Ameta, S. C., Susanto, H., Haghi, A. K., Eds.; CRC Press/Taylor & Francis; Apple Academic Press: New York, 2019b; pp 203–240.

Ameta, R.; Kapoor, N.; Pathak, B.; Bhatt, J.; Ameta, S. C. Carbon Nanotube Composites as Photocatalytic Materials. In *Carbom Nanotubes and Nanoparticles. Current and Potential Applications, Series: AAP Research Notes on Nanosciences and Nanotechnology* Vakhrushev, A. V., Kodolov, V. I., Haghi, A. K., Ameta, S. C., Eds.; CRC Press/Taylor & Francis; Apple Academic Press: New York, 2019a; pp 39–64.

Ameta, S. K.; Rai, A. K.; Hiran, D.; Ameta, R.; Ameta, S. C. Use of Nanomaterials in Food Science. In *Biogenic Nano-Particles and Their Use in Agro-ecosystems*, 2020; pp 457–488.

Anzar, N.; Hasan, R.; Tyagi, M.; Yadav, N.; Narang, J. Carbon Nanotube—A Review on Synthesis, Properties and Plethora of Applications in the Field of Biomedical Science. *Sens. Int.* **2020,** *1* (1). DOI: 10.1016/j.sintl.2020.100003.

Ardelean, I.; Sarchizian, I.; Manea, M.; Damian, V.; Apostol, I.; Cârnu, M. Armăşelu, A. CdSe/ZnS Quantum Dots Cytotoxicity against Phototrophic and Heterotrophic Bacteria. In *Proceeding of NANOCON 2011*; TANGER Ltd., Ed., 21–23 September 2011; Brno, Czech Republic, 2011; pp 608–617.

Armăşelu, A.; Popescu, A.; Apostol, I.; Ardelean, I.; Damian, V.; Iordache, I. et al. Passive Nonspecific Labeling of Cyanobacteria in Natural Samples Using Quantum Dots. *Optoelectron. Adv. Mat. Rapid Commun.* **2011a,** *5* (10), 1084–1090.

Armăşelu, A.; Popescu, A.; Damian, V.; Ardeleanu, I.; Apostol, D. Fluorescence Properties of Quantum Dots Used in the Study of Microorganisms. *J. Optoelectron. Adv. Mater.* **2011b,** *4* (13), 439–443.

Aydin, M.; Akins, D. L. Vibroelectronic Properties of Functionalized Single-Walled Carbon Nanotubes and Double-Walled Boron Nitride Nanotubes. In *Physical and Chemical Properties of Carbon Nanotubes*; Suzuki, S., Ed.; Intech: Rijeka, Croatia, 2013; pp 53–85.

Banerjee, A. N. Graphene and Its Derivates as Biomedical Materials: Future Prospects and Challenges. *Interf. Focus.* **2018,** *8* (3). DOI: 10.1098/rsfs.2017.0056.

Basu, B.; Mehta, G. K. Carbon Nanotubes: A Promising Tool in Drug Delivery. *Int. J. Pharma. Biol. Sci.* **2014,** *5* (1), 533–555.

Beachley, V.; Wen, X.; Stem Cells and Nanostructured Materials. In *NanoScience in Biomedicine*; Shi, D., Ed.; Springer-Verlag: Berlin Heildelberg, 2009; pp 1–18.

Beg, S.; Rizwan, M.; Sheikh, A. M.; Hasnain, M. S.; Anwer, K.; Kohli, K. Advancement in Carbon Nanotubes: Basics, Biomedical Applications and Toxicity. *J. Pharm. Pharmacol.* **2010,** *63* (2), 141–163.

Bhalla, G.; Sharma, A.; Kumar, V.; Bhalla, B.; Kumar, S.; Kumar, H. *Use of Carbon Nanomaterials as Potential Ion-Exchange*. In *Environmental Remediation and Technology. Through Carbon Based Nano Composites; Green Energy*; Jawaid, M.,; Ahmad, A., Ismail, N., Rafatullah, M., Eds.); Springer: Singapore, 2021; pp 231–247.

Bhatt, J.; Benjamin, S.; Ameta, R.; Ameta, S. C. Enhanced Visible Light Photocatalytic Activity of TiO_2 Nanospheres/SnO_2 Quantum Dots Composite for Degradation of 4—Nitrophenol. *J. Nanosci. Technol.* **2019,** *5* (4), 748–750.

Bhatt, J.; Jat, K. K.; Rai, A. V.; Ameta, R.; Ameta, S. C. *Photodegradation of 2-Nitrophenol, an Endocrine Disruptor, using TiO_2 Nanosphere/SnO_2 Quantum Dots*. In *Green Chemistry and Biodiversity: Principles, Techniques, and Correlations*; Aguilar, C. N., Ameta, S. C., Haghi, A. K., Eds.; CRC Press/Taylor & Francis; Apple Academic Press: New York, 2020; pp. 1–8.

Bhatt, J.; Jat, K. K.; Rai, A. V.; Ameta, R.; Ameta, S. C. *Quantum Dots and Their Sensing Applications*. In *Chemistry and Industrial Techniques for Chemical Engineers, Innovations in Physical Chemistry: Monograph Series*; Pogliani, L., Ameta, S. C., Haghi, A. K., Eds.; CRC Press/Taylor & Francis; Apple Academic Press: New York, 2021; pp 49–66.

Blais-Ouellette, S.; Verhaegen, M.; Lessard, S. Photon Etc. Applications: Carbon-Based Materials, 2020. www.photonetc.com/en/carbon-based-materials.html (accessed Sep 20, 2020).

Bleeker, E. A. J.; Cassee, F. R.; Re, G.; De John, W. H. Interpretation and Implications of the European Commission Recommendation on the Definition of Nanomaterial. *RIVM Lett. Rep*. 601358001, **2012**.

Boherhof, D. R.; Bramante, C. M.; Butala, J. H.; Clancy, S. F.; Lafranconi, M.; West, J. et al. Comparative Assessment of Nanomaterial Definitions and Safety Evaluation Considerations. *Regul. Toxicol. Pharmacol*. **2015,** *73* (1), 137–150.

Bratovcic, A. Different Applications of Nanomaterials and Their Impact on the Environment. *Int. J. Mater. Sci. Eng*. **2019,** *5* (1). DOI: 10.14445/23948884/IJMSE-V5I1P101.

Camilli, L.; Passacantando, M. Advances on Sensors Based on Carbon Nanotubes. *Chemosensors*. **2018,** *6* (4). DOI: 10.3390/chemosensors6040062.

Cardano, F.; Frasconi, M.; Giordani, S. Photo-Responsive and Carbon Nanotubes to Control and Tackle Biological Systems. *Front Chem*. **2018,** *6*, DOI: 10.3389/fchem.2018.00102.

Casais-Molina, M. L.; Cab, C.; Canto, G.; Medina, J.; Tapia, A. Carbon Nanomaterials for Breast Cancer Treatment. *J. Nanomater*. **2018,** *2018*. DOI: 10.1155/2018/2058613.

Cha, C.; Shin, S. R.; Annabi, N.; Dokmeci, M. R.; Khademhosseini, A. Carbon-Based Nanomaterials: Multi-Functional Materials for Biomedical Engineering. *ACS Nano*. **2013,** *7* (4), 2891–2897.

Chaturvedi, V. K.; Kushwaha, A.; Maurya, S.; Tabassum, N., Chaurasia, H.; Singh, M. P. *Wastewater Treatment through Nanotechnology: Role and Prospects*. In *Restauration of Wetland Ecosystem: A Trajectory Towards a Sustainable Environment*; Upadhyay, A. K., Singh, R., Singh, D. P., Eds.; Springer: Singapore, 2020; p 227.

Chatzimitakos, T.; Stalikas, C. Carbon-Based Nanomaterials Functionalized with Ionic Liquids for Microextraction in Sample Preparation. *Separations* **2017,** *4* (2). DOI: 10.3390/separations4020014.

Chaudhary, R.; Ameta, S. C.; Am eta, R. Photocatalytic Degradation of Azure B under Visible Light Irradiation by Reduced Graphene Oxide—NiS Composite. *Macedon. J. Chem. Chem. Eng*. **2019,** *38* (1), 107–114.

Chaudhary, R.; Rai, A.; Ameta, R.; Ameta, S. Reduced Graphene Oxide/CuS Nanocomposite: An Efficient Photocatalyst for Degradation of Crystal Violet. *J. Nanosci. Tech*. **2019,** *5* (2), 673–675.

Chen, B. J. Lu, G. *Carbon Nanotube Nanoparticle Hybrid Structures*. In *Carbon Nanotubes*; Marulanda, J. M., Ed.; Intech: Rijeka, Croatia, 2010; pp 611–634.

Chen, Y.; Yu, P.; Shu, Z.; Wu, M.; Wang, L.; Zhang, S. Multifunctional Graphene Oxide—Based Triple Stimuli—Responsive Nanotheranostics. *Adv. Funct. Mater*. **2014,** *24* (28), 4386–4396.

Cherukuri, P.; Bachilo, S. M.; Litovsky, S. H.; Weisman, R. B. Near-Infrared Fluorescence Microscopy of Single-Walled Carbon Nanotubes in Phagocytic Cells. *J. Am. Chem. Soc*. **2004,** *126*, 15638–15639.

Costa, P. M.; Bourgognon, M.; Wang, J. T.-W.; Al-Jamal, K. T. Functionalized Carbon Nanotubes: From Intracellular Uptake and Cell-Related Toxicity to Systemic Brain Delivery. *J. Control. Release*. **2016,** *241*, 200–219.

Damian, V.; Ardelean, I.; Armăşelu, A.; Apostol, D. Fourier Transform Spectra of Quantum Dots. *Proc. SPIE*. **2010,** 7469. DOI: 10.1117/12.862053.

Das, A.; Hall, E.; Wai, C. M. Noncovalent Attachment of PbS Quantum Dots to Single- and Multiwalled Carbon Nanotubes. *J. Nanotechnol*. **2014,** *2014*. DOI: 10.1155/2014/285857.

Diez-Pascual, A. M.; Sainz-Urruela, C.; Vallés, C.; Vera-Lopez, S.; San Andres, M. P. Tailorable Synthesis of Highly Oxidized Graphene Oxides via an Environmentally–Friendly Electrochemical Process. *Nanomater*. **2020,** *10* (2). DOI: org/10.3390/nano10020239.

Dinarvand, M.; Elizarova, S.; Daniel, J.; Kruss, S. Imaging of Monoamine Neurotransmitters with Fluorescent Nanoscale Sensors. *Chem. Plus. Chem*. **2020,** *7* (85), 1465–1480.

Dinda, S.; Kakran, M.; Zeng, J.; Sudhaharan, T.; Ahmed, S.; Das, D. et al. Grafting of ZnS: Mn—Doped Nanocrystals and an Anticancer Drug Onto Graphene Oxide for Delivery and Cell Labeling. *Chem. Plus. Chem*. **2016,** *81* (1), 100–107.

Dong, L.; Craig, M. M.; Khang, D.; Chen, C. Applications of Nanomaterials in Biology and Medicine. *J. Nanotechnol.* **2012,** *2012*. DOI: 10.1155/2012/816184.

Dudina, A.; Frey, U.; Hierlemann, A. Carbon–Nanotube–Based Monolithic CMOS Platform for Electrochemical Detection of Neurotransmitter Glutamate. *Sensors* **2019,** *19* (14). DOI: 10.3390/s19143080.

Eatemadi, A.; Daraee, H.; Karimkhanloo, H.; Kouhi, M.; Zarghami, N.; Akbarzadeh, A.; et al. Carbon nanotubes: Properties, synthesis, purification, and medical applications. *Nanoscale. Res. Lett*. **2014,** *9*. DOI: 10.1186/1556-276X-9-393.

Fujisawa, K.; Kim, H. J.; Go, S. H.; Muramatsu, H.; Hayashi, T.; Endo, M. et al. A Review of Double-Walled and Triple-Walled Carbon Nanotube Synthesis and Applications. *Appl. Sci.* **2016,** *6* (4). DOI: 10.3390/app6040109.

Gazzi, A.; Fusco, L.; Orecchioni, M.; Ferrari, S.; Franzoni, G.; Yan, J. S. et al. Graphene Other Carbon Nanomaterials and the Immune System: Toward Nanoimmunity-by-Design. *J. Phy. Mater*. **2020,** *3* (3). DOI: 10.1088/2515-7639/abc931.

Gonzales-Garcia, P.; Urones-Garrote, E.; Gomez-Herrero, A.; Avila-Brande, D.; Otero-Diaz, L. C. Graphene Production from Chlorination of Metallocenes. In *Graphene Science Handbook, Six-Volume Set*; Aliofkhazraci, M., Ali, N., Milne, W. J., Ozkan, C. S., Mitura, S., Gervasoni, J. L., Eds.; CRC Press/Taylor & Francis: Boca Raton, 2016; pp 387–404.

Gu, Z.; Zhu, S.; Yan, L.; Zhao, F.; Zhao, Y. Graphene–Based Smart Platforms for Combined Cancer Therapy. *Adv. Mater*. **2019,** *31* (9), DOI: 10.1002/adma.201800662.

Gubala, V.; Johnston, L. J.; Liu, Z.; Krug, H.; Moore, C. J.; Ober, C. K. et al. Engineered Nanomaterials and Human Health: Part 1. Preparation, Functionalization and Characterization (IUPAC Technical Report). *Pure Apple. Chem*. **2018,** *90* (8), 1283–1324.

Guerrero-Contreras, J.; Caballero-Briones, F. Graphene Oxide Powders with Different Oxidation Degree, Prepared by Synthesis Variations of the Hummers Method. *Mater. Chem. Phys*. **2015,** *153*, 1–12.

Hardy, A.; Benford, D.; Halldorsson, T.; Jeger, M. J.; Knutsen, H. K.; More, S. et al. Guidance on Risk Assessment of the Applications of Nanoscience and Nanotechnologies in the Food and Feed Chain: Part 1, Human and Animal Health. *EFSA J.* **2018,** *16* (7). DOI: 10.2903/J.EFSA.2018.5327.

Harrison, B. S.; Atala, A. Carbon Nanotube Applications for Tissue Engineering. *Biomaterials* **2007,** *28* (2), 344–353.

He, H.; Pham-Huy, L. A.; Dramou, P.; Xiao, D.; Zuo, P.; Pham-Huy, C. Carbon Nanotubes: Applications in Pharmacy and Medicine. *Biomes. Res. Int.* **2013,** *2013*. DOI: 10.1155/2013/578290.

Hochella, Jr. M. F.; Mogk, D. W.; Ranville, J.; Allen, I. C.; Luther, G. W.; Marr, L. C. et al. Natural, Incidental and Engineered Nanomaterials and Their Impacts on the Earth System. *Science* **2019,** *363* (6434). DOI: 10.1126/science.aau8299.

Hong, H.; Gao, T.; Cai, W. Molecular Imaging with Single—Walled Carbon Nanotubes. *Nano Today* **2009,** *4* (3), 252–261.

Hoshino, A.; Hanaki, K.; Suzuki, K.; Yamamoto, K. Applications of T-lymphoma Labeled with Fluorescent Quantum Dots to Cell Tracing Markers in Mouse Body. *Biochem. Biophys. Res. Commun.* **2004,** *314* (1), 46–53.

Iijima, S. Helical Microtubules of Graphitic Carbon. *Nature* **1991,** *354* (6348), 56–58.

Ion, I.; Culetua, A.; Gherase, D.; Sirbu, F.; Ion, A. C. Environmental Applications of Carbon-Based Nanomaterials. Acetylcholinesterase Biosensors for Organophosphate Pesticide Analysis. In *New Applications of Nanomaterials*; Bucharest: Academiei Române, 2014; pp 33–51.

Jain, K. K. Nanomedicine: Application of Nanobiotechnology in Medical Practice. *Med. Princ. Pract.* **2008,** *17*, 89–101.

Jat, K. K.; Bhatt, J.; Ameta, S. C. Photodegradation of Fast Green by Using SnO_2 Quantum Dots/TiO_2 Nanoparticles Composite. *J. Appl. Chem.* **2019,** *8* (1), 139–145.

Katz, E. Willner, I. Biomolecule—Functionalized Carbon Nanotubes: Applications in Nanobioelectronics. *Chem. Plus. Chem.* **2004,** *5* (8), 1084–1104.

Kim, D. Y.; Yang, C. M.; Park, Y. S.; Kim, K. K.; Jeong, S. Y.; Han, J. H. et al. Characterization of Thin Multi-Walled Carbon Nanotubes Synthesized by Catalytic Chemical Vapor Deposition. *Chem. Phys. Lett.* **2005,** *413* (1–3), 135–141.

Kim, G. B.; Bommulu, R.; Yoon, K. R. A New Strategy to Assemble CdSe/ZnS Quantum Dots with Multi-Walled Carbon Nanotubes for Potential Application in Imaging and Photosensitization. *Appl. Surf. Sci.* **2011,** *258* (3), 1027–1032.

Komane, P. P.; Choonara, Y. E.; du Toit, L.; Pradeep, K.; Kondiah, P. P. D.; Modi, G. et al. Diagnosis and Treatment of Neurological and Ischemic Disorders Employing Carbon Nanotube Technology. *J. Nanomater.* ***2016,*** 2016; DOI: 10.1155/2016/9417874.

Kostaelos, K.; Bianco, A.; Prato, M. Promises, Facts and Challenges for Carbon Nanotubes in Imaging and Therapeutics. *Nat. Nanotechnol.* **2009,** *4* (10); 627–633.

Lacerda, L.; Raffa, S.; Prato, M.; Bianco, A.; Kostarelos, K. Cell -Penetrating CNTs for Delivery of Therapeutics. *Nano Today*. **2007,** *2* (6), 38–43.

Leon, V.; Parret, R.; Almairac, R.; Alvarez, L.; Babaa, M.-R. et al. Spectroscopic Study of Double-Walled Carbon Nanotubes Functionalization for Preparation of Carbon Nanotube/ Epoxy Composites. *Carbon, Elsevier* **2012,** *50* (14), 4987–4994.

Li, R.; Guiney, L. M.; Chang, C. H.; Mansukhani, N. D.; Ji, Z.; Wang, X. et al. Surface Oxidation of Graphene Oxide Determines Membrane Damage, Lipid Peroxidation, and Cytotoxicity in Macrophages in a Pulmonary Toxicity Model. *ACS Nano*. **2018,** *12* (2), 1390–1402.

Lim, S. J.; Zahid, M. U.; Le, P.; Ma, L.; Enterberg, D.; Harney, A. S. et al. Brightness-Equalized Quantum Dots. *Nat. Commun.* **2015,** *6*. DOI: 10.1038/ncomms9210

Liu, J.; Cui, L.; Losic, D. Graphene and Graphene Oxide as New Nanocarriers for Drug Delivery Applications. *Acta. Biomater*. **2013,** *9* (12), 9243–9257.

Liu, X. Q.; Tang, R. Z. Biological Responses to Nanomaterials: Understanding Nano-Bio Effects on Cell Behaviours. *Drug Deliv*. **2017,** *24* (2), 1–15.

Liu, Z.; Winters, M.; Holodniy, M.; Dai, H. siRNA Delivery Into Human T Cells and Primary Cells with Carbon-Nanotube Transporters. *Angew. Chem. Int. Ed. Engl.* **2007,** *46* (12), 2023–2027.

Liu, Z.; Yang, K.; Lee, S.-T. Single-Walled Carbon Nanotubes in Biomedical Imaging. *J. Mater. Chem.* **2011,** *2011*, 586–598.

Lu, F.; Gu, L.; Meziani, M. J.; Wang, X.; Luo, P. G.; Veca, L. M. et al. Advances in Bioapplications of Carbon Nanotubes. *Adv. Mater*. **2009,** *21*, 139–152.

Lyu, J.; Kudiiarov, V.; Lider, A. An Overview of the Recent Progress in Modifications of Carbon Nanotubes for Hydrogen Adsorption. *Nanomaterials* **2020,** *10* (2). DOI: 10.3390/nano10020255.

Maiti, D.; Tong, X.; Mou, X.; Yang, K. Carbon-Based Nanomaterials for Biomedical Applications: A Recent Study. *Front Pharmacol*. **2018,** *9*. DOI: 10.3389/fphar.2018.01401.

Mann, F. A.; Herrmann, N.; Opazo, F.; Kruss, S. Quantum Defects as a Toolbox for the Covalent Functionalization of Carbon Nanotubes with Peptides and Proteins. *Angew. Chem. Int.* **2020,** *59* (40), 17732–17738.

Manzetti, S.; Gabriel, J.-C. P. Methods for Dispersing Carbon Nanotubes for Nanotechnology Applications: Liquid Nanocrystals, Suspensions, Polyelectrolytes, Colloids and Organization Control. *Int. Nano Lett*. **2019,** *9*, 31–49.

Makharza, S. A.; Cirillo, G.; Bachmatiuk, A.; Ibrahim, I. Graphene Oxide—Based Drug Delivery Vehicles: Functionalization, Characterization, and Cytotoxicity Evaluation. *J. Nanoparticle Res*. **2013,** *15* (12). https://doi.org/10.1007/s11051-013-2099-y.

Marques, R. R. N.; Machado, B. F.; Faria, J. L.; Silva, A. M. T. Controlled Generation of Oxygen Functionalities on the Surface of Single-Walled Carbon Nanotubes by HNO_3 Hydrothermal Oxidation. *Carbon*. **2010,** *48* (5), 1515–1523.

Mishra, V.; Patil, A.; Thakur, S.; Kesharwani, P. Carbon Dots: Emerging Theranostic Nanoarchitectures. *Drug Discov. Today*. **2018,** *23* (6), 1219–1232.

Modugno, G.; Menard-Moyon, C.; Prato, M.; Bianco, A. Carbon Nanomaterials Combined with Metal Nanoparticles for Theranostic Applications. *Br. J. Pharmacol*. **2015,** *172* (4), 975–991.

Mohajeri, M.; Behnam, B.; Sahebkar, A. Biomedical Applications of Carbon Nanomaterials: Drug and Gene Delivery Potentials. *J. Cell. Physiol*. **2018,** *233* (12), 298–319.

Mousavi, S. M.; Low, F. W.; Hashemi, S. A.; Lai, C. W.; Ghasemi, Y.; Soroshnia, S. et al. Development of Graphene Based Nanocomposites Towards Medical and Biological Applications. *Artif. Cells Nanomed. Biotechnol*. **2020,** *48* (1), 1189–1205.

Navya, P. N.; Daima, H. K. Rational Engineering of Physicochemical Properties of Nanomaterials for Biomedical Applications with Nanotoxicological Perspectives. *Nano. Converg*. **2016,** *3* (1), 1–14.

Neves, V.; Heister, E.; Costa, S.; Tîlmaciu, C.; Flahaut, E.; Soula, B. et al. Design of Double-Walled Carbon Nanotubes for Biomedical Applications. *Nanotechnol*. **2012,** *23* (36), 365102.

Pan, M.; Yin, Z.; Liu, K.; Du, X.; Liu, H.; Wang, S. Carbon-Based Nanomaterials in Sensors for Food Safety. *Nanomaterials (Basel)*. **2019,** *9* (9). https://doi.org/10.3390/nano9091330.

Pandey, K. S. Nanostructural Solar Cells. In *Recent Advances in Photovoltaic*; Ramrakhiani, M., Ed., Vol. 17; Materials Research Forum LLC: Millersville, 2017; p 58.

Park, S.-J.; Deshmukh, M. A.; Kang, B. C.; Jeon, J. Y., Chen, C.; Ha, T.-J. A Review of Advanced Electronic Applications Based on Carbon Nanomaterials. *ECS J. Solid State Sci. Technol*. **2020,** *9* (7). https://orcid.org/0000-0002-0935-795X.

Perez-Luna, V.; Moreno-Aguilar, C.; Arauz-Lara, J. L.; Aranda-Espinoza, S.; Quintana, M. Interactions of Functionalized Multi-Wall Carbon Nanotubes with Giant Phospholipid

CHAPTER 6

Carbon Nanotubes for Photosplitting of Water

NEELU CHOUHAN[1*] and SHOBHANA SHARMA[2]

[1]*Department of Pure and Applied Chemistry, University of Kota, MBS Road, Kota 24005, Rajasthan, India*

[2]*Department of Chemistry, S. S. Subodh P. G. (Autonomous) College, Jaipur 302005, Rajasthan, India*

[*]*Corresponding author. E-mail: neeluchouhan@uok.ac.in*

ABSTRACT

The carbon nanotubes (CNTs) are 1D-allotrope of carbon with appropriate electronic band positioning with fascinating conductive, electronic, and mechanical chattels. CNTs are found mainly in two forms, that is, single-walled CNTs and multiwalled-CNTs with their unique properties that have advantages over the materials of the same class. This chapter summarizes the latest progress related to the design and construction of CNT-based materials characterization with their applications focused on cleavage of water by using photocatalytic and photoelectrochemical methods. The CNT-based materials used for photoelectrochemical water splitting are divided into five main parts, that is, primary, binary, ternary, quaternary, and Z-scheme-based systems. These are described in this chapter with specific examples. This chapter includes the main properties, synthesis methods, characterization, and usages of CNTs. Stances on future prospective of CNTs are also discussed.

The Science of Nanomaterials: Basics and Applications. Suresh C. Ameta & Rakshit Ameta (Eds.)

6.1 INTRODUCTION

The fourth most copious material on earth is carbon in addition to the other elements, such as, hydrogen, helium, and oxygen. Carbon is a basic element present in all living things of the earth. The physicists and material scientists are greatly fascinated with the new allotrope of carbon, that is, CNTs that changed the whole paradigm of the basic three forms of carbon (diamond, graphite, and amorphous carbon) forever. After the invention of the CNT, a lot of thrust, experienced in the area of catalyst, electronics, medical science, energy storage devices, and so on. That boosted the research activities devoted to exploiting the profound characteristics of CNTs, that is, light weight, electronic, thermal, mechanical, tensile strength, flexibility, high aspect ratio, charged carrier mobility, etc., for their application in the field of electronics, tubular flow heat exchangers (Walvekar et al., 2016), humidity sensors (Han et al., 2012) and biosensors (Hui-Wen et al., 2005), transistors, batteries (Sung-Wook et al., 2010), supercapacitors (Ting et al., 2012), catalysts, antibacterial air filter (Hee et al., 2011), orthopedic (Debrupa et al., 2012), hydrogen storage cells (Nguyen et al., 2016), construction (Makar and Beaudoin, 2004), etc., applications. The single-walled carbon nanotube's (SWCNT's)/multiwalled carbon nanotube's (MWCNT's) properties can be adjusted by means of controlling their basic characteristics, such as structure/shape, length purity, along with their chemistry, energy, size distribution, charge of their surface and their agglomerate states. CNTs are made of sp^2 hybridized cylindrical rolled honeycomb structures of the graphitic carbon and also known as bucky tubes. These quasi-one-dimensional tubular structures of CNTs are made up of either one or tens or hundreds of concentrical CNTs cylinders that are separated by 0.34 nm distance (Popov, 2004).

6.2 HISTORICAL BACKGROUND

The presence of the CNTs (as a strengthening agent) was discovered in the swords made up of Damascus steel in the 17th century (Kochmann et al., 2004; Reibold et al., 2006; Lau and Hui, 2002; Grobert, 2007).The vapor-grown carbon nanofibers (VCNFs) of carbon nanomaterials were first explored by Hughes and Chambers (1889) that attracted many researchers due to their low production cost and their good availability with outstanding qualities (inferior to the contemporary CNTs). The allotrope of carbon, that is, CNT was first reported by Radushkevich and Lukyanovich (1952) in the form of "worm-like" hollow graphitic carbon fibers of 50 nm diameter, fabricated as a product through the carbon monoxide (CO) deposited on iron particles at 600°C temperature heating (*Oberlin et al., 1976*), they also reported the rolled-up form of the MWCNTs. Later on, in 1982, Lapidus et al. investigated the physicochemical and structural backgrounds of the 1D-carbon nanoparticles made up of the thermocatalytical disproportion of carbon monoxide. Then, a Iijima (1990) first time synthesized the MWCNTs out of the soot of arc discharge by using an arc-evaporation method. Few years later, Odom et al. (2000) reported the synthesis of the SWCNTs that consist of an enfold single faultless graphite sheet as a cylindrical tube as shown by Figure 6.1.

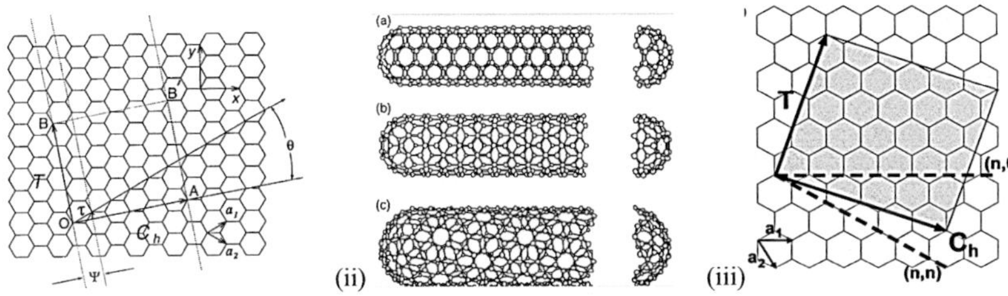

FIGURE 6.1 (i) 2D Arrangement with honey comb lattice of graphene sheet with chiral unit vectors a1 and a2 and chiral angle θ Folding directions of the 2D-graphitic sheets for making of the 1D-CNTs and (ii) a general direction B with 0 <0< 30" (a "chiral" tubule). The actual tubules shown are correspond to (n, m) values of: (a) (5,5), (b) (9,0), and (c) (10,5) and (iii) Schematic theoretical model for a 2D graphene sheet of single-wall carbon illustrating lattice vectors a1 and a2, and the roll-up vector Ch = na1 + ma2 and translation vector T along the CNT axis. The achiral cases of zigzag (n,0) and armchair (n,n)arrangement indicated by black dashed lines.

Source: i and ii: Adapted with permission from Dresselhaus et al., 1995. iii: Adapted with permission from Odom et al., 2000.

Both of the carbon networks of the shell of the CNTs are closely linked to the graphite-like honeycomb lattice structure of the carbon, rolled in their quasi-one-dimensional (1D) pattern. They show the amazing mechanical and electronic properties (Bernholc et al., 2002). CNTs are mainly classified in two forms: (1) SWCNTs (Figure 6.2a) and (2) MWCNTs (Figure 6.2b). Accordingly, the SWCNTs are a one atom thick layer of graphene rolled into a flawless cylinder of a high diameter: length ratio (1:132,000,000), shell diameter 0.5–1.5 nm, and bandgap from 0.00 to 2.00 eV (show either metallic or semiconductor nature) (Jamal et al., 2017). While the MWCNTs usually exhibit the conducting property and consist of the multiple rolled layers of graphene in concentric tubes with interlayer distance 3.4 Å and diameter more than 100 nm in two basic forms, that is, (1) Parchment model (Figure 6.2c) and (2) Russian doll model (Figure 6.2d).

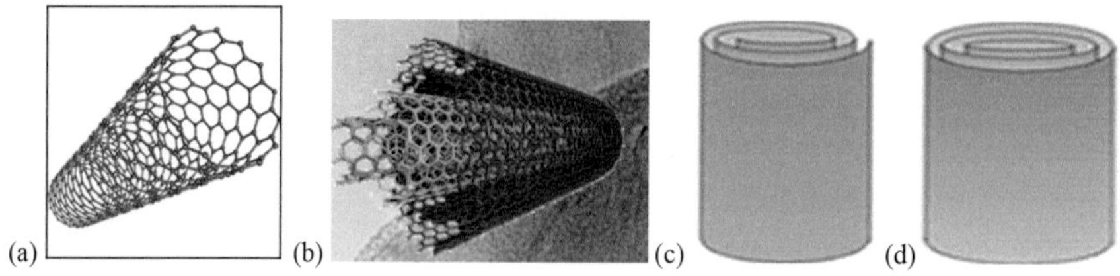

FIGURE 6.2 Systematic representation of the (a) SWCNT, (b) MWCNT and types, (c) Parchment, and (d) Russian doll models of MWCNT.

Source: Adapted from Alsharefa et al. (2017). With permission.

The electronic, thermal, and structural properties of the CNTs were governed by the changes (chirality, diameter, twist, and length), depending on the different properties of the CNTs. Few of the prominent properties of the CNTs are discussed here:

6.3 UNIQUE PROPERTIES OF CNTS

6.3.1 *ELECTRICAL CONDUCTIVITY*

CNTs exhibit the unique electrical conductivity, that is, almost similar or better (approximately,1000 times) than that of copper (Ebbesen et al., 1996; Aryasomayajula and Wolter, 2013), because each carbon atom of CNT is covalently bonded to three neighbor carbons via sp^2 molecular orbitals and the fourth electron remains free in each unit for delocalization that leads to improvement in the electrical behavior of CNTs. Based upon the geometric differences, such as defects, chirality, diameters, and degree of crystallinity, they are considered as either semiconductor or metals (Dai et al., 1996; Chandra et al., 2009). Usually, the SWCNTs are metals or p-type semiconductors (Tans et al., 1998) with their resistivities in the range 0.34×10^{-4} to 1.0×10^{-4} ohm·cm (Dai et al., 1996). SWCNTs behaved as quantum wires due to the smooth and fast transportation of the electrons in one dimension. However, the MWCNTs are composed of many tubes of SWCNTs and therefore, these are not likely to be strictly one-dimensional conductors. Conversely, the electronic transport in MWCNTs observed quite diffusive or quasi-ballistic (Delaney et al., 1999). Due to their unparalleled electronic nature, it can be used in the transistors, sensors, emitters, vacuum microelectronics, X-ray generations, and the switching applications in advanced electronics (Yue et al., 2002; Cheng et al., 2003; Modi et al., 2003; Bandaru et al., 2005).

6.3.2 THERMAL CONDUCTIVITY

The CNTs are thermally stand-alone structure up to 2800°C with longitudinal thermal conductivity of 2800–6000 W/m K, which is found good in comparison to the diamond (2000 W/m K) (Pitroda et al., 2016), silver (450 W/m K), copper (483 W/m K), graphite (100~400 W/m K), and carbon fibers (350–1100 W/m K) at room temperature. The reported thermal conductivity of the MWCNTs is ~3000 W/m K (Kim et al., 2001), where the SWCNTs exhibited ~2000 W/m K (Yu et al., 2005). The thermal energy in CNTs is acquired via a phonon conduction mechanism (phonons and inelastic Umklapp-scattering: an anharmonic phonon–phonon or electron–phonon scattering process).

6.3.3 MECHANICAL STRENGTH

CNTs are the strongest material with excellent potential for the stiffest and toughest structure ever synthesized by scientists due to their high Young's modulus (Y~1.4 TPa) over the carbon fibers ($Y \geq 800$ GPa). The Young's modulus values of the few known materials are: Rubber (0.1 GPa) < Al (70 GPa) < Fe(200 GPa) < SiC(440 GPa) < Fe nanoparticle of 100 nm size (600 GPa)< CNTs(1000 GPa)< Diamond (1200 GPa), where the experimental Young's modulus for the MWCNTs is ~1–1.2 TPa and for the SWCNT, the Y-values lie between 320 and 1470 GPa. Theoretically the mechanical properties of SWCNTs are diameter-dependent. The theoretical Young's modulus of CNTs (greater than 1 nm diameter) is reported as almost 0.6–0.7 TPa by Guanghua et al. in 1998. The enhancement in the tensile strength of CNTs (11–63 GPa) along with the radial direction was reported by Treacy et al. in 1996, which is approximately 100 times greater than the same of steel along the axial direction with a density of around 1.33 g/cm^3 (Salvetat et al., 1999).

6.3.4 FLEXIBILITY

Highly flexible and elastic (~18% elongation before rupture) CNTs possess highest flexibility in its class of materials. Furthermore, the flexible CNTs had highest breaking elongation (20–30%) power that make them not to break down upon bending due to the exceptionally high aspect ratios as reported by Iijima in 1991 and Collins and Avouris in 2000 with the tensile strength of 65–93 GPa (Belytschko et al., 2002). Falvo et al. (1997) observed that MWCNTs can be applied for AFM tip due to its sharp bending angle without letting any structural rupturing. Zhu et al. (1998) applied the high pressures (50 GPa) at room temperature, using shock waves to generate the arc-discharged MWCNTs, and they noted that the tubes do not break but collapse and resulted in the transformation of the outer shells into curled graphene domains, while the inner cores display structural defects.

6.3.5 MALLEABILITY

With their unique properties, CNTs are also extremely malleable. Like macroscopic tubes, they can bend in circles and form knots, and can be clasped or compressed under more than appropriate loading (Lourie et al., 1998).

6.3.6 DISPERSION

The properties of nanocomposites are highly influenced by their dispersion ability. The strong attractive van der Waals forces originated from their polarizable and delocalised π-electron system of these nanomaterials result in the agglomeration of chains (Tsuda et al., 2016). Nanocarbons have a comparatively high melting point of 4000 K or 3675°C (Jamal et al., 2017) along with their super hydrophobic (Wang et al., 2007) nature, and chemically inertness but they tend to agglomerate, and due to the attraction of van der Waals forces present in the CNTs fibers these are restricted to be dispersed in solvents. The dispersion behavior of CNFs and CNTs depends on a small number of serious factors, for example, the length, volume fraction, sonication duration, entanglement density, and attractive forces of the nanomaterials by various chemical routes (by using a surfactant, polymers, or functionalization). They cannot be directly disseminated in water. Therefore, their dispersion process used to be followed by their wetting with water that improved their dispersion stability (Sobolkina et al., 2012). Hence, for the instant dispersion of CNTs/nanomaterial, the chemical process is supported by the physical methods (such as ultrasonication). Regarding the Derjaguin–Landau–Verwey–Overbeek (DLVO) theory (as shown in Figure 6.3), particles are dispersible when electrostatic repulsion prompted by correspondingly charged electrical layers around the particles by overcoming the attraction force (van der Waals) as claimed by Rosen and Kunjappu in 2012.

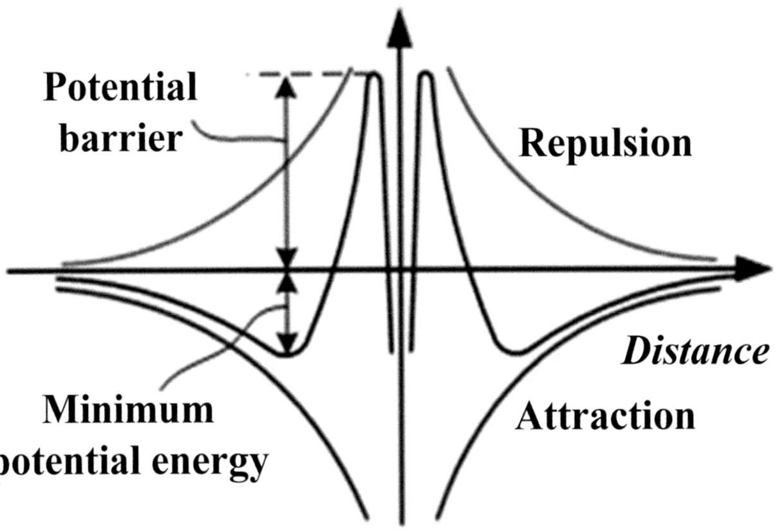

FIGURE 6.3 Derjaguin–Landau–Verwey–Overbeek (DLVO) theory of dispersion of CNTs, related with the distance, potential barrier, potential energy, and electrostatic forces.

Source: Adapted from Rosen and Kunjappu (2012). With permission.

6.3.7 HIGH ASPECT RATIOS

The term aspect ratio (λ = L/d) is associated with the high surface area-to-volume ratio that can carry forward the transmission of their superior characteristics to the associated species. It is just like the high dielectric constant of the CNT conceived composites with a higher aspect ratio of CNTs than the CNT-associated composites with a lower aspect ratio (Wang, et al., 2005). The composites with higher aspect ratio of CNTs possess higher shielding effectiveness, storage moduli, and viscosities than the composites with lower aspect ratio of CNTs (Bose et al., 2007; Jou et al., 2008). Moreover, the aspect ratio of CNTs strongly affects the electrical percolation performance of composites (Luo et al., 2007; Bauhofer et al., 2009). The above facts established that the aspect ratio of CNTs is a key parameter used to manage the final properties of the CNT-based compounds. The relationship between the aspect ratio and mesoscopic assembly of the SWCNTs/MWCNTs (high aspect ratio (HAR) and a low aspect ratio (LAR)) is well established, and schematically illustrated in Figure 6.4. A little part of LAR CNTs is distributed like stiff fibers,, while the extended part of

LAR CNTs is dispersed as bended fibers or in small bundles of fibers with decreased aspect ratio (Defeng et al., 2010).

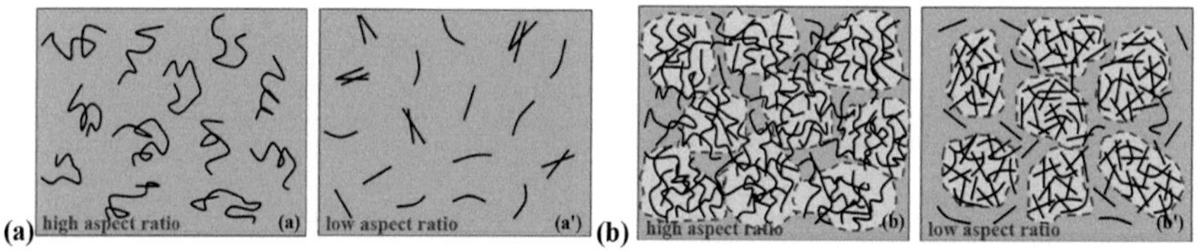

FIGURE 6.4 Schematic representation of the dispersion of (a and b) HAR CNTs and (a' and b') LAR CNTs at low loading levels (a and a') and high loading levels (b and b'). The solid black lines and curves represent CNTs; the dashed curvatures stand for flocks.

Source: Adapted from Wu et al. (2010). With permission.

6.3.8 GOOD FIELD EMISSION

The phenomena of an electron field emission (EFE) in CNTs were originally observed (Rinzler et al., 1995; Heer et al., 1995) in 1995. The CNTs can emit very large electron current. Experiments from a single MWCNT tip revealed their capability of emitting electron stably for more than 100 h at ~2 μA current flow (Bonard et al., 1998). The factors affecting the EFE include type of CNTs (i.e., MWCNTs or SWCNTs), their aspect ratio, orientation, density, interaction of CNTs with supporting surface, and uniformity.

The diameter of the SWCNTs is usually found in the range of 1–2 nm, while the diameter of the MWCNTs is observed in the range 20–100 nm. This recommended the higher field amplification factor for SWCNTs. However, they form close-packed bundles comprising 10^1–10^2 of SWCNTs with a bundle diameter of 10–100 nm.

The CNTs are excellent electron emitters due to the right combination of the properties, such as suitable morphology, nanometer-size diameter, structural integrity, high electrical and thermal conductivity, and chemical stability that lead to a low threshold field for emission and a high current density. That made them attractive as a cold cathode electron sources for a variety of vacuum electronic devices (Cheng et al., 2003). The reported work function for the SWCNTs is 4.8 eV, which is closed to the value of graphite (4.2 eV) (Suzuki et al., 2001). The aligned MWCNTs have a larger density of states (DOS) on the Fermi level and a slightly lower work function than the random MWCNTs which is attributed to the difference in the electronic state between the tip and the sidewall of the CNT (Suzuki et al., 2001). Similar to the other carbon allotropes, the electronic properties of the CNTs can be modified and tuned by chemical doping. Experimentally, it was observed that the SWCNT's function can be reduced to as low as 2.5 eV by intercalating with carbon of different types (Gayathri et al., 2010).

6.4 DISADVANTAGES OF CARBON NANOTUBES

- Comparatively a newer technology so not as much testing has been done.
- Lower lifetime (1750 compared with 6000 for silicon tips).

6.5 HOW GRAPHENES ARE DIFFERENT FROM CNTs?

Besides a lot of similar characteristics and behaviors, the diversity in their properties has categorized the graphene and CNTs as the most adaptable and potential materials for further development in the field of material science.

Both graphene and CNTs are tremendously strong materials along with their outstanding heat and electrical conductivity. They also have quite higher aspect ratio in comparison to the other contemporary materials. These two materials possessed the sp^2-bonds between the adjacent carbon atoms which make both of them extremely strong (rather stronger than the diamond) and the most stiff materials ever known. Moreover, both had the highest tensile strength due to their exceptional molecular arrangement.

6.6 SYNTHESIS OF CNTS

The vapor–solid–solid (VSS) model, which derives from the vapor–liquid–solid (VLS) model has been the most accepted growth model for CNT syntheses, which was suggested by Wagner and Ellis (1964) for the growth of silicon whiskers. The decomposition of the carbon from CNT leads to evolve the hydrocarbon gases that precipitated out of the surface of the catalyst and dispersed into the catalyst, and the structure of the CNT starts to nucleate from the boundaries of the catalyst. Nickel catalyst, in the presence of carbon vapor, can procure the CNTs on silica substrates as the thin carbon films of CNTs. The shape and size of the as-obtained catalyst particle depend on the experimental factors. The process of the catalytic growth of the CNTs mostly depends on the active sites of the catalytic particles dispersed in the solution. It is observed that the VLS growth process is operated as per the catalytic process. The interfacial energy of catalyst and CNTs is significantly responsible for the decrease in the melting temperature of the catalyst that may have resulted in various forms of carbon forms due to the interrupted VLS growth mechanism.

TABLE 6.1 The Characteristic Features of Graphene vs CNTs.

S. no.	Graphenes	CNTs
1.	Graphenes had a 2D plane like structure	It is the 1D rolled form of the graphene
2.	Defect-free graphene has better electrical properties compared with MWCNT	MWCNT is very cheap and has better mechanical and electrical properties in comparison to the most of the synthesized reduced graphene oxides (rGO)
3.	Graphene is semi-metal that show the quantum Hall effect	CNTs can either be semiconductor with a suitable bandgap or a purely metallic that must be subjected to their structural type, chirality, method of synthesis, etc. For example, the armchair nanotubes are truly metallic (Gayathri et al., 2008), while zigzag nanotubes as well as 1/3 chirality are narrow gap semiconductors and with 2/3 chirality wide-gap semiconductors
4.	The common synthesis approaches for the graphene are: • Epitaxial growth on substrate (SiC) • Exfoliation • Chemical vapor disposition (CVD) on metal catalysts	The popular growth methods for the CNTs are: • Catalytic chemical vapor disposition (CCVD) method • Laser ablation

According to this model, the VLS mechanism circumvents periodic instability caused by the presence of the liquified catalytic alloy phase that quickly adsorbed the carbon vapor up to supersaturation level, and started the formation of the CNT crystal growth from the nucleated seeds (catalyst) at the liquid–solid interface. The growth conditions can regulate the physical characteristics of the as-obtained carbon nanowires.

When hydrocarbon vapor comes into contact with hot metal nanoparticles, it decomposes into carbon and hydrogen through exothermic process, where hydrogen escapes out and the carbon is liquified into the metal catalyst. At the carbon-solubility limit of the metal, the dissolved carbon precipitated out in the form of cylindrical network (endothermic process). At this thermal gradient inside the metal, the crystallization process proceeds in two ways.

6.6.1 TIP GROWTH METHOD

If the catalyst and substrate interaction is weak or if the metal has an acute contact angle with the substrate, the hydrocarbon started its deposition on the top of the catalyst surface which leads to diffuse out carbon from the metal in the form of CNT at the bottom of the metal. If the catalyst's tip is exposed to the fresh hydrocarbon, then the CNT continues to grow longer. When the metal became supersaturated with the carbon, then the catalytic activity of the metal gradually decreases and the growth of CNT will be stopped. This method is identified as tip-growth model of CNT synthesis as shown in electron microscope analysis (Figure 6.5a). Growth of carbon nanofibers by using radiofrequency plasma-enhanced chemical vapor deposition method occurs at room temperature during which the decomposition of methane on the Ni catalyst also occurs (Boskovic et al., 2002).

6.6.2 BASE-GROWTH MODEL

When the interaction between catalyst and substrate or metal at an obtuse contact angle, then the decomposition of hydrocarbon, resulted in diffusion of carbon in the form of CNT that hailed out from the metal's apex (Figure 6.5b) (Boskovic et al., 2002). The carbon crystallizes in the form of the stand-alone graphitic cylinders with a semispherical-curved top, rooted with the catalyst particles at the bottom. This CNT growth process is known as base-growth model.

6.6.3 CZ PROCESS

Both VSS and VLS models neglect some factors of the growth (i.e., the rotation of catalyst and the chirality-related advancement in CNT), which was considered by CZ process. This process of CNT synthesis was discovered by Jan Czochralski, in 1916 (Lu and Miao, 2012) that can be used for the large-scale production of the high-quality single CNTs crystal of the diameter

~300 mm. The CZ process is based on the principle of the single crystal growth in a crucible from the melt and seed crystal as demonstrated in Figure 6.5a.

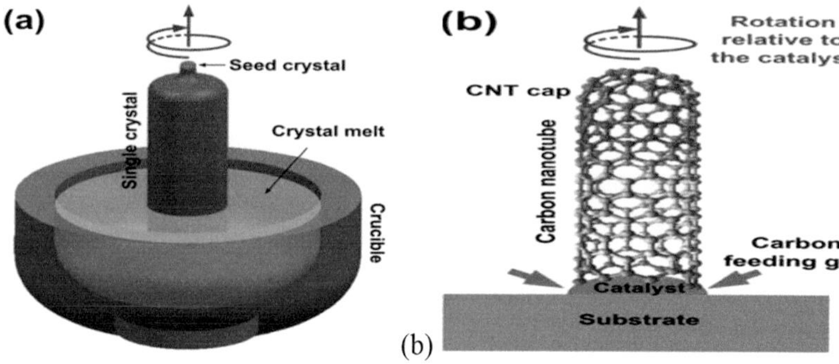

(a) (b)

FIGURE 6.5 Schematic presentation of the classical Czochralski growth models. (a) single seed crystal-oriented CZ method and (b) catalyst-based growth model of a CNT preparation.

Source: Reprinted with permission from Lu and Miao (2012). https://creativecommons.org/licenses/by/2.0.

6.6.4 ARC DISCHARGE OR EVAPORATION TECHNIQUE

This technique was first used for the synthesis of CNT (Iijima, 1991). It is one of the most common, simple, oldest, and easiest way to create CNT. It was used for developing both single-walled as well as multwalled nanotubes. In this method, electric dc arc discharge was generated between two graphite electrodes in an inert environment of helium or argon gas. If pure graphite rods are used, fullerenes are left as soot inside the synthesis chamber, the MWCNTs are deposited on the cathode surface (Dresselhaus et al., 2005). The mass production of MWCNTs by this dc arc discharge evaporation method was first achieved by Ebbesen and Ajayan (1992), as produced CNTs are illustrated by the scanning electron micrograph. These nanoparticles are easily removed by infrared irradiation or heating in air at 500°C as fabricated and thermally purified MWCNTs (Mahajan et al., 2013).

If a graphite anode contains a mixed-metal catalyst (Fe or Co) then SWCNTs are generated in the form of soot (Ando and Zhao, 2006). The quantity and quality of CNTs (Ando and Zhao, 2006) can be influenced by changing the synthetic parameters, such as the concentration of carbon vapor, the carbon vapor dispersion in inert gas, the temperature in the reactor, the composition of catalyst, the addition of promoters, and the presence of hydrogen.

6.6.5 LASER ABLATION

This technique was developed by Smalley et al. (Hafner et al., 1998) for the mass production of SWCNTs. In this technique, the laser is focused upon a carbon target (i.e., 98.8% of graphite composite) that consists of 1.2% of cobalt/nickel catalyst kept in a quartz tube furnace at 1200°C under the argon atmosphere at 500 Torr pressure. Thess et al. (1996) and Rao et al. (1997) made improvement in the above method by introducing a double-beam laser.

There are several factors (Chrzanowska et al., 2015), such as temperature, wavelength of laser power, pressure, and amount of catalyst that affect the quantity and quality of CNTs formation (Thess et al., 1996).

6.6.6 CHEMICAL VAPOR DEPOSITION

The chemical vapor deposition (CVD) technique is used to prepare CNT. This was first reported by Endo et al. (1993). In this method, the vapor deposition of a solid material on a surface results in the reactions of the precursor gases in the reaction chamber. It is a controllable process for the selective production of CNT and economically viable process for large-scale production. CNTs in desired shape can be produced by controlling their alignment, length, wall number and diameter of vessel, temperature, operation pressure, pretreatment of metallic catalyst (nature and size), concentration of hydrocarbon, nature of support and reaction time (Seo et al., 2007). Smalley's group developed a high-pressure carbon monoxide decomposition technique (known as HiPco) for mass production of SWCNTs in which iron pentacarbonyl was used as a catalyst under the long reaction time with a temperature range of 800°C–1200°C (Nikolaev et al.,1999).

Kumar and Ando (2010) reported two main growth CVD methods for the synthesis of SWCNTs and MWCNTs using the catalyst over the substrate, that is, tip-growth and base-growth methods. As-synthesized CNTs are characterized by the analytical techniques, such as XRD, EDX, SEM, TEM, ESR, Raman, and FTIR.

6.6.7 GREEN METHODS

The physical and chemical methods involve toxic solvents, high-pressure and high-temperature conditions, and therefore to overcome the various limitations of conventional methods, an alternative method suggested for the fabrication of CNTs is a green chemical method. Green chemical method is unconventional, efficient, and eco-friendly. In the green method, the catalysts are derived from the different parts of the plants and the preparation of CNT proceeds under the comparatively mild conditions and by using less infrastructure. As grown CNTs from the green method are free from toxic metal catalyst, require low growth temperature (575°C), produce high yield and do not require expensive and complex systems for the synthesis.

Zhan et al. (2019) reported a synthesis of polydopamine modified multiwalled carbon nanotube (MWCNT-PDA)/GO hybrid aerogels through a green route that is used for highly efficient heavy metal ions separation through adsorption (Figure 6.6a). The MWCNT-PDA possesses a high adsorption capacity which was used to separate heavy metals from water after adsorption. Kumar and Ando (2005) produced the SWCNTs and MWCNTs by using camphor (Figure 6.6b) upon a high-silica zeolite support (impregnated with Fe–Co catalyst) by varying the concentration of catalyst, growth temperature, and camphor vapor pressure. High yields of metal-free MWCNTs of fairly uniform diameter (~10 nm) grown at 600°C–700°C had been obtained in a noteworthy amount (~30%), where SWCNTs were formed at 850°C–900°C within a narrow diameter range of 0.86–1.23 nm (Kumar et al., 2005). Further, CNTs were prepared from coconut shell wastes (Hakim et al., 2018), by catalytic decomposition of turpentine oil (Ghosh et al., 2008), and from many other green methods.

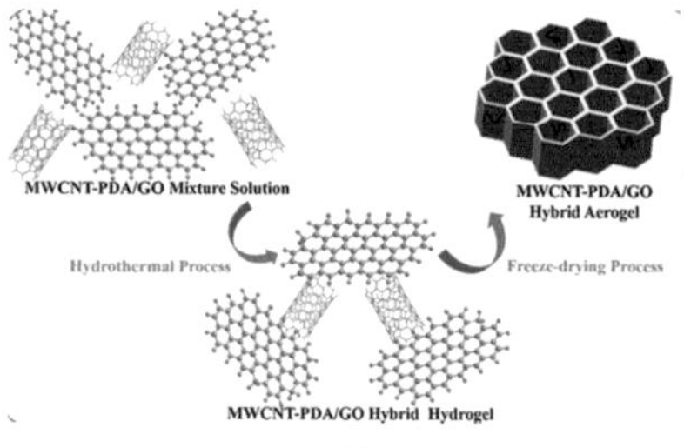

(a)

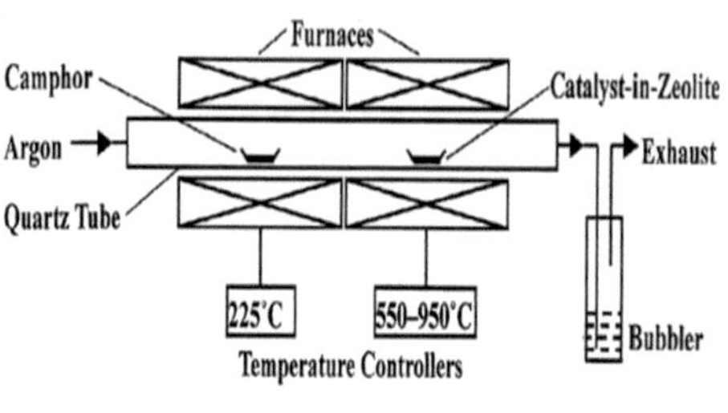

(b)

FIGURE 6.6 (a) Adsorption capacity of polydopamine modified MWCNTs / GO hybrid and (b) preparation method of SWCNT and MWCNT grown by camphor.

Source: Reprinted with permission from a) Zhan et al. (2019) © Elsevier; b) Kumar and Ando (2005) © Elsevier.

6.7 CHARACTERIZATION OF CNTS

XRD patterns for the SWCNTs and MWCNTs samples exhibited (002), (100), and (004) peaks (Zhang et al., 2008a; Cao et al., 2011), indicated the hexagonal ring structure of graphite sheets forming the CNT as the main product. The energy-dispersive X-ray spectrum (EDS) was used to calculate the elemental composition in the CNTs that contain mainly the carbon (~91.24%), which is represented by the high-intensity peak at 2.5 kV energy in EDX spectrum. TEM images of atypical low-diameter SWCNT (normal and defective) and MWCNT that exhibit the nanotubes of diameter ~0.5 nm without any visible defects, while the lower image exhibited the SWCNT with kinks like defects in the structure, known as Stones-Walls defects. A few nano-onion like structures were also observed in the images that are related to different forms of carbon in these samples. Raman spectra of the SWCNTs and MWCNTs samples recorded at radial breathing modes (RBM) and intermediate frequency modes (IFM) show reflections at 694, 791, 928, and 984 cm^{-1} along with the well-known G and D bands (Dresselhaus et al., 2005) around 1568.2 and 1350.3 cm^{-1} positions. In SWCNT samples, G/D ratio is ~10, which indicate the presence of carbonaceous impurities, residual metal catalysts. In MWCNTs, the G band is relatively weak and the intensity of the IFM modes is stronger than D-band of the sample. This indicates the structural defects in the nanotubes. The PL spectral peaks of SWCNTs found around 2.33 and 2.10 eV, respectively and their corresponding full -width at half-maxima (FWHM) values are 0.22 eV and 0.40 eV. In MWCNT, PL emission peaks appeared at 2.34 and 2.0 eV with FWHM of 0.25 and 0.25 eV, respectively (Singh et al., 2009). Differentiated TGA curves indicate the no measurable content of amorphous carbon in both the samples. Some EDX studies of MWCNT samples exhibited the catalyst (iron) traces in the samples (1.2%) and the presence of Co and Mo (~9.3%) in SWCNT samples along with carbon. FTIR spectra of the samples show C—C stretching (1625 cm^{-1}), C—H deformation (1381 cm^{-1}), C—H stretching (2912 cm^{-1}), C—O stretching (1016 cm^{-1}), CO_2 stretching (2355 cm^{-1}), and moisture-related bands (3440 cm^{-1}) (Singh et al., 2009). The ESR studies demonstrated that the samples contained a significant amount of metallic impurity (Singh et al., 2009). The ESR spectra of CNT are characterized by primarily two broad peaks with g-values 3.774 and 2.026 in the range of 60–600 mT range (Singh et al., 2009). At higher magnetic field (> 600 mT), a peak with g value of 0.748 is found, which is accredited to the wetness that was absorbed at the surface of the samples, which is consistent with the FTIR data.

6.8 PHOTOCATALYTIC WATER SPLITTING WITH CNTS

Clean energy is the need of the hour and hydrogen is the cleanest fuel ever known. Water is a zero-carbon source of hydrogen. However, water is the most abundant, renewable, and the clean source of hydrogen on planet and Sun is a well-known source of energy to mankind from ages. So, one can split water by utilizing the direct energy of sunlight in the presence of a semiconductor. Despite the pure metals (used for only reduction) and oxidizing agents (used for only oxidation), the semiconductors create electron (used for reduction) and hole (used for oxidation) by acquiring the energy from sunlight. The semiconductors can handle the overall water splitting process for the generation of hydrogen and oxygen by reduction and oxidation of water. The stability, efficiency, and cost of the photocatalytic material are the decisive factors for their practical applications. The photocatalytic water splitting emerges as one of the significant tools for the production of clean and green hydrogen. The cleavage of water into H_2 and O_2 gases is an energetical demanding reaction that required change in the standard Gibbs free energy (ΔG) of +237 kJ/mol (corresponding to 1.23 eV) (Maeda and Domen, 2007). The process is depicted in the following chemical reactions (eqs 6.1–6.3).

Anode (oxidation reaction): $2\ H_2O \rightarrow O_2 + 4\ H^+ + 4\ e^-$ E = 1.23 eV (6.1)

Cathode (reduction reaction): $4\ H^+ + 4\ e^- \rightarrow 2\ H_2$ E= 0.00 eV (6.2)

Overall reaction: $2\ H_2O \rightarrow 2\ H_2 + O_2$ E= 1.23eV (6.3)

Pure water (neutral) cannot break even under electrolysis. Consequently, the water electrolysis process takes place either in the presence of either acidic or alkaline electrolytes or in the presence of semiconductor. Various transition metal chalchogenides (where MX = S, Se, Te), nitrides, oxides, and carbides had been developed as the good photocatalytic material including the elite class of carbonaceous materials, such as the CNTs, graphenes (GO/rGO), porous carbons, carbon nitrides for hydrogen production via photoelectrochemical and photocatalytic water splitting. In the process of the development of metal-free photocatalysts/electrocatalysts, the CNTs based materials/systems play crucial role in the generation of hydrogen and oxygen through electrochemical, photocatalytic, or photoelectrochemical water splitting. Spataru et al. (2004) employed an ab initio theory of poly-electron Greene's function to look into the effect of e–h interaction on the optical spectra of small-diameter nanotubes. They found that in the semiconducting tubes, the excitons are bound by ~1 eV energy, while the metallic tube's excitons are bound with ~100 meV energy. The decay of these excitons occurred via radiative-relaxation (emission) or by creating e–h pairs within the band structure. Ma et al. (2004) concluded that the photoexcitation leads to the charge separation/trapping. These photo-induced charge carriers lead to generate photocurrent (Barazzouk et al., 2004).

In this chapter, we discuss the various functionalized forms of the CNT as an effective catalytic material for water splitting. Pristine form of CNTs show quite low catalytic activity toward hydrogen evolution reaction (HER)/ oxygen evolution reaction (OER) via water splitting. Therefore, the suitable modification of CNTs could be executed by changing the covalent and noncovalent bonding of the systems, where the covalent bonding used to be formed by devastating the graphitic skeleton of CNTs and established between the functional molecule and the carbon atom of the CNTs. The noncovalent bonding is facilitated by the loading of CNTs on supramolecules through various adsorption interactions (i.e., Van der Waals force, p-stacking interaction, electrostatic force, and hydrogen bond) without affecting the basic skeleton of CNT. The DFT calculations along with the experimental measurements proposed the 4-electron OER reaction mechanism at the oxidized surface of MWCNTs (Lu et al., 2015) and are exhibited by following eqs 6.4–6.7.

$$H_2O\ (l) \rightarrow OH\ (ad) + H^+\ (aq) + e \tag{6.4}$$

$$OH\ (ad) \rightarrow O\ (ad) + H^+\ (aq) + e \tag{6.5}$$

$$O\ (ad) + H_2O\ (l) \rightarrow OOH\ (ad) + H^+\ (aq) + e \tag{6.6}$$

$$OOH\ (ad) \rightarrow O_2\ (g) + H^+\ (aq) + e \tag{6.7}$$

The DFT calculations for the CNTs suggested the Volmer–Heyrovsky mechanism for the HER (Holmberg and Laasonen, 2015a) reaction depending upon the effects of heteroatom doping, chirality, and coverage of hydrogen over the pristine open-ended CNTs on the HER activity. The enhanced activity of CNTs for the fixed hydrogen-covered sites is attributed to the position of the valence band location close to the highest occupied molecular orbital level that facilitates the electron transfer to protons (Holmberg and Laasonen, 2015b). These findings suggested that the pristine CNT is not as inert as carbon taken for granted and the introduction of different ring structures in CNT leads to enhance the HER activity in open-ended CNTs. It is found that the CNTs ended with the five-membered carbon ring are more active over to the CNT ended with the six-membered carbon rings. Furthermore, DFT calculations performed by Holmberg and Laasonen (2015a) exhibited the N-doping in CNTs, could not improve their electrocatalytic activity toward HER reactions (Figure 6.13b and c) These two-electron HER reaction mechanism depends upon the pH of the reaction medium also (Schuldiner, 1954; Bockris and Koch 1961; Sheng et al., 2010). The following is the HER mechanism in acidic medium, shown by eqs 6.8–6.10.

$$H^+ + e^- + * \rightarrow H_{ads} \ldots.. \quad (6.8)$$

$$H_{ads} + H_{ads} \rightarrow H_2 + 2* \quad (6.9)$$

$$H^+ + H_{ads} + e^- \rightarrow H_2 + * \quad (6.10)$$

where * is a surface, 2* is a surface that was left by H_2 molecule. Similarly, the HER mechanism can be presented in alkaline media (Zheng et al., 2018), as shown by eqs 6.11–6.13:

$$H_2O + e^- + * \rightarrow H_{ads} + OH^- \quad (6.11)$$

$$H_{ads} + H_{ads} \rightarrow H_2 + 2* \quad (6.12)$$

$$H_2O + H_{ads} + e^- \rightarrow H_2 + OH^- + * \quad (6.13)$$

6.8.1 TYPES OF THE CNT-BASED SYSTEMS FOR PHOTOCATALYTIC/ PHOTO ELECTROCHEMICAL WATER SPLITTING

There are four basic categories of the systems used for photocatalytic water splitting:

- CNT-based primary systems
- CNT-based binary systems
- CNT-based ternary systems
- CNT-based quaternary systems
- CNT-based Z-scheme systems

6.8.2 CNT-BASED PRIMARY SYSTEMS

6.8.2.1 SWCNT FILMS

The SWCNT films deposited on optically transparent electrodes by electrophoretic method became photoelectrochemically active to generate photocurrent under visible light irradiation. The system represents the low photon-to-current conversion efficiency, that is, 0.15% at 400 nm suggested the recombination of the most of the photogenerated charge carriers. The system shows the maximum photocurrent and photovoltage 8 μA/cm^2 and 12 mV, respectively. The charge separation phenomena was confirmed by the time-resolved transient absorption spectroscopy that followed by the laser-pulse excitation of SWCNTs. The value of the relaxation time of the photo-assisted carriers for SWCNT is~1 ps that supported the lowest energy gap (Barazzouk et al., 2004).

6.8.2.2 N, S-doped CNT

The N, S-doped CNT can act as an active bifunctional electrocatalyst for the complete water cleavage reactions (HER/OER). The bifunctional catalytic performance was investigated by using a rotating ring-disk electrode (RRDE) technique. The electrocatalytic HER performances of carbon-based catalyst were observed in terms of the cathodic current density of 5 mA/cm^2 (HER) at the operating potential, that is, −0.40 V for N, S-CNT with Tafel slope (133 mV/decade), which is quite inferior to the reported systems, that is, CNT-PDA (−0.68 V and 178 mV/decade), oxidized CNT (ox-CNT; −0.61 V and 165 mV/decade) and N-CNT in 1 M KOH solution (−0.54 V and 187 mV/decade). This has underlined the unique role of N, S-doping in electrocatalytic process and indicates the favorable HER kinetics. The electrochemical double layer capacitance *C*dl of N,S-CNT is recorded as 4.75 mF/cm^2, which is much higher than that of N-CNT (2.02 mF/cm^2) and ox-CNT (1.78 mF/cm^2). The OER performance of the electrocatalytic reaction delivered a 10 mA/cm^2 current density (E_j= 10, that is a critical metric for the solar fuel production) at the operating potential of 1.59 V for N, S-CNT with a small Tafel slope value of 56 mV/decade, which are better than those of CNT-PDA (1.80 V, 111 mV/decade), ox-CNT (1.66 V, 97 mV/decade), and N-CNT (1.64 V, 87 mV/decade), which exhibits the better exposure and utilization of the active sites on the N, S-CNT surface for contributing effective HER and OER activities (Zhang et al., 2015).

6.8.2.3 Co-NCNT-T

Cobalt nanoparticles loaded and nitrogen-doped CNTs (Co-NCNT-T, where T = Pyrolysis temperature) are made by a simple pyrolysis method at different temperatures and used for photocatalytic oxygen and hydrogen evolution via overall water splitting (Meng et al., 2019). The system Co-NCNT-800 revealed a significantly high rate of hydrogen generation, that is, 14.7 mmol/h/g with 51% oxygen yield using cocatalyst $Ru(bpy)_3(ClO4)_2$. Under optimized condition, the system attained a turnover number (TON) of 25.8 and 0.86 s^{-1} as turn over frequency (TOF) for water oxidation reaction. The outstanding activity of the compound is attributed to the effective electron transfer process between the catalyst and photosensitizer, which was also verified by the photoluminescence emission (PLE) spectra and transient photocurrent responses, the lifetimes of EY and Co-NCNT-800/TEOA/EY are 1.48 and 1.25 ns, respectively. The low lifetime of the photogenerated charge carriers required the effectual interfacial charge transfer responsible for good hydrogen and oxygen production through water splitting (Meng et al., 2019).

6.8.3 CNT-BASED BINARY SYSTEMS

6.8.3.1 TiO_2-REINFORCED MWCNT

A series of TiO_2 NPs loaded with MWCNTs was synthesized by using the sonothermal supported hydrothermal treatment. The TiO_2-enforced MWCNT photocatalyst (2.0CNT–TiO_2 sample) showed good H_2 production rate from water splitting, that is, 69.41 μmol/g/h under visible light irradiation at pH 2 in the presence of a sacrificial electron donor triethanolamine (Olowoyo et al., 2019). In comparison to the bare TiO_2, as-synthesized sample not only showed the better photocatalytic activities for hydrogen production but also showed enhanced CO_2 reduction ability (methanol, hydrogen, and formic acid production from the reduction of CO_2 with yield rates of 2360.0, 3246.1, and 68.5 μmol/g/h under UVA radiation and 1520.0 μmol/g/h methanol production under visible light) under UVA as well as under visible light at ambient condition (Olowoyo et al., 2019). The computational study of the primary photocatalytic reactions of CNT–TiO_2 composites demonstrated the binding of CNT to the (101) surfaces of TiO_2 than at (001) facets. Visible light excitation of the CNT–TiO_2 composites leads to the charge transfer between TiO_2 and CNT, whereas UV light excitation can cause the charge transfer either from CNT to TiO_2 or from TiO_2 to CNT.

Photocatalyst 2.0, CNT–TiO_2-based photocatalyst was used for H_2 production via water cleavage at different values of pH, that is, 2, 4, 6, 7, 8, 10, and 12, in the absence of a sacrificial agent under visible light illumination during the time course of 8 h. Their corresponding H_2 gas production rates are: 69.41, 40.05, 26.35, 24.90, 23.85, 15.75, and 7.66 μmol/g/h at pH 2, 4, 6, 7, 8, 10, and 12, respectively (Olowoyo et al., 2019). Under the acidic medium, the higher performance of 2.0 CNT–TiO_2 for the H_2 evolution was observed because H^+ ions produced in the acidic solution are adsorbed on the photocatalyst surface and interact with as-generated/stored electrons, leading to produce the reactant to increase the rate of molecular hydrogen formation. The recyclability of H_2 production from H_2O over 2.0 CNT–TiO_2 catalysts at pH 2 under visible light illumination were tested for three cycles in 3 days for 8 h each day. Computational quantum study of CNT and TiO_2 interaction was reported with one-go modified CNT and two-go modified CNT attached to (101) and (001) facets of TiO_2. The electrons of CNT belong to the boundary orbital with a minimal contribution of p_O orbital of the TiO_2 (001) surface in HOMO and both p_O and d_{Ti} orbitals in LUMO.

6.8.3.2 MWCNT-TiO_2 NANOCOMPOSITE

The MWCNT–TiO_2 nanocomposite was developed by hydrothermal process and Pt loading was done to composite by adding $H_2PtCl_6 \cdot 6H_2O$ to 175 mL aqueous methanol (v/v 1:6) kept in a Pyrex glass container, under 500 W Hg lamp exposure for 5 h. The sample was centrifuged after photocatalytic activity measurement, then washed and dried out at 80°C, resulted in Pt-loaded photocatalyst (Pt/MWCNT–TiO_2). Its activity was checked in photocatalysis (hydrogen production) of water under visible light or full spectra irradiation of a Xe lamp.

The composite Pt/MWCNT–TiO_2 was used to produce the H_2 gas (up to 8 mmol/g/h) but no capacity toward splitting of water was observed for Pt-loaded pristine TiO_2 loaded (optimization of the Pt load was also done) over MWCNTs sample under visible light irradiation ($\lambda > 420$ nm). The H_2 evolution capacity of the MWCNT, MWCNT+TiO_2, TiO_2 and MWCNT–TiO_2 systems was recorded as 35, 6090.5, 7298.8 and 8092.5 μmol/g/h, respectively (Dai et al., 2009). The photocatalytic activity of the nanocomposite was measured in terms of hydrogen generation and was attributed to the synergetic effect of the inherent properties of its component which include the excellent light absorption and charge separation capacity at MWCNTs and TiO_2 interface due to the direct growth of TiO_2 nanoparticles on the MWCNTs surface in hydrothermal process.

6.8.3.3 $CoSe_2$/CNT SYSTEM

The ultrathin films of the nanocomposite $CoSe_2$/CNTs were made up by the facetious two-step scheme. In the first step, ultrathin Co-MOF (metal organic framework) garnished with cable-like CNTs to form Co-MOF/CNT–composite, and in the second step, selenization process of the same was performed at low temperature. The well synergy among the ultrathin nanosheets of $CoSe_2$ and the superior conductive CNTs resulted in the abundant number of active spots and increased conductivity for their electrocatalytic activity in the nanocomposite as observed from the voltammetry (CV) experiments performed with different scan rates in the range of potential from 0.1 to 0.2 V. The as-reported electrocatalyst $CoSe_2$/CNTs displayed the low value of overpotential, that is, 190 and 300 mV relating to the standard reversible hydrogen electrode (RHE) at a current density of 10 mA cm^2 for the HER and OER in 1 mol/L KOH solution, respectively, with the better stability (Wei et al., 2020). Furthermore, as-synthesized bifunctional electrocatalytic film of $CoSe_2$/CNTs acts both as the cathode and anode in an electrolyser with the cell voltage of 1.75 V for the current 10 mA cm^2 for the overall water splitting.

6.8.3.4 $CuWO_4$/MWCNT

The photoelectrochemical performances of a photoactive nanocomposite material $CuWO_4$/MWCNT was observed in photoelectrochemical cell. The MWCNT was added to the $CuWO_4$ thin film to minimize the bulk resistance of $CuWO_4$ material by making a composite MWCNT (light absorber):$CuWO_4$ (charge collector) with adjustable electronic transport properties. Porous, crack-free thin polycrystalline films of $CuWO_4$ with thickness of 0.1–2.0 mm were fabricated by means of spray pyrolysis. These films exhibited the bandgap ~2.20 (0.05) eV (calculated from Tauc plots). Electrochemical impedance spectroscopy confirmed the 30% bulk resistance reduction compared with the unmodified $CuWO_4$ control (Gaillard et al., 2013) at pH 10 buffer solution and irradiation of light source (AM1.5G) at 0.8 V versus saturated calomel electrode (1.63 V vs. reversible hydrogen electrode). The decrease in bulk resistance was attained due to the tremendously low weight ratio of MWCNT: $CuWO_4$ (1:10,000) composite, in which MWCNT absorbed the incident light of less than 2%. The linear scan voltametric (LSV) study of the composite exhibited the increase in the photocurrent density of 26% at 0.8 V with respect to the standard calomel electrode as compared with the control. The LSV study and incident photon-to-current efficiency (IPCE) measurements demonstrated that the MWCNT served as an effective electron accumulator that distributed the electron in the entire bulk $CuWO_4$ film.

6.8.3.5 HETEROJUNCTION SWCNT (CORE)/C_{60} (SHELL)

The SWCNTs have a tunable absorption bands, and therefore, they can act as strong optical absorber, which can be controlled by changing their chiral indices (*n*, *m*) that can be done by creating heterojunction, which is the key factor for the high-power conversion efficiency. But achieving the high solar conversion efficiency for the SWCNTs is a difficult task due to the huge binding energy (BE>100 meV) of the electron–hole pairs (excitons) produced by optical absorption. Difficulty in construction of the well-ordered SWCNT surface restricted their uses in photocatalytic reactions as compared with the thin film devices (Murakami et al., 2017)-loaded fullerodendron shell on SWCNT core to form heterojunction SWCNT (core)/C_{60} (shell), where the dendron shell act as a co-catalyst to support the H_2 production from water. The band offset counter balanced the LUMO level of (8, 3) SWCNT and $C_{60,}$ allowed the exciton detachment by shattering the high exciton binding energy of SWCNTs, resulted in H_2 generation activity (0.083 μmol^{-1} h) with quantum yield of 0.015 under 680 nm light exposure without any initiation period or reduction in the catalytic activity of the studied time for 6 h light irradiation. The respective H_2 generation efficiencies for the (6, 5) and (7, 5) SWCNTs are 0.022 and 0.0065 μmol^{-1} h under the monochromatic light irradiation of wavelength 570 and 650 nm (Murakami et al., 2017).

6.8.4 CNT-BASED TERNARY SYSTEMS

6.8.4.1 CdS/ CNT/NiS

It was found that CdS photocatalysts prepared in alkaline (0.6 mol NaOH) environment showed higher hydrogen production activity in aqueous sacrificial electrolyte Na_2S and Na_2SO_3 under 350 W Xe light source irradiation for 5 h, that is, 794.6 μmol/h/g. CdS/CNTs made by two-step hydrothermal method. (CdS was prepared by hydrothermal method for 12 h with the addition of 0.06 mol NaOH, then CNTs were added, ultrasonicated for 1 h and then followed by hydrothermal treatment for 12 h. In the second step, first CdS was hydrothermally prepared in 24 h, then CNTs were mixed with CdS, ultrasonicated for 1 h and then followed by washing and drying) show the hydrogen production rate of 1.771 mmol/h/g, which was further enhanced to 12.13 mmol/h/g when the composite CdS/CNT was loaded with cocatalyst NiS under the same condition.

6.8.4.2 HYBRID CATALYST NIFE–N-CNT–rGO

The fabrication of noble metal-free catalysts is vital for the generation of low cost and clean hydrogen through water splitting. A successful try was made with the transition metal encapsulated nitrogen-doped CNTs (M–N-CNTs, M = Ni, Co, Fe) that can play important role in the production of low-cost electrocatalysis by direct and easy pyrolysis of bulk g-C_3N_4. As-generated M–N-CNTs are further modified by anchoring with rGO surface, and it was found that the catalyst NiFe–N-CNT/rGO exhibited tremendously the small overpotential, that is, 270 mV with glassy carbon electrode for the OER reaction at the standard current density of 10 mA/cm^2, which is superior to most of the noble-metal-free catalysts (Wan et al., 2019). The metallic nanoparticles (NPs) after thermal reduction, provides the nucleation sites for the carbon species to facilitate the N-CNT growth process. Modification in electronic structure of the CNT was observed from the calculations of the density functional theory (DFT). The process of HER involved either one-step or two-electron transfer reactions, which was governed by the number of N-centers than the OER process with the four-electron transfer. The hybrid catalyst NiFe–N-CNT–rGO can serve as a scalable and straightforward synthesis of large-scale photocatalytic water splitting.

6.8.4.3 HETEROJUNCTION COMPOSITE MWCNTS/$TiO_{2)}$1D/0D

Heterojunction catalyst of MWCNTs/TiO_2(1D/0D) composite distributed on the 2D surface of the montmorillonite (MMT) clay, which was used to extend the ability of the visible light absorption. The synergistic effect between MWCNTs/TiO_2/MMT in 1D/0D/2D heterojunctions shorten the migration distance for charges and abundant active sites. The composite exhibited the good photocatalytic H_2 evolution (158, 310, 446, 210, 1157 and 947 ppm/h) rate with the different sacrificial reagents (pure water, ethanol, methanol, propanol, glycerol, and ethylene glycol with 5% (v/v) concentration), respectively under visible light irradiation, that is, 1888 ppm/h, which is almost 1.49, 3.07 and 7.12 times larger than the composite MWCNTs/TiO_2, MMT/TiO_2 and pristine TiO_2, respectively (Umer et al., 2018). Due to the α-hydrogen atoms attached to carbon atoms, glycerol–water mixture shows highest H_2 production among the different sacrificial reagents. The photocatalytic turn-over productivity (PTOP) of the composite MWCNTs/TiO_2/MMT heterojunction was observed (6.15×10^{-5}) 5.94 times higher than the pure TiO_2 NPs. This ternary composite exhibited excellent stability and activity of H_2 generation. The study recommended the construction of the heterojunction between the constituents MWCNTs, TiO_2, and MMT of the composite that improve the efficiency of H_2 production over the sunlight exposure.

*6.8.4.4 MOF/NANO TiO_2 MATRIX/CNT (*IN SITU *GENERATED)*

The CNTs prepared from pyrolysis of the solid wastes through recrystallization and also prepared through catalytic decomposition of methane at high temperatures and prolong heating. But both these methods are not considered economic. Therefore, research was focused on facile and low-cost methods for in situ CNTs generation along with the production of hydrogen gas through water splitting. Precisely, the gas production from the metal organic framework (Cr-MIL-101) (153.6 mmol/h of H_2 and CO_2) and TiO_2 nanoparticles (192.0 mmol/h of H_2 and CO_2) had been demonstrated. The binary photocatalyst was prepared by physically mixing Cr-MIL-101 and TiO_2 (144.3 mmol/h of H_2 and CO) that could produce nanocarbon species, where the coupling of MOF(Cr-MIL-101) with TiO_2 nanoparticles performed by sonication in the presence of methanol resulted in a novel matrix that produces pure hydrogen gas (107.0 mmol/h with 99.3% H_2) along with in situ CNT formation that was free from CO_2 or CO under visible light irradiation (Mohamed et al., 2020). The new advancement in the field of water splitting is made through introduction of this promising candidate for pure hydrogen production that can be used as a clean fuel.

6.9 QUATERNARY COMPOSITE OF CNT

6.9.1 TiO_2/WO_3/Au/MWCNT

The quaternary composite of CNT, TiO_2/WO_3/Au/MWCNT was experienced as a good photocatalytic material for the hydrogen production by Pap et al. (2012). Furthermore, the effect of change in the various crystalline phases of titania and their combination with MWCNT were also studied for the hydrogen production. Photocatalytic hydrogen production by using the series of the nanocomposites, that is, AR-ATO-Au, AA-ATO-Au, P25-ATO-Au, P25-ATO-Au-MWCNT (0.1%), P25-ATO-Au-MWCNT (1.0%) and P25-ATO-Au-MWCNT (5%) has been recorded as 1.99, 2.82, 47. 56, 30.78, 20.58, and 5.22

μmol (where Aldrich Rutile—AR, Aldrich Anatase—AA, Aldrich Tungsten Oxide—ATO, 14% WO_3, 1%Au). The hydrogen production study revealed that the anatase TiO_2 nanoparticles were found effective in the H_2 evolution but rutile TiO_2 was found to be more competent regarding the normalized surface area, where the P25-based composites showed a maximum hydrogen production compared with the pure anatase- and rutile-incorporated samples. The addition of the MWCNTs to the nanocomposite lowered the hydrogen productivity. Among the all MWCNT-loaded P25-ATO-Au composites, the MWCNT (0.1%) shows the small improvement in hydrogen production rate over the 1%- and 5%-loaded MWCNT samples.

6.9.2 CNT-BASED Z-SCHEME SYSTEMS

6.9.2.1 SWCNTs-C_{60}/$RuCl_3$ /t $BiVO_4$ IN MEDIATOR $[Co(bpy)_3]^{3+/2+}$

The SWCNTs modified with fullerodendrons (C_{60}) and complexed with $RuCl_3$ are treated as CNT photocatalysts in a Z-scheme photocatalytic system with another photocatalyst $BiVO_4$, and Co-complex $[Co\ (bpy)_3]^{3+/2+}$ as an electron mediator to split water. The system exhibited the high quantum yields (QY = 12.8% in light of λ_{irr} = 1005 nm) for the production of H_2 (17 μmol/h) and O_2 (5.5 μmol/h) from water splitting under illumination from a solar simulator (AM 1.5) with 0.089% (Izawa et al., 2019). Solar-to-hydrogen (STH) efficiency and system also worked fine under illumination with NIR light (λ_{irr} = 1005 nm) in the presence of sacrificial donor. This particulate Z-scheme photocatalytic system can progressively split water for 12 h in the absence of any sacrificial agent and buffer solution. The observed TON for the system is 1.6 for the reacted electrons with respect to the mediator $[Co(bpy)_3]^{3+/2+}$, indicated the H_2 and O_2 gas evolution reactions for the Z-scheme.

6.9.2.2 $Zn_{0.5}Cd_{0.5}S$-MWCNT-TiO_2 TERNARY CORE-SHELL NANOCOMPOSITES

The effectiveness of the tailored MWCNTs (i.e., ternary core-shell nanocomposites $Zn_{0.5}Cd_{0.5}S$-MWCNT-TiO_2) was demonstrated for facilitating the Z-scheme water splitting. This anisotropic solid-state electron mediator had been fabricated by coating and hydrothermal route. This system rendered the greater photocatalytic activity than the respective one- and two-component systems of above constituents, where MWCNTs are used as a mediator to transfer the photo generated electrons from the photosystems PS II (TiO_2) to PS I ($Zn_{0.5}Cd_{0.5}S$) of the Z-scheme systems, via MWCNT. This sole system has been empowered with the efficient spatial charge isolation with apt band positions and strong redox ability for the Z-scheme system. This Z-scheme system evolved H_2 gas with the rate of 21.9 μmol/h, which is almost 4.5 and 2.8 times greater than the pristine $Zn_{0.5}Cd_{0.5}S$ and MWCNT-$Zn_{0.5}Cd_{0.5}S$ systems, correspondingly (Ng et al., 2017) under solar stimulator (1.5AMG). The plausible mechanism for charge transfer in the Z-schematic system was also proposed by the researchers.

Few more state-of-the-art examples of all four prominent categories (primary, binary, ternary, quaternary, and Z-scheme) of the photocatalytic- and photoelectrochemical-type materials are listed in Tables 6.2 and 6.3, respectively for their water splitting capacity under different experimental conditions.

6.10 APPLICATIONS OF CNT-BASED MATERIALS

The recent escalation of electronic products (computer, TV, laptops, mobile phones, e-gadgets, etc.) show the widespread application of CNTs in the field of electronic and electrical industry. CNTs are used due to their extraordinary characteristics (such as light weight, low cost, nontoxicity, mechanical strength, thermal- and electrical conductivity, etc.) as a promising material for on-chip and off-chip interconnects (Zhang et al., 2008). Microelectronics had expanded the benefits of CNTs with growing in situ CNTs in fulfillment of the basic requirement of microelectronics. Furthermore, semiconducting single-walled tubes can be easily operated as CNT field-effect transistors (CNTFETs) (Hoenlein et al., 2003).

CNTs market has been expected to draw extraordinary attention of the medical field. CNTs prove themselves as ideal for developing innovative biomimetic materials in the bone biomedical field. CNTs have been used in biomedicine for making them nanocarriers for the targeted-drug delivery in human body, cancer diagnosis and therapy (Chen et al., 2017) and offered the remarkable features like site-specificity, better stability, encapsulation efficiency, enhanced skin penetration, prolonged action, and high drug-loading capability (Prakash and Kulamarva, 2007). The nanotoxicological research has raised question over the excessive use of nanomaterials/nanoparticles in cosmetics, because there is a possibility that their penetration into skin might cause health hazards (Bilal and Iqbal, 2020). Various nanomolecules, that is, antigen, peptide, and nucleic acid can be directly delivered to cancer cells by using the CNTs of high efficacy. CNT-based materials and their composites

TABLE 6.2 Photoelectrochemical Materials for Water Splitting in Terms of Current Density at Different Applied over Potential.

SL. no.	Sample	Amount	Sacrificial agent	Cocatalyst	Light source	Current density	Quantum efficiency	References
Pure CNT								
1.	SWCNT	–	0.5 M LiI + 0.01 M I_2 in acetonitrile	Pt	Visible light source (> 400 nm)	8 mA cm^2at η=12 mV	0.15% (IPCE) at 400 nm	Barazzouk et al. (2004)
2.	N,S-CNT	2 mg	1.0 M KOH	–	–	10 mA/cm^2at η=2.03 V	–	Zhang et al. (2015)
CNT-based binary system								
3.	Cysteamine and 2-thioethane sulfonic acid) protected CdS NPs (5 nm) with SWCNT	1 mg/mL	Phosphate buffer (pH 10) with triethanol- amine	Au	300 W Xe lamp of wavelength 390 nm	0.83 mA at	25% P-EQE	Sheeney-Haj-Ichia et al. (2005)
4.	MWCNT $CuWO_4$-based PEC electrode	MWCNT:$CuWO_4$ (1: 10,000)	Water	Pt foil	1000 W Xe arc lamp	0.5 mA/cm^2at η=0.8 V	13%	Gaillard et al. (2013)
5.	TiO_2 dispersed on SWCNT	–	1 M KOH	–	150 W Xe lamp with Intensity 50 mW/cm^2	52 mAcm^2 at η=0.00 V	16%	Kongkanand et al. (2007)
6.	M-CNTs shell $NiCo_{0.16}Fe_{0.34}$ metal core	0.1 mg/cm^2	1 M KOH	–	–	500 A/g at η=0.37 V	FE66%	Cheng et al. (2014)
7.	MWCNT over ZnO nanorod	–	0.5 M Na_2SO_4	Ag/ AgCl	100 W Xe Lamp with AM 1.5G	1.45 mA/cm^2at η=1.00 V	0.646%	Wei et al. (2014)
8.	NiFe-LDH/ CNT	0.25 mg/cm^2	1 M KOH	–	–	10 mA/cm^2 at η=1.538 mV	FE 99.3%	Gong et al. (2013)
9.	ZnPc/CNT-4	–	1 M KOH	–	UV and visible light	0.32 mA/cm^2 at η= 1.20 V	–	Cheng et al. (2016)
10.	ZnO-NWs/ MWCNTs	–	0.10 M Na_2SO_4	Ag/ AgCl	400 W UV and Hg lamp with solar simulator	1.68 mA/cm^2 at η=1.00 V	–	Zhang et al. (2009)
11.	Bilayer Fe_2O_3-MWCNT	–	1 M NaOH	Ag/ AgCl	100 mW/cm^2	0.22 mA/cm^2at η=0.40 V	–	Kim et al. (2011)

TABLE 6.2 *(Continued)*

SL. no.	Sample	Amount	Sacrificial agent	Cocatalyst	Light source	Current density	Quantum efficiency	References
12.	MWCNTs with α-Fe_2O_3	0.2 wt.%	Water	Pt	Solar light	2.8 mA/cm^2 at η=0.70 V	2%	Rai et al. (2015)
13.	CNTs/Co	–	Water	Pt	–	3.00 mA/cm^2, at η=1.75 V	–	Cozzarini et al. (2017)
14.	Co_3O_4/CNTs	153 µg/cm^2	0.1 M KOH solution	–	–	22 mA/cm^2 at η=270 mV	–	Ahmed et al. (2018)
15.	CoS_2 encap- sulated in self-catalyzed CNTs	0.2 mg/cm^2	1.0 mol/L KOH	Nickel foam	–	10.0 mA/cm^2 at η= 0.112 V &10.0 mA/cm^2 at η= 1.57 V		Wang et al. (2018)
16.	PdMn-N_4/CNTs	–	1 M KOH and 2 M ethanol	–	–	56.3 mA/cm^2	–	Begum et al. (2019)
CNT—Based ternary system								
17.	Ru@MWCNT	–	0.5 M H_2SO_4 and 1 M KOH solution	IrO_2 Electrode	–	10 mA/cm^2 at η= 17 mV	FE (92.28%)	Kweon et al. (2020)
18.	Fe_2O_3-CNT/ Graphene	–	1 MNaOH	–	100 mW/cm^2 intensity	325 mA/cm^2 at η=1.23 V	7% IPCE at 400 nm	Kim et al. (2012)
19.	NH_2-CNTs co-doped Ti-Fe_2O_3	0.10 mg/mL	1.0 M NaOH	Ag/ AgCl	300 W Xe lamp of 150 m/Wcm^2 intensity	4.5 mA/cm^2 at η=1.23 V	–	Liu et al. (2012)
20.	(Ti-Fe_2O_3)/ (MWCNTs / Cu_2O)	–	0.1 M NaOH	Pt	150 W Xe lamp	5.1 mA/cm^2		Verma et al. (2018)
21.	Fe_2O_3/MWCNTs/RGO	–	1 M NaOH	–	Visible light	495 mm/cm^2	–	Pawar et al. (2015)
22.	Ag/TiO_2/CNT	–	0.5 M Na_2SO_4	Pt	Visible region of intensity 20 mW/cm^2	0.91 mA/cm^2	–	Chaudhary et al. (2016)

Photon-to-electron Quantum efficiency = P-EQE; Faradaic efficiency = FE; η = Overpotential; IPCE= Incident photon-to-current efficiency.

TABLE 6.3 Photocatalytic Materials for Water Splitting In Terms of Hydrogen and Oxygen Gas Generation Under Exposure of Light Source.

SL. no.	Sample	Amount	Sacrificial agent	Cocatalyst	Light source	H_2 production	O_2 production	Quantum efficiency	References
1.	CNT/Pt and *f*-CNT/Pt (20 wt.% Pt NPs)	10 mg catalyst in 100 mL DIW	None	Pt	UV lamp with 4400 mWcm^2	45.4 and 193.9 µmol/h/g	–	–	Mercado-Zúñiga et al. (2020)
Pure CNTs									
2.	CdS/CNTs	–	0.06 M NaOH	NiS	λ > 500 nm	794.6 m mol/h/g	–	–	Wang et al. (2013)
CNT-based binary system									
3.	Ta_2O_5/CNT	–	Methanol	Pt	450 W mercury lamp	32 mmol/h/g	–	–	Cherevan et al. (2014)
4.	MWCNT-TiO_2	40 mg	85 mL Deionized water 15 mL TEA	Pt	Visible light (420 nm)	15.1 µmol/h	–	4.4%	Dai et al. (2014)
5.	CNT-TiO_2–δ-15	7.2 wt.%	Methanol	Pt	UV light	9.7 mmol/h	–	–	Chen et al. (2015)
6.	(CNTs)/$ZnFe_2O_4$	50 mg	Methanol	Pt	Solar light	18 µmol/h/g	–	–	Dang et al. (2016)
7.	CNT–TiO_2	2.0 mol	Water	CNTs	Visible light	69.41 µmol/h/h	–	–	Olowoyo et al. (2019)
8.	Co–NDCNTs	–	1.0 M KOH	Co	Visible light	598.5 µmol/g/h	–	–	Zheng et al. (2019)
9.	CNTs@TNRs	0.35 g	0.3 M KOH	–	Visible light	300 mmol/h/cm^2	–	–	Ahmed et al. (2020)
10.	CNTs/MnO_2–C_3N_4	–	–	MnO_2	Visible light-	122 µmol/h	–	–	Song et al. (2018)
CNT-Based Ternary System									
11.	TiO_2–$Ni(OH)_2$/CNT/CdS	50 mg	Lactic acid	$Ni(OH)_2$ and CNT	visible light with λ>400 nm	12 mmol/h/g			Wang et al. (2017)
12.	Pt/C_f-CNTs / CdS nanorods	0–10 wt.%	20 vol.% of Lactic acid	Pt	Visible light (≥425 nm) irradiation	120.1 mmol/h/g	–	29.10%	Gopannagari et al. (2018)

TABLE 6.3 *(Continued)*

SL. no.	Sample	Amount	Sacrificial agent	Cocatalyst	Light source	H_2 production	O_2 production	Quantum efficiency	References
13.	CNT-GR- TiO_2	5 mg	Methanol	Graphene	UV–vis irradiation	29 mmol/h/g	–	14.6%.	Bellamkonda et al. (2019)
14.	SWCNT/ fullerodendron/ Ru(III) and $[Co(bpy)_3]^{2+}$	0.08 wt.%	Water	–	solar light	17 μmol/h	5.5 μmol/h	12.8%	Izawa et al. (2019)
15.	TiO_2/CdS/ CNT	0.1 g	Methanol	–	100 W Vis sun lamp & 8 W UV lamp	3502 μmol/h	–	–	Sampath et. al. (2019)
16.	CuO/Cr_2O_3–MWCNTs–eosin Y	60 mg of CuO/Cr_2O_3–MWCNTs and 60 mg of eosin Y	15 vol.% of triethanolamine	CuO/Cr_2O_3	1000 W tungsten lamp and 8 W UV lamp	67.2 μmol w	–	5.43% at 450 nm.	Xiangqing et al. (2014)

due to their unique physicochemical and biomedical properties possesses the promising potential to revolutionize the design and integration of bone tissue regeneration and engineering, as well as used for drug therapeutic systems (Pei et al., 2019).The nanosystems including CNTs have excellent potential in targeted-drug delivery system due to the following notable properties, such as site-specificity, better stability, encapsulation efficiency, enhanced skin penetration, prolonged action, and high drug-loading capability. The nano toxicological research has raised their apprehension on the extreme practice of nanomaterials in cosmetic industries, which might make possible for the nanoparticles to enter into the skin which might cause the health-related problems (Kaul et al., 2018).

Carbon fiber-based polymer composites are used in high-performance applications particularly in aerospace, automotive, energy, and sports sectors due to their lighter, stronger, and durable nature. This technology has huge potential for the forthcoming materials and it also has a noteworthy influence on the manufacturing industry (Anvesh and Madhu 2015; Das et al., 2019).

The detoxification of wastewater as well as the technologies for the desalination of seawater are using nano-structured CNT sponge comprising sulfur and iron elements, which are quite effective in soaking up water-contaminants like oil, pesticides, fertilizers, and pharmaceuticals products. Magnetic properties of CNTs make them easier to retrieve once their cleanup job is done (Wang et al., 2019; Ihsanullah etal., 2019).

6.11 FUTURE PROSPECTIVES

Efficient solar energy conversion to chemical energy is possible only when the photocatalyst with suitable band gap and band positions is able to absorb all the major portions of the sunlight including the visible and infrared spectrum of solar insolation. In addition, it should utilize all the photogenerated charge carriers effectively without any recombination to produce hydrogen. This needs the novel materials which can have suitable narrow band gap for the effective solar light harvesting capacity suitable for the redox reactions of water. CNTs had great potential to meet out all requirements. Moreover, the layered and one-dimensional structure of CNT with ready to tune surface confirmed its candidature for a good photocatalyst that can be used to cleave water. Surface-active sites of the CNT, which can trap the charge carriers, are also equally important so that the desired photocatalyst should be made with higher surface-active sites. There is a lot of space for studying this wonderful material to understand all electronic and morphological aspects that influence the interface junction power and charge transfer ability for the photocatalytic hydrogen production. Most of the studies devoted to wide band oxide materials which had light harness ability toward UV part of the sunlight (only 6% of incident photons) are involving in the photocatalysis. There is a need to find out ways to prepare narrow band gap photocatalyst with effective electron transfer to the surface-active sites for redox reactions without charge carrier recombination. As CNTs and their modified versions are available and few more better forms of the CNTs are in the pipeline, which can be used as excellent option over the available classes of the photocatalytic and photoelectrochemical materials to revolutionize the field of artificial photosynthesis, which is the potential method to replace the fossil fuel by efficient production of hydrogen.

KEYWORDS

- **carbon nanotubes**
- **photosplitting**
- **hydrogen generation**
- **CNT-composites**
- **MWCNT**
- **SWCNT**

REFERENCES

Ahmed, A. M.; Mohamed, F.; Ashraf, A. M.; Shaban, M.; Khan, A. A. P.; Asiri, M. A. Enhanced Photoelectrochemical Water Splitting Activity of Carbon Nanotubes @ TiO_2 Nanoribbons in Different Electrolytes. *Chemosphere* **2020,** *238*, 124554.

Ahmed, M. S.; Choi, B.; Kim, Y. -B. Development of Highly Active Bifunctional Electrocatalyst Using Co_3O_4 on Carbon Nanotubes for Oxygen Reduction and Oxygen Evolution. *Sci. Rep.* **2018,** *8*, 1–10.

Alsharef, J. M.; Taha, M. R.; Khan, T. A. Physical Dispersion of Nanocarbons in Composites–A Review. *J. Teknol.* **2017,** *79,* 69–81.

Ando, Y.; Zhao, X. Synthesis of Carbon Nanotubes by Arc-Discharge Method. *New Diamond Front. Carbon Technol.* **2006,** *16*, 123–137.

Anvesh, P.; Madhu, S. Applications of Carbon Nano Tubes (CNT) in Automobiles—A Review. *Inter. J. Appl. Eng. Res.* **2015,** *10*, 25450–25461.

Aryasomayajula, L.; Wolter, K. Carbon Nanotube Composites for Electronic Packaging Applications: A Review. *J. Nanotechnol.* **2013**. DOI: 10. 1155/2013/296517.

Bandaru, P. R.; Daraio, C.; Jin, S.; Rao, A. M. Novel Electrical Switching Behaviour and Logic in Carbon Nanotube Y-Junctions. *Nat. Mater.* **2005,** *4*, 663–666.

Barazzouk, S.; Hotchandani, S.; Vinodgopal, K.; Kamat, P. V. Single-Wall Carbon Nanotube Films for Photocurrent Generation. A Prompt Response to the Visible-Light Irradiation. *J. Phys. Chem. B* **2004,** *108*, 17015–17018.

Bauhofer, W.; Kovacs, J. Z. A Review and Analysis of Electrical Percolation in Carbon Nanotube Polymer Composites. *Compos. Sci. Technol.* **2009,** *69*, 1486–1498.

Begum, H.; Ahmed, M. S.; Lee, D-W.; Kim, Y-B. Carbon Nanotubes-Based PdM Bimetallic Catalysts through N4-System for Efficient Ethanol Oxidation and Hydrogen Evolution Reaction. *Sci. Rep.* **2019,** *9*, 1–10.

Bellamkonda, S.; Thangavel, N.; Hafeez, H. Y.; Neppolian, B.; Rao, G. R. Highly Active and Stable Multi-Walled Carbon Nanotubes-Graphene- TiO_2 Nanohybrid: An Efficient Non-Noble Metal Photocatalyst for Water Splitting. *Catal. Today* **2019,** *321–322*, 120–127.

Belytschko, T.; Xiao, S.; Schatz, G.; Ruoff, R. Atomistic Simulations of Nanotube Fracture. *Phys. Rev. B* **2002,** *65*, 235430.

Bernholc, J.; Brenner, D. W.; Nardelli, M. B.; Meunier, V. Mechanical and Electrical Properties of Nanotubes. *Annu. Rev. Mater. Res.* **2002,** *32*, 347–375.

Bilal, M.; Iqbal, H. M. N. New Insights on Unique Features and Role of Nanostructured Materials in Cosmetics. *Cosmetics* **2020,** *7,* 24.

Bockris, J. O. M.; Koch, D. F. A. Comparative Rates of the Electrolytic Evolution of Hydrogen and Deuterium on Iron, Tungsten and Platinum. *J. Phys. Chem.* **1961,** *65*, 1941–1948.

Bonard, J. M.; Maier, F.; Stökli, T.; Chäelain, A.; Heer, W. A. D.; Salvetat, J. -P.; Forr, L. Field Emission Properties of Multiwalled Carbon Nanotubes. *Ultramicroscopy* **1998,** *73*, 7–15.

Bose, S.; Bhattacharyya, A. R.; Bondre, A. P.; Kulkarni, A. R.; Pötschke, P. Rheology Electrical Conductivity, and the Phase Behavior of Cocontinuous PA6/ABS Blends with MWCNT: Correlating the Aspect Ratio of MWCNT with the Percolation Threshold. *J. Polym. Sci. Part B: Polym. Phys.* **2008,** *46*, 1619–1631.

Boskovic, B. O.; Stolojan, V.; Khan, R. U.; Haq, S.; Silva, S. R. Large-Area Synthesis of Carbon Nanofibres at Room Temperature. *Nat. Mater*. **2002,** *1*, 165–168.

Cao, A.; Xu, C.; Liand, J.; Wu, D.; Wei, B. X-ray Diffraction Characterization on the Alignment Degree of Carbon Nanotubes. *Chem. Phys. Lett*. **2011,** *344*, 13–17.

Chandra, B.; Bhattacharjee, J.; Purewal, M.; Son, Y. W.; Wu, Y.; Huang, M. et al. Molecular-Scale Quantum Dots from Carbon Nanotube Heterojunctions. *Nano Lett.* **2009,** *9*, 1544–1548.

Chaudhary, D.; Singh, S.; Vankar, V. D.; Khare, N. A Ternary Ag/TiO_2/CNT Photoanode for Efficient Photoelectrochemical Water Splitting under Visible Light Irradiation. *Int. J. Hydrogen Energy* **2016,** *41*, 7826–7835.

Chen, P.; Wang, L.; Wang, P.; Kostka, A.; Wark, M.; Muhler, M. et al. CNT-$TiO_{2-\delta}$ Composites for Improved Co-Catalyst Dispersion and Stabilized Photocatalytic Hydrogen Production. *Catal* **2015,** 5, 270–285.

Chen, Z.; Zhang, A.; Wang, X.; Zhu, J.; Fan, Y.; Yu, H. et al. The Advances of Carbon Nanotubes in Cancer Diagnostics and Therapeutics. *J. Nanomater.* **2017**. DOI: 10. 1155/2017/3418932.

Cheng, Y.; Liu, C.; Cheng, H-M.; Jiang, S. P. One-Pot Synthesis of Metal–Carbon Nanotubes Network Hybrids as Highly Efficient Catalysts for Oxygen Evolution Reaction of Water Splitting. *ACS Appl. Mater. Interf.* **2014,** *6*, 10089–10098.

Cheng, Y.; Memar, A.; Saunders, M.; Pan, J.; Liu, C.; Gale, J. D. et al. Dye Functionalized Carbon Nanotubes for Photoelectrochemical Water Splitting—Role of Inner Tubes. *J. Mater. Chem. A* **2016,** *4*, 2473–2483.

Cheng, Y.; Zhou, O. Electron field Emission from Carbon Nanotubes. *C. R. Phys*. **2003,** 4, 1021–1033.

Cherevan, A. S.; Gebhardt, P.; Shearer, C. J.; Matsukawa, M.; Domen, K.; Eder, D. Interface Engineering in Nanocarbon–Ta_2O_5 Hybrid Photocatalysts. *Energy Environ. Sci.* **2014,** *7*, 791–796.

Chrzanowska, J.; Hoffman, J.; Malolepszy, A.; Mazurkiewicz, M.; Kowalewski T. A.; Szymanski, Z. et al. Synthesis of Carbon nanotubes by the Laser Ablation Method: Effect of Laser wavelength. *Phys. Status Solidi* B **2015,** *252*, 1860–1867.

Collins, P. G.; Avouris, P. Nanotubes for Electronics. *Sci. Am.* **2000,** *283*, 62–69.

Cozzarini, L.; Bertolini, G.; Suran-Brunelli, S. T.; Radivo, A.; Bracamonte, M. V.; Tavagnacco, C. et al. Metal Decorated Carbon Nanotubes for Electrocatalytic Water Splitting. *Int. J. Hydrog. En*ergy. **2017,** *42*, 18763–18773.

Dai, H.; Wong, E. W.; Lieber, C. M. Probing Electrical Transport in Nanomaterials: Conductivity of Individual Carbon Nanotubes. *Science* **1996,** *272*, 523–526.

Dai, K.; Peng, T.; Ke, D.; Wei, B. Photocatalytic Hydrogen Generation Using a Nanocomposite of Multi-Walled Carbon Nanotubes and TiO_2 Nanoparticles under Visible Light Irradiation. *Nanotechnology* **2009,** *20*, 125603.

Dai, K.; Zhang, X.; Fan, K.; Zeng, P.; Peng, T. Multiwalled Carbon Nanotube-TiO_2 Nanocomposite for Visible-Light-Induced Photocatalytic Hydrogen Evolution. *J. Nanomater*. **2014**. DOI: 10. 1155/2014/694073.

Dang, H.; Qiu, Y.; Cheng, Z.; Yang, W.; Wu, H.; Fan, H. Hydrothermal Preparation and Characterization of Nanostructured CNTs/$ZnFe_2O_4$ Composites for Solar Water Splitting Application. *Ceram. Int.* **2016,** *42*, 10520–10525.

Das, T. K.; Ghosh, P.; Das, N. C. Preparation, Development, Outcomes, and Application Versatility of Carbon Fiber-Based Polymer Composites: A Review. *Adv. Compos. Hybrid Mater.* **2019,** *2*, 214–233.

Debrupa, L.; Ghosh, S.; Agarwal, A. Carbon Nanotube Reinforced Hydroxyapatite Composite for Orthopedic Application: A Review. *Mater. Sci. Eng. C* **2012,** *32*, 1727–1758.

Delaney, P.; Di, V. M.; Pantelides, S. T. Quantized Conductance of Multiwalled Carbon Nanotubes. *Appl. Phys. Lett.,* **1999,** *75*, 3787–3788.

Dresselhaus M. S.; DresselhausG.; Saita R.; JorioA. Raman Spectroscopy of Carbon Nanotubes. *Phys. Rep.* **2005,** *409*, 47–99.

Dresselhaus, M. S.; Dresselhaus, G.; Saito, R. Physics of Carbon Nanotubes. *Carbon* **1995,** *33*, 883–891.

Ebbesen T. W.; Ajayan P. M. Large-Scale Synthesis of Carbon Nanotubes. *Nature* **1992,** *358*, 220–222.

Ebbesen, T. W.; Lezec, H. J.; Hiura, H.; Bennett, J. W.; Ghaemi, H. F.; Thio, T. Electrical Conductivity of Individual Carbon Nanotubes. *Nature* **1996,** *382*, 54–56.

Endo, M.; Takeuchi, K.; Igarashi, S.; Kobori, K.; Shiraishi, M.; Kroto, H. W. The Production and Structure of Pyrolytic Carbon Nanotubes. *J. Phys. Chem. Solids* **1993,** *54*, 1841–1848.

Falvo, M. R.; Clary, G. J.; Taylor, R. M.; Chi, V.; Brooks, F. P.; Washburn, S.; Superfine, R. et al. Bending and Buckling of Carbon Nanotubes under Large Strain. *Nature* **1997,** *389*, 582–584.

Gaillard, N.; Chang, Y.; Angelis, A. D.; Higgins, S.; Braun, A. A Nanocomposite Photoelectrode Made of 2. 2 eV Band Gap Copper Tungstate ($CuWO_4$) and Multi-Wall Carbon Nanotubes for Solar-Assisted Water Splitting. *Inter. J. Hyd. Energy* **2013,** *38*, 3166 -3176.

Gayathri, V.; Devi, N. R.; Geetha, R. Hydrogen storage in coiled carbon nanotubes. *Int. J. Hydrog. Energy* **2010,** 35, 1313–1320.

Ghosh, P.; Soga, T.; Afre, R. A.; Jimbo, T. Simplified Synthesis of Single-Walled Carbon Nanotubes from a Botanical Hydrocarbon: Turpentine Oil. *J. Alloys Compd.* 2008, 462, 289–293.

Gong, M.; Li, Y.; Wang, H.; Liang, Y.; Wu, J. Z.; Zhou, J.; Wang, J. et al. Advanced Ni–Fe Layered Double Hydroxide Electrocatalyst for Water Oxidation. *J. Am. Chem. Soc.* **2013,** *135*, 8452–8455.

Gopannagari, M. D.; Kumar, P.; Park, H.; Kim, E. H.; Bhavani, D. P.; Reddy, A. et al. Influence of Surface-Functionalized Multi-Walled Carbon Nanotubes on CdS Nanohybrids for Effective Photocatalytic Hydrogen Production. *Appl. Catal. B: Environ.* **2018,** *236*, 294–303.

Grobert, N. Carbon Nanotubes–Becoming Clean. *Mater. Today* **2007,** *10*, 28–35.

Guanghua, G.; Tahir, C.; William, A. G.; Energetics, Structure, Mechanical and Vibrational Properties of Single-Walled Carbon Nanotubes. *Nanotechnology* **1998,** *9*, 184–191.

Hafner, J. H.; Bronikowski, M. J.; Azamian, B. R.; Nikolaev, P.; Rinzler, A. G.; Colbert, D. T. et al. Catalytic Growth of Single-Wall Carbon Nanotubes from Metal Particles. *Chem. Phys. Lett.* **1998,** *296*, 195–202.

Hakim, Y. Z.; Yulizar, Y.; Nurcahyo, A.; Surya, M. Green synthesis of Carbon Nanotubes from Coconut Shell Waste for the Adsorption of Pb (II) Ions. *Acta Chim. Asiana* **2018,** *1*, 6–10.

Han, J-W.; Kim, B.; Li, J.; Meyyappan, M. Carbon Nanotube Based Humidity Sensor on Cellulose Paper. *J. Phys. Chem. C* 2012, *116*, 22094–22097.

Hee, J. J.; Hwang, G. B.; Lee, J. E.; Bae, G. N. Preparation of Airborne Ag/CNT Hybrid Nanoparticles Using an Aerosol Process and Their Application to Antimicrobial Air Filtration. *Langmuir* **2011,** *27*, 10256–10264.

Heer, W. A. D.; Chatelain, A.; Ugarte, D. A carbon nanotube field-emission electron source, *Science* **1995,** 270, 1179–1180.

Hoenlein, W.; Kreupl, F.; Duesberg, G. S.; Graham, A. P.; Liebau, M.; Seidel, R. et al. Carbon Nanotubes for Microelectronics: Status and Future Prospects. *Mater. Sci. Eng. C* **2013,** *23*, 663–669.

Holmberg, N.; Laasonen, K. *Ab Initio* Electrochemistry: Exploring the Hydrogen Evolution Reaction on Carbon Nanotubes. *J. Phys. Chem. C* **2015a,** *119*, 16166–16178.

Holmberg, N.; Laasonen, K. Theoretical Insight Into the Hydrogen Evolution Activity of Open-Ended Carbon Nanotubes. *J. Phys. Chem. Lett.* **2015b,** *6*, 3956–3960.

Hughes, T. V.; Chambers, C. R. Manufacture of Carbon Filaments. *US Patent* 405, 480, 1886.

Hui-Wen, C.; Wu, R-J.; Chan, K-H.; Sun, Y-L.; Su, P-G. The Application of CNT/Nafion Composite Material to Low Humidity Sensing Measurement. *Sens. Actuat. B-Chem.* **2005,** *104*, 80–84.

Ihsanullah. Carbon Nanotube Membranes for Water Purification: Developments, Challenges, and Prospects for the Future. *Sep. Purif. Technol.* **2019,** *209*, 307–337.

Iijima, S. Helical Microtubules of Graphitic Carbon. *Nature* **1991,** *354*, 56–58.

Izawa, T.; Kalousek, V.; Miyamoto, D.; Murakami, N.; Miyake, H.; Tajima, T. et al. Carbon-Nanotube-Based Photocatalysts for Water Splitting in Cooperation with $BiVO_4$ and $[Co(bpy)_3]^{3+/2+}$. *Chem. Lett.* **2019,** *48*, 410–413.

Jou, W. S.; Cheng, H. Z.; Hsu, C. F. The Electromagnetic Shielding Effectiveness of Carbon Nanotubes Polymer Composites. *J. Alloys Comp.* **2007,** *434–435*, 641–645.

Kaul, S.; Gulati, N.; Verma, D.; Mukherjee, S.; Nagaich, U. Role of Nanotechnology in Cosmeceuticals: A Review of Recent Advances. *J. Pharm.* 2018. DOI: 10. 1155/2018/3420204.

Kim, J. Y.; Jang, J. W.; Youn, D. H.; Kim, J. Y.; Kim, E. S.; Lee, I. S. Graphene Carbon Nanotube Composite as an Effective Conducting Scaffold to Enhance the Photoelectrochemical Water Oxidation Activity of a Hematite Film. *RSC Adv.* **2012,** *2*, 9415–9422.

Kim, J. Y.; Jun, H.; Hong, S. J.; Kim, H. G.; Lee, J. S. Charge Transfer in Iron Oxide Photoanode Modified with Carbon Nanotubes for Photoelectrochemical Water Oxidation: An Electrochemical Impedance Study. *Int. J. Hydrog. Energy* **2011,** *36*, 9462–9468.

Kim, P.; Shi, L.; Majumdar, A.; McEuen, P. L. Thermal Transport Measurements of Individual Multiwalled Nanotubes. *Phys. Rev. Lett.* **2001,** *87*, 215502.

Kongkanand, A.; Domínguez, R. M.; Kamat, P. V. Single Wall Carbon Nanotube Scaffolds for Photoelectrochemical Solar Cells. Capture and Transport of Photogenerated Electrons. *Nano Lett.* **2007,** *7*, 676–680.

Kochmann, W.; Reibold, M.; Goldberg, R.; Hauffe, W.; Levin, A. A.; Meyer, D. C. et al. Nanowires in Ancient Damascus Steel. *J. Alloys Comp.* **2004,** *372*, 1–2.

Kumar, M. *Carbon Nanotube Synthesis and Growth Mechanism. Carbon Nanotubes-Synthesis, Characterization*; Applications Intechopen, Limited: London, 2011; pp 147–170.

Kumar, M.; Ando, Y. Controlling the Diameter Distribution of Carbon Nanotubes Grown from Camphor on a Zeolite Support. *Carbon* **2005,** *43*, 533–540.

Kumar, M.; Ando, Y. Chemical Vapor Deposition of Carbon Nanotubes: A Review on Growth Mechanism and Mass Production. *J. Nanosci. Nanotechnol.* **2010,** *10*, 3739–3758.

Kweon, D. H.; Okyay, M. S.; Kim, S. J.; Jeon, J. P.; Noh, H. J. et al. Ruthenium Anchored on Carbon Nanotube Electrocatalyst for Hydrogen Production with Enhanced Faradaic Efficiency. *Nat. Commun.* **2020,** *11*. DOI: 10.1038/s41467-020-15069-3.

Lapidus, A. L.; Jem, H. C.; Krylova, A. Y. Effect of the Type of Carrier on the Properties of Cobalt Catalysts in the Synthesis of Aliphatic Hydrocarbons from Carbon Monoxide and Hydrogen. *Bull. Acad. Sci. USSR, Div. Chem. Sci.* **1982,** *31* (10), 1953–1956.

Lau, K. T.; Hui, D. The revolutionary creation of new advanced materials-carbon nanotube composites. *Compos. B. Eng.* **2002,** 33, 263–277.

Liu, Y.; Wang, D-P.; Yu, Y-X.; Zhang, W. D. Preparation and Photoelectrochemical Properties of Functional Carbon Nanotubes and Ti Co-doped Fe_2O_3 Thin Films. *Int. J. Hydrog. Energy* **2012,** *37*, 9566–9575.

Lourie, O.; Cox, D. M.; Wagner, H. D. Buckling and Collapse of Embedded Carbon Nanotubes. *Phys. Rev. Lett.* **1998,** *81*, 1638–1641.

Lu, J.; Miao, J. Growth Mechanism of Carbon Nanotubes: A Nano Czochralski Model. *Nanoscale Res. Lett.* **2012,** 7. DOI: 10. 1186/1556-276X-7-356.

Lu, X.; Yim, W. L.; Suryanto, B. H. R.; Zhao, C. Electrocatalytic Oxygen Evolution at Surface-Oxidized Multiwall Carbon Nanotubes. *J. Am. Chem. Soc.* **2015,** *137*, 2901–2907.

Luo, D. M.; Wang, W. X.; Takao, Y. Effects of the Distribution and Geometry of Carbon Nanotubes on the Macroscopic Stiffness and Microscopic Stresses of Nanocomposites. *Compos. Sci. Technol.* **2007,** *67*, 2947–2958.

Ma, Y. Z.; Stenger, J.; Zimmermann, J.; Bachilo, S. M.; Smalley, R. E.; Weisman, R. B. et al. Ultrafast Carrier Dynamics in Single-Walled Carbon Nanotubes Probed by Femtosecond Spectroscopy. *J. Chem. Phys.* **2004,** *120*, 3368–3373.

Maeda, K.; Domen, K. New Non-Oxide Photocatalysts Designed for Overall Water Splitting under Visible Light. *J. Phys. Chem. C* **2007,** *111*, 7851–7861.

Mahajan, A.; Kingon, A. I.; Kukovecz, A.; Kónya, Z. Studies on the Thermal Decomposition of Multiwall Carbon Nanotubes under Different Atmospheres. *Mater. Lett.* **2013,** *90*, 165–168.

Mahmood, N.; Mohammad, I.; Mahmood, A. *Handbook of Carbon Nanotubes-Polymer Nanocomposites*; LAP Lambert Academic Publishing, 2014.

Makar, J. M.; Beaudoin, J. J. Carbon Nanotubes and Their Application in the Construction Industry. *Spec. Publ.—R. Soc. Chem.* **2004,** *292*, 331–342.

Meng, X.; Dong, Y.; Hu, Q.; Ding, Y. Co Nanoparticles Decorated with Nitrogen Doped Carbon Nanotubes for Boosting Photocatalytic Water Splitting. *ACS Sustain. Chem. Eng.* **2019,** *7*, 1753–1759.

Mercado-Zúñiga, C.; Oliva, J.; Vargas-García, J. R.; Diaz-Torres, L. A.; Gomez-Solis, C. Effect of Pt Loading on the Hydrogen Production of CNT/Pt composites Functionalized with Carboxylic Groups. *Int. J. Hydrog. Energy* **2020,** *45* (51), 27012–27025.

Modi, A.; Koratkar, N.; Lass, E.; Wei, B.; Ajayan, P. M. Miniaturized Gas Ionization Sensors Using Carbon Nanotubes. *Nature* **2003,** *424*, 171–174.

Mohamed, R. S.; Ekahlawy, A. A.; Naggar, A. M. A. E.; Gobara, H. M. Innovative Approach for Production of Carbon Nanotubes (CNTs) and Carbon Nano-Sheets through Highly Efficient Photocatalytic Water Splitting Into Hydrogen Using Metal Organic Frameworks (MOF)-Nano TiO_2 Matrix as Novel Catalyst. *New J. Chem.* **2020,** *44*, 5097–5108.

Murakami, N.; Tango, Y.; Miyake, H.; Tajima, T.; Nishina, Y.; Kurashige, W. et al. SWCNT Photocatalyst for Hydrogen Production from Water Upon Photoexcitation of (8, 3) SWCNT at 680-nm Light. *Sci. Rep.* **2017,** *7*. DOI: 10. 1038/srep43445.

Ng, B. J.; Putri, L. K.; Tan, L. L.; Pasbakhsh, P.; Chai, S. P. All-Solid-State Z-Scheme Photocatalyst with Carbon Nanotubes as an Electron Mediator for Hydrogen Evolution under Simulated Solar Light. *Chem. Eng. J.* **2017,** *316*, 41–49.

Nguyen, V. H.; Kang, C.; Roh, C.; Shim, J. J. Supercritical CO_2-Mediated Synthesis of CNT@ Co_3O_4 Anocomposite and Its Application for Energy Storage. *Ind. Eng. Chem. Res.* **2016,** *55*, 7338–7343.

Nikolaev, P.; Bronikowski, M. J.; Bradley, R. K.; Rohmund, F.; Colbert, D. T.; Smith, K. A. et al. Gas-Phase Catalytic Growth of Single-Walled Carbon Nanotubes from Carbon Monoxide. *Chem. Phys. Lett.* **1999,** *313*, 91–97.

Oberlin, A.; Endo, M.; Koyama, T. Filamentous Growth of Carbon through Benzene Decomposition. *J. Cryst. Growth* **1976,** *32,* 335–349.

Odom, T. W.; Huang, J. -L.; Kim, P.; Lieber, C. M. Structure and Electronic Properties of Carbon Nanotubes. *J. Phys. Chem. B* **2000,** *104*, 2794–2809.

Olowoyo, J. O.; Kumar, M.; Jain, S. L.; Babalola, J. O.; Vorontsov, A. V.; Kumar, U. Insights Into Reinforced Photocatalytic Activity of the CNT−TiO_2 Nanocomposite for CO_2 Reduction and Water Splitting. *J. Phys. Chem. C* **2019,** *123*, 367−378.

Pap, Z.; Karácsonyi, E.; Baia, L.; Pop, L. C.; Danciu, V.; Hernadi, K.; Mogyorosi, K. et al. TiO_2/WO_3/Au/MWCNT Composite Materials for Photocatalytic Hydrogen Production: Advantages and Draw-Backs. *Phys. Status Solidi B* **2012,** *12*, 2592–2595.

Pawar, R. C.; Choi, D-H.; Lee, C. S. Reduced Graphene Oxide Composites with MWCNTs and Single Crystalline Hematite Nanorhombohedra for Applications in Water Purification. *Int. J. Hydrog. Energy* **2015,** *40*, 767–778.

Pei, B.; Wang, W.; Dunne, N.; Li, X. Applications of Carbon Nanotubes in Bone Tissue Regeneration and Engineering: Superiority, Concerns, Current Advancements, and Prospects. *Nanomaterials* **2019**. DOI: 10.3390/nano9101501.

Pitroda, J. K.; Jethwa, B.; Dave, S. K. A Critical Review on Carbon Nanotubes. *Int. J. Constr. Res. Civ. Eng.* **2016,** 2, 36–42.

Popov, V. N. Carbon Nanotubes: Properties and Application. *Mater. Sci. Eng. R Rep.* **2004,** *43*, 61–102.

Prakash, S.; Kulamarva, A. G. Recent Advances in Drug Delivery: Potential and Limitations of Carbon Nanotubes. *Recent Pat. Drug Deliv. Formul.* **2007,** *1*, 214–221.

Radushkevich, L. V.; Lukyanovich, V. M. The Structure of Carbon Forming in Thermal Decomposition of Carbon Monoxide on an Iron Catalyst. *Russ. J. Phys. Chem. A* **1952,** *26*, 88–95.

Rai, S.; Ikram, A.; Sahai, S.; Dass, S.; Shrivastav, R.; Satsangi, V. R. Photoactivity of MWCNTs Modified α-Fe_2O_3 Photoelectrode towards Efficient Solar Water Splitting. *Renew Energy* **2015,** *83*, 447–454.

Ramesh, P.; Okazaki, T.; Sugai, T.; Kimura, J.; Kishi, N.; Sato, K. et al. Purification and Characterization of Double-Wall Carbon Nanotubes Synthesized by Catalytic Chemical Vapor Deposition on Mesoporous Silica. *Chem. Phys. Lett.* 2006, *418*, 408–412.

Rao, A. M.; Richter, E.; Bandow, S.; Chase, B.; Eklund, P. C.; Williams, K. A. et al. Diameter-Selective Raman Scattering from Vibrational Modes in Carbon Nanotubes. *Science* **1997,** *275*, 187–191.

Reibold, M.; Paufler, P.; Levin, A. A.; Kochmann, W.; Patzke, N.; Meyer, D. C. Materials: Carbon Nanotubes in an Ancient Damascus Sabre. *Nature* **2006,** *444*. DOI: 10.1038/444286a.

Rinzler, A. G.; Hafner, J. H.; Nikolaev, P.; Lou, L.; Kim, S. G.; Tomanek, D. et al. Unraveling Nanotubes: Field Emission from an Atomic Wire. *Science* **1995,** *269*, 1550–1553.

Rosen, M. J.; Kunjappu, J. T. *Surfactants and Interfacial Phenomena*; John Wiley & Sons, Hoboken: New Jersey, 2012.

Salvetat, J. P.; Bonard, J. M.; Thomson, N. H.; Kulik, A. J.; Forro L.; Benoit, W. et al. Mechanical Properties of Carbon Nanotubes. *Appl. Phys. A.* **1999,** *69*, 255–260.

Sampath, S.; Sellappa, K. Visible-Light-Driven Photocatalysts for Hydrogen Production by Water Splitting. *Energ. Source Part A* **2019,** *42*, 719–729.

Schuldiner, S. Hydrogen Overvoltage on Bright Platinum II. pH and Salt Effects in Acid, Neutral, and Alkaline Solutions. *J. Electrochem. Soc.* **1954,** *101*, 426–432.

Seo, J. W.; Magrez, A.; Milas, M.; Lee, K.; Lukovac, V.; Forro, L. Catalytically Grown Carbon Nanotubes: From Synthesis to Toxicity. *J. Phys. D: Appl. Phys.* **2007,** *40*, 109–120.

Sheeney-Haj-Ichia, L.; Basnar, B.; Willner, I. Efficient Generation of Photocurrents by Using CdS/Carbon Nanotube Assemblies on Electrodes. *Angew. Chem. Int. Ed.* **2005,** 44, 78 –83.

Sheng, W.; Gasteiger, H. A.; Shao-Horn, Y. Hydrogen Oxidation and Evolution Reaction Kinetics on Platinum: Acid vs. Alkaline Electrolytes. *J. Electrochem. Soc.* **2010,** *157*, B1529–B1536.

Singh, D. K.; Iyer, P. K.; Giri, P. K. Optical Signature of Structural Defects in Single Walled and Multiwalled Carbon Nanotubes. *J. Nanosci. Nanotechnol.* **2009,** *9*, 5396–5401.

Sobolkina, A.; Mechtcherine, V.; Khavrus, V.; Maier, D.; Mende, M.; Ritschel, M. et al. Dispersion of Carbon Nanotubes and Its Influence on the Mechanical Properties of the Cement Matrix. *Cem. Concr. Compos.* **2012,** *34*, 1104–1113.

Song, L.; Kang, X.; Zhang, S. CNT/g@ C_3N_4 Photocatalysts with Enhanced Hydrogen Evolution Ability for Water Splitting Based on a Noncovalent Interaction. *Int. J. Energy Res.* **2018,** *43*. DOI: 10.1002/er.3960.

Spataru, C. D.; Ismail-Beigi, S.; Benedict, L. X.; Louie, S. G. Excitonic Effects and Optical Spectra of Single-Walled Carbon Nanotubes. *Phys. Rev. Lett.* **2004,** *92*, 077402.

Sung-Wook, K.; Seo, D-H.; Gwon, H.; Kim, J.; Kang, K. Fabrication of FeF_3 Nanoflowers on CNT Branches and Their Application to High Power Lithium Rechargeable Batteries. *Adv. Mater.* **2010,** *22*, 5260–5264.

Suzuki, S.; Watanabe, T.; Kiyokura, T.; Nath, K. N.; Ogino, T.; Heun, S. et al. Electronic Structure at Carbon Nanotube Tips Studied by Photoemission Spectroscopy. *Phys. Rev. B* **2001,** 63. DOI: 10. 1103/PhysRevB. 63. 245418.

Tans, S. J.; Verschuerenc, A. R. M.; Dekker, C. Room-Temperature Transistor Based on a Single Carbon Nanotube. *Nature* **1998,** *393*, 49–52.

Thess, A.; Lee, R.; Nikolaev, P.; Dai, H.; Petit, P.; Robert, J. et al. Crystalline Ropes of Metallic Carbon Nanotubes. *Science* **1996,** *273*, 483–487.

Ting, Z.; Xia, B.; Zhou, L.; Lou, X. W. D. Arrays of Ultrafine CuS Nanoneedles Supported on a CNT Backbone for Application in Supercapacitors. *J. Mater. Chem.* **2012,** *22*, 7851–7855.

Treacy, M. M. J.; Ebbesen, T. W.; Gibson, J. M. Exceptionally High Young's Modulus Observed for Individual Carbon Nanotubes. *Nature* **1996,** *381*, 678–680.

Tsuda, A.; Venkata, N. K. The Role of Natural Processes and Surface Energy of Inhaled Engineered Nanoparticles on Aggregation and Corona Formation. *Nano Impact.* **2016,** *2*, 38–44.

Umer, M.; Tahir, M.; Azam, M. U.; Jaffar, M. M. Metal Free MWCNTs@TiO_2@MMT Heterojunction Composite with MMT as a Mediator for Fast Charges Separation towards Visible Light Driven Photocatalytic Hydrogen Evolution. *Appl. Surf. Sci.* **2018,** *463*, 747–757.

Verma, A.; Srivastav, A.; Sharma, S.; Badami, P.; Satsangi, V. R.; Shrivastav, R. MWCNTs and Cu_2O Sensitized Ti/Fe_2O_3 Photoanode for Improved Water Splitting Performance. *Int. J. Hydrog. Energy* **2018,** *43*, 6049–6059.

Wagner, R. S.; Ellis, W. C. Vapor-Liquid-Solid Mechanism of Single Crystal Growth. *Appl. Phys. Lett.* **1964,** *4*, 89–90.

Walvekar, R.; Mohammad, K. S.; Seik, S. O.; Ahmad, F. I. Application of CNT Nanofluids in a Turbulent Flow Heat Exchanger. *J. Exper. Nanosci.* **2016,** *11*, 1–17.

Wan, W.; Wei, S.; Li, J.; Triana, C. A.; Zhou, Y.; Patzke, G. R. Transition Metal Electrocatalysts Encapsulated Into N-Doped Carbon Nanotubes on Reduced Graphene Oxide Nanosheets: Efficient Water Splitting through Synergistic Effects. *J. Mater. Chem. A,* **2019,** *7*, 15145–15155.

Wang, J.; Wang, Z.; Zhua, Z. Synergetic Effect of $Ni(OH)_2$ Cocatalyst and CNT for High Hydrogen Generation on CdS Quantum Dot Sensitized TiO_2 Photocatalyst. *Appl. Catal. B: Environ.* **2017,** *204*, 577–583.

Wang, J. Y.; Ouyang, T.; Li, N.; Ma, T.; Liu, Z. Q. S. N. Co-Doped Carbon Nanotube-Encapsulated Core-Shelled CoS_2@Co Nanoparticles: Efficient and Stable Bifunctional Catalysts for Overall Water Splitting. *Sci. Bull.* **2018,** *63*, 1130–1140.

Wang, L.; Dang, Z. Carbon Nanotube Composites with High Dielectric Constant at Low Percolation Threshold. *Appl. Phys. Lett.* **2005**. DOI: 10.1063/1.1996842.

Wang, X.; Liu, M.; Chen, Q.; Zhang, K.; Chen, J.; Wang, M. et al. Synthesis of CdS/CNTs Photocatalysts and Study of Hydrogen Production by Photocatalytic Water Splitting. *Inter. J. Hydrog. Energy* **2013,** 38, 13091–13096.

Wang, Y.; Pan, C.; Chu, W.; Kiliyankil, V. A. Environmental Remediation Applications of Carbon Nanotubes and Graphene Oxide: Adsorption and Catalysis. *Nanomaterials* **2019,** *9*, 349.

Wang, Z.; Ci, L.; Chen, L.; Nayak, S.; Ajayan, P. M.; Koratkar, N. Polarity-Dependent Electrochemically Controlled Transport of Water through Carbon Nanotube Membranes. *Nano Lett.* **2007,** *7*, 697–702.

Wei, G.; Du, K.; Zhao, X.; Wang, J.; Yan, W.; An, C. et. al. Cable-Like Carbon Nanotubes Decorated Metal-Organic Framework Derived Ultrathin $CoSe_2$/CNTs Nanosheets for Electrocatalytic Overall Water Splitting. *Chin. Chem. Lett.* **2020,** *31*, 2641–2644.

Wei, Y.; Du, H.; Kong, J.; Lu, X.; Ke, L.; Sun, X. W. Multi-Walled Carbon Nanotubes Modified ZnO Nanorods: A Photoanode for Photoelectrochemical Cell. *Electrochim. Acta.* **2014,** *143*, 188–195.

Wu, D., Wu, L., Zhou, W., Sun, Y. and Zhang, M. (2010), Relations between the aspect ratio of carbon nanotubes and the formation of percolation networks in biodegradable polylactide/carbon nanotube composites. J. Polym. Sci. B Polym. Phys., 48: 479-489. https://doi.org/10.1002/polb.21909.

Xiangqing, L.; Zhang, J.; Kang, S.; Li, G.; Mu, J. Visible Light Photocatalytic Activity of CuO/Cr_2O_3 Co-loaded Multiwalled Carbon Nanotubes Sensitized with Eosin Y for Hydrogen Evolution from Water. *Ceram. Int.* **2014,** *40*, 10171–10176.

Yu, C.; Shi, L.; Yao, Z.; Li, D.; Majumdar, A. Thermal Conductance and Thermopower of an Individual Single-Wall Carbon Nanotube. *Nano Lett.* **2005,** *5*, 1842–1846.

Yue, G. Z.; Qiu, Q.; Gao, B.; Cheng, Y.; Zhang, J.; Shimoda, H. et al. Generation of Continuous and Pulsed Diagnostic Imaging X-ray Radiation Using a Carbon-Nanotube-Based Field Emission Cathode. *Appl. Phys. Lett.* **2002,** *81*, 355–357.

Zhan, W.; Gao, L.; Fu, X.; Siyal, S. H.; Sui, G.; Yang, X. Green Synthesis of Amino-Functionalized Carbon Nanotube-Graphene Hybrid Aerogels for High Performance Heavy Metal Ions Removal. *Appl. Surf. Sci.* 2019, *467–468*, 1122–1133.

Zhang, J.; Li, J.; Cao, J.; Qian, Y. Synthesis and Characterization of Larger Diameter Carbon Nanotubes from Catalytic Pyrolysis of Polypropylene. *Mater. Lett.* **2008,** *62*, 1839–1842.

Zhang, J.; Zhao, Z.; Xia, Z.; Dai, L. A Metal-Free Bifunctional Electrocatalyst for Oxygen Reduction and Oxygen Evolution Reactions. *Nat. Nanotechnol.* **2015,** *10*, 444–452.

Zhang, W-D.; Jiang, L-C.; Ye, J-S. Photoelectrochemical Study on Charge Transfer Properties of ZnO Nanowires Promoted by Carbon Nanotubes. *J. Phys. Chem. C* **2009,** *113*, 16247–16253.

Zhang, X.; Wang, T.; Liu, J.; Andersson, C. Overview of Carbon Nanotubes as Off-Chip Interconnects. *In 2nd Electronics System-Integration Technology Conference*; Greenwich, 2008; pp 633–638.

Zheng, X.; Liu, K.; Chen, L.; He, H.; Chen, L.; Sun, C. $Co(O)_x$ Particles in polymeric N-doped carbon nanotube applied for photocatalytic h_2 or electrocatalytic O_2 evolution. *Polymers* **2019,** *11*. DOI: 10. 3390/polym11111836.

Zheng, Y.; Jiao, Y.; Vasileff, A.; Qiao, S. Z. The hydrogen evolution reaction in alkaline solution: from theory, single crystal models, to practical electrocatalysts. *Angew. Chem. Int. Ed.* **2018,** 57, 7568–7579.

Zhu, Y. Q.; Sekine, T.; Kobayashi, T.; Takazawa, E.; Terrones, M.; Terrones, H. Collapsing carbon nanotubes and diamond formation under shock waves. *Chem. Phys. Lett.* **1998,** 287, 689–693.

CHAPTER 7

Nanomaterials as Effective Tools for Detection and Microbial Diagnosis of Foodborne Pathogens

C. O. ADETUNJI[1*], A. INOBEME[2], O. T. OLANIYAN[3], O. A. ANANI[4], J. T. MATHEW[5], R. E. BODUNRINDE[6], J. B. ADETUNJI[7], and D. I. HEFFT[8]

[1]*Applied Microbiology, Biotechnology and Nanotechnology Laboratory, Department of Microbiology, Edo University Iyamho, PMB 04, Auchi, Edo State, Nigeria*

[2]*Department of Chemistry, Edo University Iyamho, Nigeria*

[3]*Laboratory for Reproductive Biology and Developmental Programming, Department of Physiology, Edo University Iyamho, Nigeria*

[4]*Laboratory of Ecotoxicology and Forensic Biology, Department of Biological Science, Faculty of Science, Edo University, Iyamho, Edo State, Nigeria*

[5]*Department of Chemistry, Ibrahim Badamasi University Lapai, Niger State, Nigeria*

[6]*Department of Microbiology, Federal University of Technology Akure, Nigeria*

[7]*Nutritional and toxicological Research Laboratory, Department of Biochemistry Sciences, Osun State University, Osogbo, Nigeria*

[8]*Edgbaston Campus, University of Birmingham, School of Chemical Engineering, Birmingham B15 2TT, UK*

[*]*Corresponding author. E-mail: adetunji.charles@edouniversity.edu.ng*

The Science of Nanomaterials: Basics and Applications. Suresh C. Ameta & Rakshit Ameta (Eds.)

ABSTRACT

Foodborne diseases have been identified as a significant factor, which have threatened the wellbeing of consumers. This has also resulted in a high mortality level around the globe, particularly in susceptible groups. These diseases normally emanate from the toxins of microbes, which are responsible for these diseases. Typical sources are poisons and toxins, which have been reported from prions, bacteria, viruses, parasites, and so forth, which are normally responsible for foodborne diseases and gastroenteritis in human being. Therefore, there is a need to search for a potent antimicrobial material that could be utilized in the food industry for processing and packaging of foods, which has led to the emergence of nanotechnology for adequate intervention. Therefore, this chapter intends to provide a comprehensive information on the application of nanomaterials that could serve as antimicrobial agents for effective food control as well as the up-to-date legislation regarding these materials.

7.1 INTRODUCTION

The application of nanomaterials in the development of nanotechnology will transform many aspects of the food industry and food science. Nanomaterials are particles containing naturally unbound materials such as metal oxides, metals, clay plus full-carbon constituents, and organic nanoncapsulates, currently utilized in many agricultural and food processing sectors.

The most commonly utilized forms of nanotechnology are the nanobiocides and nanobiotics for improving food quality and safety like packaging and processing (Suresh and Digvir, 2011). In the agricultural sector, nanotechnology plus machine coatings, which have received massive attention in the food industry against foodborne pathogens like *Salmonella, Pseudomonas aeruginosa, Campylobacter jejuni*, are recently being utilized to eliminate and filter out microorganisms, and microbial disinfection, (Nile et al., 2020).

Quite a number of nanoparticles such as nanocomposites, copper oxide, zinc oxide, silver nanoparticles, and so on, have found way into the production processes of many food processing plants, including the coatings and packaging sector for the microbial control system. They perform these protective functions through their physicochemical properties, such as binding of Ag ions to electron donor molecules of oxygen, nitrogen, and sulfur, interference of DNA replication and formation of oxidative stress, and alteration in the physiological function plus cell damage.

Many of these nanoparticles are utilized for the fortification of cereal-based foods and poultry microbial control agents for improved meat and dairy quality (Pengju et al., 2008). Zhen-Xing and Bin-Feng (2014) reported that in the agricultural sector, pathogenic contamination has become a serious growing concern responsible for many health issues across the world; thus, the development of an effective nanoparticle antimicrobial agent is of tremendous benefit. Studies have revealed that antimicrobial agents can be grouped into two groups: organic and inorganic (Syafiqah et al., 2021). The organic generally develops weak resistance during processing with the limited application; thus, inorganic antimicrobial agents are becoming more attractive in the scientific community due to their stability and safety. In line with this, nanotechnology can be utilized to develop novel inorganic antimicrobial agents for many promising applications from agriculture to medical science.

Ulatowska-Jarża et al. (2011) reported that in the nanotechnology development, many of the physiochemical properties may be explored to enhance the diagnostic and therapeutic capacity of these nanoparticles. They revealed that nanosensor gold-based nanoparticles have begun to show promising results as a diagnostic agent or tool for various pathogenic diseases and cells, glucose, urea, pesticide, and metabolites through nanotherapeutics or nanochips. Fakruddin et al. (2012) revealed that in the field of biological sciences, nanobiotechnology is a fast-growing field where nanodevices and nanoparticles are utilized for diagnosis and detection of microorganisms that affect various food processing industries.

Ramos et al. (2017) demonstrated that molecular biology has offered tremendous opportunity for investigation and search for many biomolecules with physiological functions, particularly diagnosis of disease and drug design for many foodborne diseases. The molecular structure of these materials enhances their utilization in biosensing, diagnosis, and bioimaging of disease pathogens.

It has been established that nanobiotechnology can be utilized to solve many agricultural problems, particularly in food production and identification of disease plus diagnosis due to their cost effectiveness, physiochemical properties, eco-friendly, and nontoxic antibacterial effects. It was revealed that as a result of the increasing prevalence of foodborne diseases, many safe environmental friendly nanoparticles were synthesized (Shrivastava and Dash, 2009).

A single gold nanoparticle assay could be utilized for single or multiple pathogenic detection and control through hybridization or immobilization of the nanoparticles to enhance their antimicrobial effects (Dakrong et al., 2019). Their antimicrobial effect is achieved through the reduction in oxygen supply, reduced respiration, and killing of the pathogenic organisms; thereby, producing quality, healthier, and safer functional foods. It is generally known that potent serious advantages of nanotechnology over the traditional methods are sterilization of food processing and vector control system, enhancement of food quality, prevention of contamination, food protection, nanosensors in smart food packaging, increase in shelf life, taste, bioavailability, and texture (Valavanidis and Vlachogianni, 2016).

It has been shown that due to the advancement in technology, conventional or traditional methods are now being renewed with more sophisticated ones. The traditional sterilization or conventional filtration technique in the purification of water and mycotoxins in animal feeds is continually being replaced with nanotechnology to improve food processing yields and decrease waste or depletion of nutrients (Horky et al., 2018).

7.2 ANTIMICROBIAL AGENTS FOR EFFECTIVE FOOD CONTROL

There have been a series of studies investigating the antimicrobial properties of nanomaterials. Some nanomaterials, which exhibit antimicrobial properties, include some metals such as copper, silver, zinc, and iron oxide (Seil and Websters, 2012). Though the modes of action of the nanoparticles, most especially the antimicrobial action, are not well understood; however, there are some proposed mechanisms relating to their actions, which include its physical structure and improved antibacterial metal ions released from nanoparticle surfaces. Some relevant factors affecting antibacterial activity include particle shape, size, chemistry, and zeta potential.

Recent findings by Lima et al. (2013) and El Sheekh et al. (2014) have proved the ability of nanomolecules to effectively stop the action of *Salmonella typhi, Microcystis aeruginosa,* and *Cyanobacteria. Cyanobacteria* and *Microcystis aeruginosa* are cancer-promoting microbes. Yu et al. (2014) reported the antimicrobial potential of silver nanomolecules, while Azam et al. (2012) observed significant antimicrobial activity of ZnO nanoparticles against Gram-negativity and Gram-positive bacteria; however, the effect of iron oxide nanomolecules on bacterial cells was found to be least in the same study. Gram-positive and Gram-negative bacterial strains reported for the study include *Bacillus subtilis, Escherichia coli, Pseudomonas aeruginosa, Staphylococcus aureus.* The activities of the foodborne pathogens (*P. Aeruginosa, E. coli*, and *Aspergillus niger*) were reported to be inhibited by ZnO nanoparticles at varied concentrations (Chitra and Annadurai, 2013).

This study further proved the potential of nanoparticles application in the food packaging industry.

The viability of *S. aureus* and *E.coli* was found to decrease when subjected to decreasing size of ZnO (Nair et al., 2009). Irregularity in the membrane of cells of bacteria exposed to ZnO nanoparticles was also observed. A possible mechanism suggested for the antimicrobial effect of ZnO includes cell damage is a result of increased abrasiveness of smaller nanoparticles and improved surface to volume ratios of the nanoparticles. In another investigation, lower zinc concentration was observed to have more effect on *S. Aureus* as compared to *P. aeruginosa* (McCarthy et al., 1992).

Nanoparticle prepared from some Allium species inhibited bacterial pathogens such as *Enterobacter spp, Staphylococcus spp, E. coli, Pseudomonas spp*, and *Bacillus spp*. Cell wall damage and DNA condensation were changes recorded in bacterial species exposed to silver ions concentration within a 4–12 h duration (Zeelie and McCarthy, 1998). The silver ion concentration was kept-same for the experiment as 10 μg/mL.

Some species of Gram-negative bacteria have been reportedly inhibited by silver nanoparticles of about 75 μg/mL and 21 nm in diameter (Morones et al., 2005). These Gram-negative bacteria species include *P. aeruginosa, Vibrio cholerae*, and *E. coli.* It was reported that CFU population was reduced when agar plates were incorporated with silver nanoparticles (Kim et al., 2007). Likewise, silver and selenium nanomolecules produced from bacteria isolate derived from coal mines were also found effective against the antibacterial activity of *S. aureus, Klebsiella spp, E. coli, S. aureus*, and *Klebsiella spp* (Singh et al., 2014; Deepa et al., 2013).

7.3 DIAGNOSIS AND MANAGEMENT OF PLANT PATHOGENS

Singh et al. (2019) examined the role of microbes (fungi and bacteria) enhanced with nanomaterials in drug delivery and for the management of flora pathogens. They reported that nanomaterials can be used in the production of cosmetics, in the field of agriculture and medicine. Nano-based particles have a wide range for their use in the food industries for the detection and identification of fungal and other microbial species with high sensitivity and specificity in food constituents. Nanomaterials play a significant part in the management and control of pathogenic fungi and bacteria. It was also reported that the mode of actions of nanomaterials in the inhibition of fungi involves cellular and metal ions interactions via different pathways that include cell capture, DNA, and reactive oxygen species generation. In conclusion, the therapeutic management of plant pathogens by nanomaterials has guaranteed better food quality and safety. The utilization of biomolecules in conjunction with nanosensors and nanomaterials was proposed as novel nanotechnology to further improve the interactions between the microbes and nanomaterials for efficient food safety because of the level of toxins generated by the fungi and detection of adulterant during the process of treatment.

The application of engineered nanomaterials in the management of plant and food pathogens has gained immense contribution in the field of agriculture. Chen et al. (2020) tested and evaluated the possible use of nanomaterials sourced from MgO (magnesium oxide) in the management of soilborne fungal (*Thielaviopsis basicola* and *Phytophthora nicotianae*) pathogenic activities in a greenhouse setting. The results from the in vivo evaluation showed that the magnesium oxide was able to inhibit the spore and fungal development. It was able to stop the sporangium growth to a microscale level. The direct relations between the cells of the fungi and the nanomaterials were found to provoke subsequent morphological vicissitudes, when viewed under a transmission electron microscopy and scanning electron microscopy/energy-dispersive spectrometry. Besides, the spreading of the zeta accumulation and potential of different mechanisms of oxidative stress on the nanomaterial used and these are also reported for the fundamental antifungal action. In the greenhouse setting, approximately 42.35% and 36.58% reduction in the black root rot and shank diseases, respectively, could attest to the competence by which about 500 μg/mL of magnesium oxide acted as antifungal with a final efficacy of 50.20% and 62.10%, respectively, when related to the untreated controls of the nanomaterials. It was recommended that the effective understanding of the nanomaterials (magnesium oxide) will aid in the future prevention of different fungal diseases affecting agricultural produce.

Elgorban et al. (2016) tested and evaluated the antifungal properties of AgNPs. AgNPs have been known to possess enhanced antimicrobial activity and are well used for different cleansing purposes, such as medical, materials, and water sanitization purposes. It has been documented that few kinds of literature abound in the utilization of AgNPs for the management of fungal diseases. They used the AgNPs concentration of 0.0, 0.0002, 0.0005, 0.0007, 0.0009, 0.0014, and 0.0019 mol/L in the management of the pathogenicity of six different fungi (*Rhizoctonia solani*), belonging to the anastomosis clusters on cotton plants. The results of the biological test revealed that the silver nanoparticles showed strong antifungal assets in the management of *R. solani*. It was revealed that the powerful inhibitory effects of the silver nanoparticles were observed in all the tested concentrations

Álvarez et al. (2016) looked at the application of nanotechnology as a tool in the management of fungal diseases. It is known that fungal infections cause significant damage to agricultural products globally. However, they can be controlled by fungicides, which are chemicals that can irritate their activities and kill them. Ascomycetes fungi having pathogenic properties in general are *Verticillium, Fusarium*, *Alternaria,* and *Basidiomycetes* of the genus *Rhizoctonia* and *Sclerotium*. The effect of fungicides on the various general has been recounted to be inefficient. Of recent, the application of nanotechnology using nanomaterials has been shown to improve the efficacy and the control of pathogenic fungi, when combined with fungicides. Some specific examples of these nanomaterials are chitosan, Al-silicates, silica, Ag, and carbon nanoforms. They suggested that the transportation and penetration of nanomaterials on plants need to be well understood to know their mode of actions.

The use of fungicides has been dated since the year 1940, which has also been fingered to cause several environmental problems. Their utilization not only kills the fungi but also eliminate nontargeted species and portend possible impacts on living structures. Patel et al. (2014) evaluated in a review the importance of nanotechnology in the agricultural control and management of fungal pathogens. They stated that to preserve the extant biodiversity, it is necessary to employ the services of green technology. The biocontrol messengers such as biopesticides and phytocompounds are necessary tools that can substitute the artificial chemical pesticides widely used. Currently, nanofungicides sourced from Ag emulsion have been recounted to control fungal pathogens threatening agricultural produce. The use of Ag nanomaterials could open a wide area of integrated research in integrated pest management and phytopathogen control in many ways. However, different biological aspects such as soil toxicity and the ecological fate in the terms of their life span also should be looked into. The mode of action on the fungal species should also be looked for a proper understanding of the effect of nanoemulsion on the fungal cells.

Abd-Elsalam (2012) evaluated the nanoplatforms of using nanomaterials in the management of fungi. They were of the opinion that in recent times, floral pathologists have been looking for a better and more novel solution in the management of various viral, fungal, and bacterial diseases to forestall food safety for the present and future generation. The utilization of nano-platforms as a biosensor tool in current agricultural practice has elicited significant directions in the detection and management of fungi disease in agriculture. Nanomaterial formulation with fungicides can be prepared in a way and manner so that it is cost-efficient and highly effective for its utilization. A specific example is the "Syngenta's Banner MAXX ™," which is a very systemic and potent nanofungicide with a broad spectrum in the management of diseases in the soil and ornamental floras. Nanomaterial application in agriculture has a place in the protection and detection of plants against specific fungal pathogens. Hence, it can be applied as a next generational nanophytopathology.

Atiq et al. (2020) reviewed the application of nanomaterials against fungal pathogens in plants. Different techniques using chemicals such as fungicides have proved to be less efficient and not environmentally friendly. Currently, a novel "bionanotechnology" has offered a more efficient and environmentally friendly alternative means for the effective control of fungi diseases in agriculture. This advanced method has opened a novel era in the agricultural sector in combating fungal invasion on the plant as compared to the traditional method.

Recent studies in nanotechnology have led to the fabrication of nanoparticles that have unique chattels toward the drug delivery process. Sharma et al. (2017) looked into in vitro and in vivo mechanisms of action of nickel and cobalt nanoferrites ($NiFe_2O_4$ and $CoFe_2O_4$) in managing fungi pathogens in plants. It was reported that fungal pathogens are the major leading plant diseases that destroy most of the agricultural produce. The effective utilization of nanomaterials such as nickel and cobalt ferrites on infected plants with fungal strains like *Dematophora necatrix, Colletotrichum gloeosporioides*, and *Fusarium oxysporum* have yielded positive antifungal influence. Besides, it has also been recounted that these nanomaterials are also effective on wilt in capsicum plants. The potencies of the nanomaterials demonstrated the phytopathogenic response toward different fungal strains.

Vargas-Hernandez et al. (2020) evaluated the potentials of nanomaterials as an antiviral tool. It was reported that about 50% of plant disease is caused by a virus. Sometimes, plant diseases are very difficult to manage or control. So, it is important to develop a novel technique to manage plant viruses. Nanomaterials derived from metals have been identified as one of the novel technologies that has been employed in different industrial sectors, especially in agriculture. There have been positive reports on the influence of nanomaterials as antiviral, antifungal, and antibacterial tools. The antiviral properties of nanoparticles derived from metals have been known to possess strong inhibitory and biostimulant modes of activity against plant viruses.

Young et al. (2008) evaluated the biotemplates of nanomaterials as phytoantiviral. It was observed that the capsids of phytovirus possess protein covering (shell) that emerged as an important material for the synthesis of nanoscale structures (3D) from subunits codes of the virus. The capsids are opened to both; chemical as well as genetic modifications, permitting the incorporation of nanomaterials to interface of the superficial region to pathogenic plants as an antiviral tool. This technology can serve as future bionanotechnology and can be applied to combat several strains of plant viruses.

Marwal and Gaur (2019) discussed the current management of virus threats to plant growth by the application of nanotechnology. It was reported that nanophytovirology is a novel aspect of nanotechnology that aids in the control of virus infections in the plant. The nanomaterials act as a delivery messenger to hit strains of plant viruses. These nanomaterials not only provide the speed of detection but also the power of inhibitions toward strains of viruses to better the growth and wellbeing of the plant infected. In conclusion, it was concluded that finding nanophytovirology-based remedies will help stakeholders, agriculturists, and researchers to discover a healthier management technique against challenges such as antimicrobial invasion and natural factors impeding plant growth and productivity.

Cao et al. (2015) evaluated the potentials of virus nanomaterials in resolving phytoinfection caused by parasitic roundworms by manipulating the physical and chemical structures of the soil. It was reported that parasitic roundworm infection in the plant is one of the largest farm pests producing excessive damage of about $157 billion globally to farm produce yearly. A biological nematicides Abm (Abamectin) has been widely used in the control and management of parasitic roundworm affecting plants for last so many years. However, its mode of action has been recently impeded by unfortunate movement in the soil; thus, compromising its efficiency in the soil in safeguarding the nascent developing root part of the plant. It was indicated that the capsid of the virus in conjunction with the nanoparticles showed good integrity in the improvement of the Abamectin in combating the soil nematodes as well as enhancing the soil movement for the nematicides to be readily bioavailable to the roundworm as evident from the electron microscope. As a consequence, the plant virus nanomaterial Abamectin, increased the region of protection against root-knot roundworm—*Meloidogyne hapla* in the soil, when compared to the treatment with only Abamectin. It was observed that the young tomato plants treated with the amalgam-plant virus nanomaterial, Abamectin demonstrated good growth of root and a decreased in the root irritation by the nematodes, which is an indication of an effective drug delivery process for the improved efficiency of the nematicides to control and manage plants damaged by the nematodes.

Pérez-de-Luque and Rubiales (2009) reviewed the application of nanomaterials in combating and managing plant parasites. It was stated that nanotechnology is a new aspect of science and engineering that can be applied in agriculture, medicine, and pharmacology. Moreover, special interest has been driven into its applications in crop management as delivery structures (nanocarriers) to target specific regions, where traditional chemicals cannot reach. Nanocapsules (a nanocarrier) can be used to enhance extant plant management techniques in a specific range of time via the introduction of herbicides in controlling parasitic plants/weeds. Besides, nanoencapsulation can also enhance the persistent release of active materials, excellent infiltration via tissues and cuticles, and application of herbicide efficiently when combined in the control of plant parasites. They have also reported that nanomaterials with herbicides have the countless possibility as "magic bullets" for the drug delivery into specific target regions, where only herbicides cannot reach, penetrating the nucleic acids and tissues contents of the parasites. The application of viral capsid with nanomaterials can also achieve various configuration purposes and transport specific antimicrobial enzymes, nucleic acids, and peptides against the plant parasites. It was concluded that several issues such as the toxicology of the nanomaterials, herbicides, and viral capsid as well as the reduction of the economic cost of the nanomaterials should be looked at to proffer sustainable management and control of plant parasites.

7.4 MANAGEMENT OF DIFFERENT FOODBORNE PATHOGENS

The antibacterial effectiveness of nanoparticles is dependent on the administered dose and the native NPs. It is known to have more cidal effects on Gram-negative bacteria as compared to bacteria Gram-positive due to their cellular makeup. The potency of nanoparticle is due to their ability to produce reactive species, which damages cellular components. Silver nanoparticles are globally used as antibacterial agent. It was reported that Ag NPs inhibited *P. aeruginosa* and *E. coli* by disrupting the chain in their respiration and hindering the process of division of cells. Ag NPs act together concurrently with proteins that have sulfur in their bacterial cell wall and complexes that have phosphorus-like DNA in other to block replication (Patel et al., 2014). Mohanraj and Chen (2006) revealed that Ag NPs are a substitute antibacterial representative to conventional antibiotics and, thus, cannot experience the problem of resistance on like antibiotics. Ag NPs were found to be most effective against bacteria due to its presence of increased surface area, there was pill up of Ag NPs in the cell wall of *E. coli* after treatment with the NPs and the cell was perforated, which later caused the death of the cell (Dastjerdi and Montazer, 2010).

Ag NPs also showed antibacterial potency against methicillin-resistant *S. aureus,* which is known to cause food poisoning. This antibacterial approach was possible due to its ability to synthesize different saccharides (Liu et al., 2017). Before the advent of NPs, a lot of organic and inorganic materials have been used in other to control food disease/infection (Mujeebur and Tanveer, 2014). The application of NPs has a distinct way of helping to manage the rate at which plant is being affected by pathogen. This could be done via straightforward application of the desired nanoparticles to the soil before plant commences. This method has helped in the reduction of the harm caused by pathogen as compared to the application of chemicals, which could even pose a threat to human life after consumption of the plant produce. The effect of direct application of NPs to soil is usually greater than other means in the elimination of mineral fixing pathogenic microorganisms (Alghuthaymi et al., 2015). The results of the application of NPs in controlling plant pathogenic microbes could be divided into two, straightforward results upon application of NPs to the soil and the utilization of the NPs in the production of nanopesticides/nanocidal agents (Rai and Kratosova, 2015).

Silver nanoparticles were found to have over 15 microelements like zinc, which has helped to stop the spoilage of sunflower. Upon the experimentation of Ag NPs on *Candida albicans, Candida glabrata,* and *Candida tropicalis,* there was a great antifungal effect against the pest (Ouda, 2014). Polgoma et al. (2011) showed that when tiny pieces of DNA are joined in close association to gold NPs, the nanoparticles are gathered on the surface, where the signal is received in the presence of a substitutionary target. When this is used, it can help the Au NPs to detect over millions of various DNA sequence and acting on them appropriately (Brigger et al., 2012).

Research has shown that various pathogenic microbes have formed the mechanism of resistance to conventional microbial agents. Nanoparticles are very effective against microorganism, due to the lack of the microorganisms to resist the potency of the nanoparticles (Zhang et al., 2010). Silver nanoparticles were synthesized from the leaves of *Protiun serratum* and used against pathogens that are actively involved in foodborne like *P. earuginosa, B. Subtilis*, and *E. coli* were found to be susceptible with varying zones of inhibition with a great reduction in the viable number of cells (Mohanta et al., 2017). *Jatropha curcas* aqueous extract was also used to synthesize Ag NPs, which was tested against *E. coli*, *S. enterica,* and gave different zones of inhibition.

7.5 SPECIFIC BIOSENSOR FABRICATED WITH NANOMATERIAL FOR EFFECTIVE DETECTION AND MICROBIAL DIAGNOSIS OF FOODBORNE PATHOGENS

The loss of food as a result of crop contagions from pathogens, which includes viruses, fungi, and bacteria are tenacious issues in food production for years throughout the globe. Food-borne diseases are affected through bacteriological pathogens, which are dangers to human wellbeing. There are conservative principles established microbes techniques for the investigation of microbial food infection produce, but these are time-consuming and laborious in addition to entail precise services.

Ruyuan et al. (2020) used several outstanding nanomaterials for the production of new biosensors for foodborne pathogen detection. They were able to establish that the novel biosensors applied have numerous advantages such as specificity, rapidity, accuracy, sensitivity, as well as simplicity using nanomaterials properties. They also accounted for an innovative biosensors, which include colorimetric, electrochemical, along with outward improved Raman scattering, as well as fluorescent biosensors for detecting key foodborne pathogens. However, they opined that the traditional methods for foodborne pathogen detection have several limitations, such as low sensitivity, low selectivity, and long detection time.

Kumar et al. (2020) reported that nanomaterials were applied for innovative skills in biosensors, which was the relocation of signals to improve their sensitivity along with efficiency. Although microbial polluted food poses a lot of health concerns as a result of epidemics of severe food-borne illnesses. For that reason, the need for specific identification and revealing of pathogenic toxins in addition to microbes in food is necessary to avoid some of these concerns. Consequently, they were able to establish the concept of biosensing, which allowed them to formed nanobiosensors by means of diverse composites along with nanomaterials to increase the specificity and the sensitivity of pathogen detection. Nanomaterials used for developing biosensors are gold and magnetic, carbon nanotubes, graphene nanomaterials, quantum dots, as well as dendrimers, which predominantly improved sensitivity and specificity for detection as a result of their limited magnetic, chemical, mechanical, physical, and optical properties. Also, they opined that new composites and nanoparticles utilized in biosensors are required to be categorized and classified for the quick detection, improved productivity, effective, as well as unobtrusive apply in foodborne investigation.

Zhaolui et al. (2017) used a modest, fast, as well as sensitive detection of *E. coli* according to urease-catalyzed signal amplification and antimicrobial peptide (AMP) functionalized magnetic nanoparticles (MNP). It was revealed that AMP functionalized MNP capture *E. coli* by means of high effectiveness through hydrophobic as well as electrostatic interfaces. So, as the microbes are netted by means of MNP-AMP, the required situates for urease on the exterior of AMP functionalized MNP, which are blocked with a resulting in the huge quantity of urease in the supernatant after attractive separation. In addition, this novel podium was effectively useful to discover *E. coli* in ground beef and pointed apple nectar samples, which can detect between 84 and 233 CFU/mL, respectively.

Neha et al. (2017) provided a general idea of the application of diverse nanomaterials using fluorescent in the improvement of nanobiosensors through distinct highlighting fundamental sensitivity, revealing scheme, capability of multiplexed analysis and specificity while polymerase chain reaction (PCR) based molecular and immunologic techniques lack specificity, costly, and might produce false results. They also established fluorescent nanobiosensors, which demonstrate alternative tools for the routine and rapid detection of foodborne microorganisms.

Fang and Ramasamy (2015) observed that in order to reduce the infection-induced destruction in yields all through growth, postharvest, harvest processing, safeguarding agricultural sustainability, maximize efficiency, and prevention in crops are imperative and advanced disease detection. It was revealed by research laboratory based methods that include immunofluorescence, enzyme-linked immunosorbent assay, PCR, gas chromatography–mass spectrometry, fluorescence *in-situ* hybridization, as well as flow cytometry (some of the direct detection methods), while incidental techniques include fluorescence imaging, hyperspectral techniques, as well as thermography. Similarly, they make available a complete impression of biosensors established on exceedingly discriminating biorecognition elements like antibody, enzyme, bacteriophage, and RNA/DNA as a novel device for the primary documentation of farm product infections.

Jia et al. (2019) studied immunofiltration band technique centered on the photothermal influence of gold nanoparticles and aimed at the rapid, simple low-cost, sensitive, as well as handheld revealing of *E. coli*. Nevertheless, the band analyze is usually sensitive and small as compared to other immunological methods. They used photothermal effect of gold nanoparticles to obtain the intensification of bands to advance the sensitivity of the strip technique. It was noted that thermal differences caused were through the photothermal effect, which were relative to the microbes' yield as well as concentrations in virtually logarithmic relation. When subjected to optimum conditions, it has detection strength up to about 1.95×10^4 CFU/mL, where the sensitivity was enhanced approximately ten times than that of the conventional optical strip technique. They opined that photothermal-based immunofiltration band might be utilized for the sensitive and rapid revealing of additional pathogens as a pronounced prospective for food quality device practice.

Umesha and Manukumar (2018) reported that *Bacillus cereus, S. aureus, Proteus vulgaris, E. coli, Listeria monocytogenes, Campylobacter, Aspergillus spp., Penicillium spp., Fusarium spp., Salmonella spp.,* and some of the other pathogens are discovered in contaminated food substances that cause continuously sicknesses in human in someone otherwise the other system. They used a short period of time to identify foodborne pathogens in food technology. Some level of pollution of food through main pathogens entails specific sensitive revealing stages, where the present work is making way forward to novel nanomolecular systems for nanomaterials, making them appropriate for the improvement of analyses with portability, response time, as well as high sensitivity.

Cho and Ku (2017) explored present analytical skills, which include magnetic and metallic nanomaterials, electrochemistry, spectroscopy, as well as optics. They based their findings on the existing techniques on the early pathogens detection by means of improved analytical specificity and sensitivity. However, in order to acquaint with the prospective comparative and combination investigation of numerous advanced techniques, they also made reference to a new specimen preparation procedure that usages recovery technologies and microbial concentration. This technology has the prospective to advance the pre-enhancement stage that paves the way for the detection process. The improvement of high-tech along with novel solutions for accurate, nonlaborious, and rapid bacterial detection techniques is imperative to increase the supplying of food globally.

Yang and Bashir (2008) as an opinion of transduction, the method of impedance has been used in the area of microbiology as a way to discover or enumerate foodborne pathogenic microorganisms. The incorporation of impedance by means of biological gratitude skill for bacteria detection has steered to the enlargement of impedance biosensors having extensive-spread practice. They were working on applications and progress of microbiology impedance for foodborne pathogenic detection of bacteria, especially the novel area, which has been added to this subject in the past years, comprising the application of interdigitated microelectrodes, the use of corresponding circuits for investigation of the impedance systems, and the improvement of chip-based impedance microbiology. The importance of improvements of impedance biosensors for detection of bacteria in the past 5 years has been discussed, focusing on microfluidic-based Faradaic electrochemical impedance biosensors and microfabricated microelectrodes-based, non-Faradaic impedance biosensors, and the incorporation of impedance biosensors through other methods, such as electropermeabilization and dielectrophoresis.

Recently, Luo et al. (2020) made the use of lateral flow strip assays, which have gained an extensive attention as a substitute technique for fast diagnosis of foodborne pathogens to the conventional culture-based analyzes, which includes PCR and plate counting. They noted some main drawbacks of the present lateral flow strip assays for the pathogenic detection of bacteria that is it is restrictive in its applied uses in POCT, which is a result of low sensitivity. However, it shows existing advances in lateral flow strip assays for the pathogenic revealing of bacteria in terms of analyte types and chromatic agents. In addition, current methods for signal modifications and enhancement of the lateral movement band analyze architecture for complex of pathogenic revealing of bacteria were also discussed.

Based on the potential dangers related through the globalization of the food production, attention has been shifted to investigation on the improvement of novel sensing methods to make available the means of pollution revealing at various steps in the supply of food chain. It was revealed that the modern developments in platforms have the highest prospective for low cost, identification, and real-time detection of foodborne pathogens. There is recent shift in emphasis on the use nano- and micro-scale technology in this regard (Carlson et al., 2018).

S. aureus is part of the most usually isolated foodborne pathogen and it is considered as a major cause of foodborne diseases globally. Nevertheless, several of these techniques are time-consuming and laborious and are not appropriate for on-site uses. So, it is extremely significant to advance fast and more approachable detection approaches. Rubab et al. (2018) emphasized the state-of-the-art biosensor techniques toward the quantification and detection of *S. aureus*.

Bonah et al. (2019) observed that foodborne diseases caused through foodborne bacterial pathogens often happen, as a result, endangering the health and safety of human. Influences such as culturing, pretreatments, intensification make the traditional routine identification, enrichment, and account of large numbers of bacteria in a complex bacterial consortium complex, time-consuming, as well as expensive. They evaluated certain essential information about hyperspectral imaging, including data acquisition, instrumentation, data analysis, and image processing—the existing application of hyperspectral imaging for the classification, discrimination, and detection of numerous foodborne pathogens.

Sharma et al. (2015) noted that current developments in nanobiotechnology show an important role possible solutions for developing affordable detection and highly sensitive tools for food examination. Nanomaterials-based aptasensors hold great prospective to overwhelming drawbacks of conservative analytical procedures. Aptamers consist of a new class of extremely precise biorecognition elements that are shaped through the efficient evolution of ligands through exponential enhancement process. Several nanomaterials including metal nanoclusters, metal nanoparticles, carbon and metal quantum dots, metal oxide nanoparticles, carbon nanotubes, nanocomposites, and graphene enable higher sensitivity through signal intensification and introduce numerous new transduction schemes, such as fluorescence, improved chemiluminescence, electrochemical signals, Raman signals, super-paramagnetic properties, as well as enhanced catalytic activity to the biosensor.

Sangeeta et al. (2019) opined that there are shortcomings related to the conventional procedures, such as the requirement of costly sophisticated devices and multistep sample processing, which limit their applications at laboratory level. To address these restrictions related through the traditional approaches, some novel techniques have been established by means of several study groups. Recent improvements in nanomaterial-based detecting of histamine in diverse food products have exposed important measurement as a result of high sensitivity, specificity, ease of operation, field deployability as well as cost. Therefore, the improvement of nanomaterials-based histamine detecting strategies/analyzes is there, where the revealing is based on optical and electrochemical. Table 7.1 indicates the various nanomaterials and their applications in the food sector.

7.6 CONCLUSION AND FUTURE RECOMMENDATIONS

Here, detailed information has been provided on several nanomaterials and their wider applications in the food industry. There is a need to provide some facts on numerous biomolecules, application of nanotechnology, and their incorporation into the system devices for adequate improvement as well as higher sensitivity for effective transmission of biological signal information

TABLE 7.1 Application of Various Nanomaterials and Their Applications in Food Sector.

Nanomaterial	Type of microorganism	References
Nanosensor gold based nanoparticles	Diagnostic agent for various pathogenic diseases, urea, and pesticide	Ulatowska-Jarża et al. (2011)
Cu, Ag, and Fe oxide NPs	Antimicrobial property	Seil and Websters (2012)
Silver NPs	Activity against *Salmonella typhi, Microcystis aeruginosa*	Lima et al. (2013)
Ag NPs	Gram-negative bacteria	Yu et al. (2014)
ZnO NPs	Gram-negative and Gram bacteria	Azam et al. (2012)
Zinc nanoparticles	*Bacillus subtilis, E. coli, Pseudomonas aeruginosa*	Chitra and Annadurai (2013)
Zinc oxide nanoparticles	*S. aureus*	McCarthy et al. (1992)
Silver nanoparticles	Gram-negative bacteria	Morones et al. (2005)
Selenium NPs	*S. aureus, Klebsiella spp, E. coli*	Singh et al. (2014)
MgO (magnesium oxide)	Soilborne fungal	Chen et al. (2020)
Ag NPs	Antifungal potential	Elgorban et al. (2016)

for adequate diagnoses of these pathogenic food spoilage microorganism. Also, there is an urgent need to provide a comprehensive report on the health implication of these nanomaterials on the gut and oral microbiota. There is also a need to provide adequate up-to-date legislation regarding these materials, while there is a need for policymakers, scientists, and government to sensitize health of the people benefits of nanomaterials most especially in the food industry.

KEYWORDS

- **foodborne diseases**
- **packaging**
- **food control**
- **nanotechnology**
- **nanomaterials**
- **antimicrobial agents**
- **consumer**

REFERENCES

Abd-Elsalam, K. A. Nanoplatforms for Plant Pathogenic Fungi Management. *Fungal Genom. Biol.* **2012,** *100*, 829–834.

Alghuthaymi, M. A.; Almoammar, H.; Rai, M.; Said-Galiev, E.; Abd-Elsalam, K. A. Myconanoparticles: Synthesis and Their Role in Phytopathogens Management. *J. Biotechnol. Biotechnol. Equip.* **2015,** *29* (2), 221–236.

Álvarez, S. P, López, N. E. L.; Lozano, J. M.; Negrete, E. A. R.; Cervantes, M. E. S. Plant Fungal Disease Management Using Nanobiotechnology as a Tool. In *Advances and Applications through Fungal Nanobiotechnology, Fungal Biology*; Prasad, R.; Ed.; Springer: Switzerland, 2016; pp 169–192.

Atiq, M.; Naeem, I.; Sahi, S. T.; Qayyum, A. Nanoparticles: A Safe Way towards Fungal Diseases. Arch. Phytopathol. *Plant Protection* **2020,** *53* (3). DOI: 10.1080/03235408.2020.1792599.

Azam, A.; Ahmed, A. S.; Oves, M.; Khan, M. S.; Habib, S. S.; Memic, A. Antimicrobial Activity of Metal Oxide Nanoparticles against Gram-Positive and Gram-Negative Bacteria: A Comparative Study. *Int. J. Nanomed.* **2012,** *7*, 6003–6009.

Bonah, E.; Huang, X.; Aheto, J. H.; Osae, R. Application of Hyperspectral Imaging as a Nondestructive Technique for Foodborne Pathogen Detection and Characterization. *Foodborne Pathog. Dis.* **2019,** *16* (10), 712–722.

Brigger, I.; Dubernet, C.; Patrick, C. Nanoparticles in Cancer Theraphy and Diagnosis. *Adv. Drug Deliv. Rev.* **2012,** *64* (12), 24–36.

Cao, J.; Guenther, R. H.; Sit, T. L.; Lommel, S. A.; Opperman, C. H.; Willoughby, J. A. Development of Abamectin Loaded Plant Virus Nanoparticles for Efficacious Plant Parasitic Nematode Control. *CS Appl. Mater. Interf.* **2015,** *7*, 9546–9553.

Carlson, K.; Misra, M.; Mohanty, S. Developments in Micro- and Nanotechnology for Foodborne Pathogen Detection. *Foodborne Pathog. Dis.* **2018,** *15* (1), 16–25.

Chen, J.; Wu, L.; Mei, L.; Shasha, L.; Ziyan, L.; Wei, D. Comparative Study on the Fungicidal Activity of Metallic MgO Nanoparticles and Macroscale MgO against Soilborne Fungal Phytopathogens. *Front. Microbiol.* **2020**. DOI: 10.3389/fmicb.2020.00365

Chitra, K.; Annadurai, G. Antimicrobial Activity of Wet Chemically Engineered Spherical Shaped ZnO Nanoparticles on Food Borne Pathogen. *Int. Food. Res. J.* **2013,** *20* (1), 59–64.

Cho, I. H.; Ku, S. Current Technical Approaches for the Early Detection of Foodborne Pathogens: Challenges and Opportunities. *Int. J. Mol. Sci.* **2017,** *30, 18* (10), 2078–2097.

Dakrong, P.; Camilla, G.; Skorn, M.; Michael, B. Cortie. Single and Multiple Detections of Foodborne Pathogens by Gold Nanoparticle Assays. *WIREs Nanomed. Nanobiotechnol.* **2019,** *12*, DOI: 10.1002/wnan.1584.

Dastjerdi, R.; Montazer, M. Colloids and Surfaces B: Biointerfaces: A Review on the Application of Inorganic Nano-Structured Materials in the Modification of Textiles: Focus on Anti-Microbial Properties. *Colloids Surf. B: Biointerf.* **2010,** *79* (1), 5–18.

Deepa, S.; Kanimozhi, K.; Panneerselvam, A. Antimicrobial Activity of Extracellularly Synthesized Silver Nanoparticles from Marine Derived Actinomycetes. *Int. J. Curr. Microbiol. App. Sci.* **2013,** *2* (9), 223–230.

El Sheekh, M. M.; El Kassas, H. Y. Application of Biosynthesized Silver Nanoparticles against a Cancer Promoter Cyanobacterium, Microcystis Aeruginosa. *Asian Pac. J. Cancer Prev.* **2014,** *15* (16), 6773–6779.

Elgorban, A. M.; El-Samawaty, A. E. R. M.; Yassin, M. A.; Shaban, R. S.; Syed, F. A.; Khaled, M. E.; Marwa, B.; Khan, M. Antifungal Silver Nanoparticles: Synthesis, Characterization and Biological Evaluation. *Biotechnol. Biotechnol. Equip.* **2016,** *30* (1), 56–62.

Fakruddin, M. D.; Hossain, Z.; Afroz, H. Prospects and Applications of Nanobiotechnology: A Medical Perspective. *J. Nanobiotechnol.* **2012,** *10*. DOI: 10.1186/1477-3155-10-31.

Fang, Y.; Ramasamy, R. P. Current and Prospective Methods for Plant Disease Etection. *Biosensors*, **2015,** *5* (3), 537–561.

Hang, X.; Liu, Z.; Shen, W.; Gurunathan, S. Silver Nanoparticles: Synthesis, Characterization, Properties, Applications, and Therapeutic Approaches. *Int. J. Mol. Sci.;* **2016,** *17* (9), 1534–1555.

Ho, K.; Wang, X. U.; Nie, S.; Shin, D. M. Therapeutic Nanoparticles for Drug Delivery in Cancer. *Clin. Cancer Res.* **2008,** *14* (5), 1310–1316.

Horky, P.; Skalickova, S.; Baholet, D.; Skladanka, J. Nanoparticles as a Solution for Eliminating the Risk of Mycotoxins Nanomaterials (Basel). **2018,** *8* (9). DOI: 10.3390/nano8090727.

Jia, M.; Liu, J.; Zhang, J.; Zhang, H. An Immunofiltration Strip Method Based on the Photothermal Effect of Gold Nanoparticles for the Detection of *Escherichia coli* O157:H7. *Analyst*. **2019,** *14, 144* (2), 573–578.

Kim, J. S.; Kuk, E.; Yu, K. N.; Kim, J. H.; Sung, J. P.; Par, J. et al. Antimicrobial Effects of Silver Nanoparticles. *Nanomedicine* **2007,** *3* (1), 95–101.

Kumar, H.; Kuca, K.; Bhatia, S. K.; Saini, K.; Kaushal, A.; Verma, R. et al. Applications of Nanotechnology in Sensor-Based Detection of Foodborne Pathogens. *Sensors*, **2020,** *20* (7). DOI: 10.3390/s20071966

Lima, E.; Guerra, R.; Lara, V.; Guzmán, A. Gold Nanoparticles as Efficient Antimicrobial Agents for *Escherichia coli* and *Salmonella. Typhi. Chem. Cent. J.* **2013,** *7*. DOI: 10.1186/1752-153X-7-11.

Liu, Z.; Zhang, X.; Wei, S.; Sangiliyandi, G. Silver Nanoparticles: Synthesis, Characterization, Properties, Applications, and Therapeutic Approaches. *Int. J. Molecul. Sci.* **2017,** *17* (9), 1534–1552.

Luo, K.; Kim, H.Y.; Oh, M. H.; Kim, Y. R. Paper-Based Lateral Flow Strip Assay for the Detection of Foodborne Pathogens: Principles, Applications, Technological Challenges and Opportunities. *Crit. Rev. Food Sci. Nutr.* **2020,** *60*, 157–170.

Marwal, A.; Gaur, R. K. Nanophytovirology: An Emerging Field for Disease Management, Plant Diseases—Current Threats and Management Trends, Snježana Topolovec-Pintarić. *IntechOpen*, **2019**. DOI: 10.5772/intechopen.86653.

McCarthy, T. J.; Zeelie, J. J.; Krause, D. J. The Antimicrobial Action of Zinc Ion/Antioxidant Combinations. *J. Clin. Pharm. Ther.* **1992,** *17* (1), 51–54.

Mohanraj, V. J.; Chen, Y.; Nanoparticles—A Review. *Trop. J. Pharm. Res.* **2006,** *5* (1), 561–573.

Mohanta, Y. K.; Panda, S. K.; Bastia, A. K.; Mohanta, T. K. Biosynthesis of Silver Nanoparticles from Protiumserratum and Investigation of Their Potential Impacts on Food Safety and Control. *Front. Microbiol.* **2017,** *6*, 626–634.

Morais, M. G. D.; Martins, V. G.; Steffens, D.; Pranke, P.; Costa1, J. A. V. D. Biological Applications of Nanobiotechnology. *J. Nanosci. Nanotechnol*, **2014,** *14*, 1007–1017.

Morones, J. R.; Elechiguerra, J. L.; Camacho, A.; Holt, K.; Kouri, J. B. et al. The Bactericidal Effect of Silver Nanoparticles. *Nanotechnology* **2005,** *16* (10), 2346–2353.

Mujeebur, R. K.; Tanveer, F. R. Nanotechnology: Scope and Application in Plant Disease Management. *Plant Pathol. J.* **2014,** *14* (13), 214–231.

Nair, S.; Sasidharan, A.; Divya Rani, V. V.; Menon, D.; Nair, S.; Manzoor, K. et al. Role of Size Scale of ZnO Nanoparticles and Microparticles on Toxicity toward Bacteria and Osteoblast Cancer Cells. *J. Mater. Sci. Mater. Med.* **2009,** *20* (Suppl 1), S235–S241.

Neha, B.; Sanjeev, B.; Manoj, K. N. Akash, D. Fluorescent Nanobiosensors for the Targeted Detection of Foodborne Bacteria. *TrAC Trend. Anal. Chem.* **2017,** *97*. DOI: 10.1016/j.trac.2017.09.010

Nile, S. H.; Baskar, V.; Selvaraj, D.; Selvaraj, D.; Nile, A.; Xiao, J.et al. Nanotechnologies in Food Science: Applications, Recent Trends, and Future Perspectives. *Nano-Micro Lett.* **2020,** *12*, 45. DOI: 10.1007/s40820-020-0383-9.

Ouda, S. M. Antifungal Activity of Silver and Copper Nanoparticles on Two Plant Pathogens, *Alternaria alternata* and *Botrytis cinerea*. *Res. J. Microbiol.* **2014,** *5* (9), 34–42.

Patel, N.; Desai, P.; Patel, N.; Jha, A.; Gautam, H. K. Agronanotechnology for Plant Fungal Disease Management: A Review *Int. J. Curr. Microbiol. App. Sci* **2014,** *3* (10), 71–84.

Pengju, G.; Luo, L.; Stutzenberger, F. J. Nanotechnology in the Detection and Control of Microorganisms. *Adv. Appl. Microbiol.;* **2008,** *63*, DOI: 10.1016/S0065-2164(07)00004-4

Pérez-de-Luque, A.; Rubiales, D. Nanotechnology for Parasitic Plant Control. *Pest. Manage. Sci.* **2009,** *65* (5), s540–s545.

Polgoma, C.; Rajasundari, K.; Ilamurugu, K. Nanotechnology and Its Applications in Medical Diagnosis. *J. Basic Appl. Chem.* **2011,** *1* (2), 26–32.

Qasim, M.; Lim, D. J.; Park, H.; Na, D. Nanotechnology for Diagnosis and Treatment of Infectious Diseases. *J. Nanosci. Nanotechnol.* **2014,** *14* (10), 7361–8128.

Rai, M.; Kratosova, G. Management of Phytopathogens by Application of Green Nanobiotechnology: Emerging Trends and Challenges. *J. Agri. Sci.* **2015,** *15* (66), 15–22.

Ramos, A. P.; Cruz, M. A. E.; Tovani, C. B.; Ciancaglini, P. Biomedical Applications of Nanotechnology. *Biophys. Rev.* **2017,** *9*, 79–89.

Rubab, M.; Shahbaz, H.M.; Olaimat, A.N.; Oh, D.H. Biosensors for Rapid and Sensitive Detection of *Staphylococcus aureus* in Food. *Biosens. Bioelectron.* 2018, *15* (105), 49–57.

Ruyuan, Z.; Tarun, B.; Li, L.; Xingyu, L.; Yangun, L.; Zisheng, L. Nanomaterial Based Biosensors for Sensing Key Foodborne Pathogens: Advances from Recent Decades. 2020, *19* (4), 1465–1487.

Sangeeta, Y.; Sheethal, S. N.; Sai, V. V. R.; Jitendra, S. Nanomaterials Based Optical and Electrochemical Sensing of Histamine: Progress and Perspectives. *Food Res. Int.* **2019,** *119*, 99–109.

Seil, T. S.; Websters, T. J. Antimicrobial Applications of Nanotechnology: Methods and Literature. *Int. J. Nanomed*, **2012,** *7*, 2767–2781.

Sharma, R. Ragavan, K. V.; Thakur, M. S.; Raghavarao, K. S. Recent Advances in Nanoparticle Based Aptasensors for Food Contaminants. *Biosens. Bioelectron.*; **2015,** *15* (74), 612–627.

Sharma, P.; Adikshita, S.; Monica, S.; Nikhil, B.; Pedro, E.; Jain, A. et al. Nanomaterial Fungicides: In *Vitro and In Vivo Antimycotic Activity of Cobalt and Nickel Nanoferrites on Phytopathogenic Fungi. Global Challenges* **2017,** *1*, 9,. DOI: 10.1002/gch2.201700041.

Shrivastava, S.; Dash, D. Applying Nanotechnology to Human Health: Revolution in Biomedical Sciences. **2009**. DOI: 10.1155/2009/184702.

Singh, J.; Kanchan, V.; Naleeni, R.; Padmaja, R.; Vivek, K. S.; Rohit, K. M. et al. Nanomaterials and Microbes' Interactions: A Contemporary Overview. 3 *Biotech*. **2019,** *9*. DOI: 10.1007/s13205-019-1576-0.

Singh, N.; Saha, P.; Rajkumar, K.; Abraham, J. Biosynthesis of Silver and Selenium Nanoparticles by Bacillus sp. JAPSK2 and Evaluation of Antimicrobial Activity. *Der. Pharm. Lett.* **2014,** *6* (1), 175–181.

Suresh, N.; *Digvir,* S. J. Nanotechnology for the Food and Bioprocessing Industries. *Food Bioprocess Technol.* ***2011,*** *4*, 39–47.

Syafiqah, S.; Mohamad, A.; Nur Ain Atiqah, M. A.; Abdullah SharafSaleh Al-Hammadia, C. Organic and Inorganic Antibacterial Approaches in Combating Bacterial Infection for Biomedical Application. *Mater. Sci. Eng.: C* **2021,** *118*. DOI: 10.1016/j.msec.2020.111382

Tang, Z. X.; Lv, B. F. MgO Nanoparticles as Antibacterial Agent: Preparation and Activity. *Braz. J. Chem. Eng*. **2014,** *31* (3), 591–601.

Tang, Z. X.; Lv, B-F. MgO Nanoparticles as Antibacterial Agent: Preparation and Activity. *Braz. J. Chem. Eng*. **2014,** *31* (3), 591–601.

Ulatowska-Jarża, A.; Pucińska, J.; Wysocka-Król, K.; Hołowacz, I.; Podbielska, H. Nanotechnology for Biomedical Applications—Enhancement of Photodynamic Activity by Nanomaterials. *Bull. Polish Acad. Sci. Tech. Sci.* **2011,** *59* (3), 253–261.

Umesha, S.; Manukumar, H.M. Advanced Molecular Diagnostic Techniques for Detection of Food-Borne Pathogens: Current Applications and Future Challenges. *Crit. Rev. Food. Sci. Nutr*. **2018,** *58* (1): 84–104.

Valavanidis, A.; Vlachogianni, T. Engineered Nanomaterials for Pharmaceutical and Biomedical Products New Trends, Benefits and Opportunities. *J. Pharma. Rep.* **2016,** *4* (1), 105, 13–24.

Vargas-Hernandez, M.; Israel, M-B.; Guevara-Gonzalez, R. G.; Rico-Garcia, E; Ocampo-Velazquez, R. V.; Avila-Juarez, L.; Torres-Pacheco, I. Nanoparticles as Potential Antivirals in Agriculture. *Agri*. **2020,** *10*, 444. DOI: 10.3390/agriculture10100444.

Yang, L.; Bashir, R. Electrical/Electrochemical Impedance for Rapid Detection of Foodborne Pathogenic Bacteria. *Biotechnol. Adv.* **2008,** *26* (2), 135–150.

Young, M.; Willits, D.; Masaki, U.; Trevor, D. Plant Viruses as Biotemplates for Materials and Their Use in Nanotechnology. *Annu. Rev. Phytopathol.* **2008,** *46*, 361–384.

Yu, L.; Zhang, Y.; Zhang, B.; Liu, J. Enhanced Antibacterial Activity of Silver Nanoparticles/Halloysite Nanotubes/Graphene Nanocomposites with Sandwich-Like Structure. *Sci. Rep.* **2014,** *4*. DOI: 10.1038/srep04551.

Zeelie, J. J.; McCarthy, T. J. Effects of Copper and Zinc Ions on the Germicidal Properties of Two Popular Pharmaceutical Antiseptic Agents Cetylpyridinium Chloride and Povidone-Iodine. *Analyst* **1998,** *123* (3), 503–507.

Zhang, L.; Pornpattananangku, D.; Hu, C. M.; Huang, C. M.; Development of Nanoparticles for Antimicrobial Drug Delivery. *Curr. Med. Chem.* **2010,** *17* (6), 585–594.

Zhaolui, Q.; Chunyang, L.; Yingchun, F.; Yanbin, L. Rapid and Sensitive Detection of *E. coli* O157:H7 Based on Antimicrobial Peptide Functionalized Magnetic Nanoparticles and *urease-catalyzed signal amplification. Anal. Methods* **2017,** *39*, 5204–5210.

CHAPTER 8

Use of Nanomaterials in Wastewater Treatment

RAKSHIT AMETA[1*], JAYESH P. BHATT[2], M. N. AVADHANULU[3], MONIKA JANGID[2], GAJANAN KOTHARU[3], and SHUBANG VYAS[2]

[1]*Department of Chemistry, J. R. N. Rajasthan Vidhyapeeth (Deemed to be University), Udaipur 313001, India*

[2]*Department of Chemistry, PAHER University, Udaipur 313001, India*

[3]*Department of Physics, Kavikulguru Institute of Technology and Science, Ramtek 441106, India*

[*]*Corresponding author. E-mail: rakshit_ameta@yahoo.in*

ABSTRACT

Water scarcity has already become a challenge for all of us with rapid population growth and industrialization. It has increased the demand for green chemical water treatment techniques so as to get clean water in sufficient amount and that too in a sustainable manner, which is affordable and reliable. Some efficient as well as low cost water purification methods are required so as to fulfill the demand of potable water for living beings in the future. The fast deteriorating quality of water and ever-increasing global demand for this less-available-but-most-desired material has reached an alarming situation and there is an urgent demand to search for better purification materials or techniques. This chapter includes the main types of water pollutants and the existing challenges of their effective removal from waste water with a particular focus on the use of nanomaterials such as photocatalyst, carbon-based materials, and composites.

The Science of Nanomaterials: Basics and Applications. Suresh C. Ameta & Rakshit Ameta (Eds.)

8.1 INTRODUCTION

Clean water is quite vital for life as it is required not only for drinking purpose but also for preparation of food and maintaining proper hygiene. Unfortunately, many countries of the world are facing a severe scarcity of water. Scientists are working hard to develop different technologies to decontaminate and purify water. Nanomaterials-based technology offers a low-cost and effective technology for purification and decontamination of water (Figure 8.1). Nanomaterials have remarkable antimicrobial activity apart from their ability to remove different (organic and inorganic) pollutants from water effectively. They also have the ability to adsorb toxic metals and organic contaminants from polluted water. Use of nanoparticles (NPs) has certain problems such as aggregation, seepage into drinking water, and also some adverse effects on human health. Nowadays, nanocomposites are being developed to overcome these problems by combining two or more compounds. Time is not far off when the use of nanomaterials for water purification will not remain a dream but becomes a reality in the coming years.

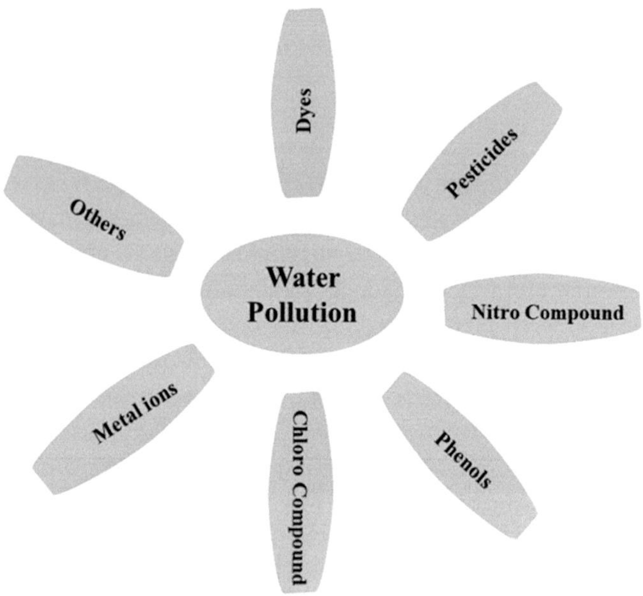

FIGURE 8.1 Nanomaterials and water pollution.

Water pollutants have major toxic effects on the complete living systems including aquatic and aerial flora and fauna. Newly emerging micro /

nanopollutants increase global warming and as a result, climate changes are posing major threats to the availability of fresh water for agriculture, household, and industries. Therefore, an appropriate water treatment technology is required, which will desalinate water to a desired extent apart from removing macro-, micro-, and nanopollutants. Antifouling and self-cleaning functions of CNT membranes with low energy consumption make them excellent for water purification over the conventional techniques.

2D nanoporous material membranes are considered suitable for the elimination of pollutant from waste water purification. This is all due to their thin structure, large surface area, and mechanical strength. Recently, some more 2D materials have been developed such as 2D boron nitride, molybdenum disulfide, tungsten chalcogenides, titanium carbide, and so on.

8.2 DYES

Discharge of dyeing, textile, and other industry waste is a growing concern from environment point of view. Effluents of dyeing, printing, and textile industries throw a large amount of dyes in water, which adversely affects its physical, chemical, and biological properties. As a result, nearby agricultural fields are deteriorated, when such polluted water is directly reach to these agricultural lands.

Majority of organic pollutants are present in the industrial and domestic waste water effluents, which should be removed before discharging these into the water bodies. Such organic pollutants are dominant in surface water and groundwater. It may be hazardous to human as well as aquatic lives. Nanotechnology played a major role in degrading such pollutants.

Mahmoodi and Arami (2006) immobilized titanium (IV) oxide NPs to degrade acid red 14 (AR 14) photocatalytically. They investigated effects of operational parameters such as H_2O_2 and dye concentrations, anions (NO_3^-, Cl^-, SO_4^{2-}, HCO_3^-, and CO_3^{2-}), and pH. They could detect acetate and oxalate anions as dominant aliphatic intermediates, which ultimately slowly oxidized to CO_2.

The photodegradation of methyl orange (MO) was investigated by Dai et al. (2007) in an aqueous suspension containing mesostructured titania NPs (m-TiO_2) under ultraviolet (UV) irradiation. It was observed that 98% MO can be mineralized in 1.0 g L^{-1} m-TiO_2 suspension, pH 2.0 on 45 min illumination. It was reported that photodegradation of the quinonoid form of MO was involved as the dominated process.

Anandan et al. (2008) degraded acid red 88 (AR88) in presence of Ag–TiO_2 NPs and visible (vis) light. Some oxidative reagents were also added to the photocatalytic system, such as hydrogen peroxide, peroxomonosulfate (PMS), and peroxodisulfate. These may serve as an electron acceptor and resulted in enhanced degradation of acid red 88, which is almost sevenfold higher as compared to PMS as an oxidant. It was observed that 65% mineralization could be obtained in 7 h in this case. The present system was found to be relatively inexpensive, reproducible, stable, and efficient.

Bokare et al. (2008) used Fe-Ni bimetallic NPs for degradation of orange G in aqueous solutions. Complete dye degradation was observed after 10 min with nanocatalyst loading of 3 g L^{-1}. The efficiency was found to increase on decreasing pH and increasing Fe-Ni concentration. It was also found to decrease on increasing the dye concentration.

The sonolytic degradation of basic blue 41 in aqueous solution was performed by Abbasi and Asl (2008) using ultrasonic power of 160 W within 180 min. The dye removal rate was decreased at lower pH values. But it increased with the increase in H_2O_2 concentration and decreasing initial dye concentration. During the decolorization, it was revealed that all nitrogen atoms and aromatic groups of basic blue 41 were transformed to urea, nitrate, formic acid, acetic acid, oxalic acid, etc.

Undoped ZnO and manganese-doped ZnO photocatalysts were synthesized by Ullah and Dutta (2008) via wet-chemical technique. The photoreduction activities of these photocatalysts were evaluated using methylene blue (MB) as organic contaminant in presence of vis light. It was reported that manganese-doped ZnO ($ZnO:Mn^{2+}$) degrades MB much faster as compared with undoped ZnO on exposure to vis light.

Zhang et al. (2008) embedded ZnO nanorods in TiO_2 nanotube arrays (ZnO/TiO_2 NR/Ts) and then electrodes were prepared in two steps:

- TiO_2 nanotube arrays (TiO_2 NTs) were electrosynthesized by anodization method in HF solution
- Then ZnO (cathodic electrodeposition) was embedded in the TiO_2 nanotube arrays

Photocatalytic and photoelectrocatalytic activity of ZnO/TiO_2 NR/Ts electrode in degradation of MO was also evaluated by them.

A series of polyaniline/TiO_2 (PANI/TiO_2) composites were prepared by Salem et al. (2009) using aniline and TiO_2 under various polymerization conditions. It was found that there was an increase in activity on increasing concentration of aniline and persulfate, but a decreasing trend with increasing concentration of sodium dodecyl sulfate, HCl, and amount of TiO_2. The activity of these composites followed the order in the presence of different acids used:

$$H_2SO_4 > H_3PO_4 > HCl > HNO_3$$

It was reported that rate of interaction of composite/dye was increased with increasing UV-irradiation, composite loads, as well as temperature, but it was adversely affected with the increasing dye concentration and pH.

Pouretedal et al. (2009) prepared NPs of undoped zinc sulfide and ZnS doped with copper, nickel, and manganese (photocatalyst) and used in the photodegradation of MB and safranin. It was reported that maximum degradation efficiency could be achieved in the presence of $Zn_{0.98}Mn_{0.02}S$, $Zn_{0.94}Ni_{0.06}S$, and $Zn_{0.90}Cu_{0.10}S$ as nanophotocatalysts. It was observed that 150.0 mg L^{-1} was the optimum dose of photocatalyst. The degradation efficiency decreases for higher concentrations of dye above 5.0 mg L^{-1} for MB and safranin dyes. The 87.3–95.6 and 85.4–93.2 degradation efficiency was obtained for MB and safranin, respectively.

The photocatalytic degradation of acid orange 10 (AO10), acid orange 12 (AO12), and acid orange 8 (AO8) was investigated by Khataee et al. (2009) using glass supported TiO_2 photocatalyst under UV light irradiation. It was indicated that the photocatalytic decolorization was in the order:

$$AO10 > AO12 > AO8$$

It was found that dyes were completely decolorized and their average overall TOC removal was more than 94% in 6 h.

CeO_2 nanocrystals were synthesized by Pouretedal and Kadkhodaie (2010) via simple precipitation method and calcination at 600°C. They used $(NH_4)_2Ce(NO_3)_6$ and ammonia as precursors. The photodegradation of MB by CeO_2 NPs was studied under both UV and sunlight irradiation. The highest degradation was achieved with 1.0 g L^{-1} CeO_2 at pH 11 within 125 min. It was indicated that photogenerated holes played a role in the degradation of the dye.

Biosynthesis of silver NPs was reported by Kumar et al. (2013) using *Ulva lactuca* (sea weed). These were used for photocatalytic degradation of MO. It was indicated that the presence of bioactive functional groups acted as capping and stabilizing agents. The silver NPs were found to be spherical (48.59 nm). The presence of silver ions was considered responsible for spherical shape of Ag NPs.

Assembled coatings of Co_3O_4 NPs were prepared by Warang et al. (2013) via physical as well as chemical methods. Degradation of MB (model system) was observed to compare the catalytic activities of Co_3O_4 heterogeneous catalyst coatings. The NPs prepared by pulsed laser deposition process were having smaller average size (18 nm), perfect spherical shape, narrow size distribution, mixed amorphous-nanocrystalline phase, and low degree of agglomeration. The coating synthesized in this manner exhibited the best photodegradation of methylene in presence of hydrogen peroxide (4 h) or oxone (8 min).

Highly crystalline ZnO NPs (ZnO NPs) were synthesized by Rahman et al. (2013). They used zinc acetate as precursor. As-synthesized ZnO NP was found to possess monodispersity with the average size 20–30 nm. The photocatalytic activity of ZnO NPs was evaluated for the degradation of rhodamine-B dye on UV exposure. It was reported that about 95% rhodamine B dye was degraded in 70 min. Mahmoodi (2013) synthesized magnetic zinc ferrite ($ZnFe_2O_4$) NPs and used for photocatalytic degradation of reactive red 198 (RR198) and reactive red 120 (RR120). Different dyes could be degraded in wastewater using $ZnFe_2O_4$ (magnetic photocatalyst).

The photocatalytic degradation of MB dye was investigated by Ahmed et al. (2013) under UV irradiation over Fe_2O_3/TiO_2 NPs. It was revealed that the incorporation of Fe_2O_3 (up to 7%) was associated with a significant increase in surface area, reduction in particle size, stabilization of anatase phase, and shifting the photoexcitation response in vis region.

Ameen et al. (2013) used a simple solution method to synthesize ZnO-flowers. They used zinc acetate as precursor in basic medium and used for the degradation of crystal violet (CV) dye. It was observed that ZnO-flower was comprised of well-defined petals with the average length of ~300 ± 50 nm. As-synthesized ZnO-flowers showed very rapid degradation of CV (~96%) within 80 min.

A rapid and high yield, but simple synthesis of Ce-doped ZnO NPs was reported by Kumar et al. (2015) using solution combustion method. Several concentrations of Ce (0.5, 1.39, 2.55, 3.28, 3.71, and 4.14%) were used for doping of ZnO. It was revealed that the prepared Ce-doped ZnO NPs were well-crystalline having good optical properties. As-prepared Ce-doped ZnO NPs were used for the photocatalytic degradation of direct red-23 (DR-23). The degradation percentage was found to increase with increasing concentration of Ce up to 3.28%; but it decreases on increasing Ce concentration further. It was found that as-prepared Ce-doped ZnO NPs exhibited remarkable photocatalytic degradation (almost 99.5%) in 70 min only.

Saikia et al. (2015) synthesized ZnO nanomaterials of different morphologies. As-prepared photocatalysts were evaluated for photodegradation of malachite green (MG) in aqueous solutions, which is self-sensitized. The effects of various parameters were also investigated on degradation efficiency such as pH, catalyst loading, dye concentration, and light source. It was reported that flower-shaped homocentric-pencil-like ZnO nanorod bundles, which were hydrothermally synthesized, exhibited highest photocatalytic activity in presence of sunlight. A comparative study was also carried out to determine photodegradation efficiency of different ZnO NPs (hydrothermally synthesized ZnO NPs and flower-shaped homocentric-pencil-like ZnO nanorod bundles). It was reported that semiconductor photocatalysis leads to degradation of organic pollutants (complete mineralization, and of course; sometimes partial also).

Flower-like hierarchical zinc oxide nanostructure were synthesized by Hussain et al. (2016) via coprecipitation method, while was hydrothermally functionalized with 8 nm Au and 15 nm Ag NPs. The composite exhibited a strong interaction between noble metal NPs and zinc oxide nanoflowers (NFs). It was reported that presence of Au NPs on ZnO surface significantly increased photocatalytic activity as compared to Ag–ZnO and pure ZnO in degradation of rhodamine B dye.

Raliya et al. (2017) observed degradation of MO in simulated wastewater using three different oxide nanomaterials. These were titanium dioxide (TiO_2), zinc oxide (ZnO), and graphene oxide (GO) as photocatalysts under vis light. Effects of different factors such as initial concentration of dye and size of NPs were monitored to achieve optimum conditions for dye removal. Then, nanocomposites of all these three materials (GO–TiO_2–ZnO) were also synthesized and tested for its photocatalytic activity.

Safardoust-Hojaghan and Salavati-Niasari (2017) prepared N-doped graphene quantum dots (N-GQDs)/titanium dioxide nanocomposites. This N-GQDs/TiO_2 was used as catalyst for photodegradation of MB and it was observed that presence of N-GQDs had an excellent effect on activity of TiO_2. It was also revealed that photodegradation efficiency of TiO_2 toward MB under UV light was increased from 40 to 85% by introducing N-GQDs.

A rapid, convenient, and one-pot microwave method was used by Hemraj-Benny et al. (2018) to synthesize Ru NPs supported onto unfunctionalized single-walled carbon nanotubes (SWCNTs). It was found that small ruthenium NPs of 2.0 ± 0.5 nm or 3.5 ± 0.5 nm were evenly distributed and stabilized onto SWCNTs support without agglomeration and it depends on the reaction temperature. It was revealed that SWCNT-Ru NP composites exhibited excellent catalytic properties in the degradation and decolorization of Congo red (CR) dye within minutes at ambient temperature.

The photocatalytic degradation of reactive blue 19 (RB-19) and reactive red 76 (RR-76) dyes (pollutant) has been studied by Helmy et al. (2018) using TiO_2, C-doped TiO_2(C-TiO_2), S-doped TiO_2 (S-TiO_2), and C,S co-doped TiO_2 (C,S-TiO_2) NPs as photocatalysts. All these NPs were synthesized by sol–gel process. The degradation of dyes was investigated with various parameters such as catalyst load, pH, dye concentration, shaking speed, and irradiation time so as to achieve optimum conditions. It was observed that when photocatalyst dose of 1.6 g L^{-1} was kept then 100% efficiency of RB-19 and RR-76 photocatalytic degradation could be achieved in 1 h under vis light. They also investigated the toxicity and biological activity of the treated and untreated wastewater on different marine aquatic organisms (*Vibrio parahaemolyticus,* rotifer, and artemia).

Chauhan et al. (2019) reported synthesis of zinc oxide NFs/GO composite. This composite was uniformly grown over silicon (Si) substrate. The ZnO nanoseeds were obtained using zinc acetate and isopropanol, while GO was prepared by Hummer's method. The nanocomposite was prepared by mixing GO and ZnO nanoseed solutions, and it was followed by vigorous ultrasonication and spin-coating over the Si substrate. These NF-like structures were grown from ZnO nanoseeds by putting the coated Si substrate in a solution containing Zn^{2+} environment. The photocatalytic degradation of MB was observed in presence of this nanocomposite under sunlight exposure. It was revealed that composite (GO/ZnO) showed ~33% faster decolorization of MB (20 mg L^{-1}) as compared to pure ZnO. This enhanced performance was due to the prevention of recombination of electron hole. The GO provided excess electrons for increasing the photocatalytic reaction by generating more hydroxyl radicals.

Cruz et al. (2019) synthesized magnetic cobalt nanoparticles (Co NPs) and used it for removing remazol golden yellow RNL (RGY) from textile wastewater. The CoNPs contained 60.38% cobalt. It was observed that Co NPs have size smaller than 10 nm with high agglomeration capacity and spherical morphology. The nanomaterial has a mesoporous structure with low specific surface area (15.70 m^2 g^{-1}) and a pore volume and pore diameter of 0.072 cm^3 g^{-1} and 3.64 nm, respectively. It was reported that Co NPs completely removed RGY in 30 min. It was revealed that the reductive degradation prevailed in acidic conditions but in neutral or basic conditions, two processes occur and these are reductive degradation and adsorption. The Co NPs were also tried with textile wastewater and the results showed that there was almost 88% discoloration, but only 32% decrease in chemical oxygen demand (COD).

Kapoor et al. (2019) investigated photocatalytic degradation of Evans blue using composite of barium chromate and multiwalled carbon nanotubes (MWCNTs). Here, MWCNTs will act as adsorbent of Evans blue, while barium chromate will act as photocatalyst. This composite (MWCNTs-$BaCrO_4$) is a promising material for treatment of wastewater.

Rupa et al. (2019) used coprecipitation method for synthesis of flower-shaped zinc oxide NF from sea buckthorn fruit (SBT). The ZnO/NF was further used to degrade cationic and anionic dyes, such as MG, CR, MB, and eosin Y in presecence of ultraviolet light. It was reported that as-prepared photocatalyst degraded ~99% of the MG, MB, CR, and EY dyes (concentration = 15 mg L^{-1}) within 70, 70, 80, and 90 min, respectively. The 5 mg L^{-1} SBT-ZnO/NF degraded almost 100% of these dyes within 23, 25, 28, and 30 min, respectively. It was indicated that SBT-ZnO/NFs is a low cost, nontoxic, rapid, and reusable photocatalyst.

Pathak et al. (2019) carried out photocatalytic degradation of bismarck brown. They used calcium molybdate–multiwalled carbon nanotube ($CaMoO_4$–MWCNT) composite for this purpose. It was reported that this composite showed better photocatalytic activity as compared with its individual components ($CaMoO_4$ and MWCNT). Rate affecting parameters were optimized as pH = 8.0; dye concentration = 2.0×10^{-4} M; amount of composite = 0.08 g; and light intensity = 40.0 mW cm^{-2}.

The photocatalytic degradation of fast green (FG) was studied by Jat et al. (2019) under vis light using SnO_2 quantum dots decorated TiO_2 nanospheres. This composite (SnO_2-TiO_2) was prepared hydrothermally and stannic chloride (hydrate) was used as precursor for SnO_2 quantum dots. This composite showed more photocatalytic activity for degradation of FG as compared to titania nanopowder. The effect of various parameters like pH, amount of catalyst, concentration of dye, and light intensity on the rate of degradation was also monitored to attain optimum conditions.

Gupta et al. (2019) explored the photocatalytic activity of GO and zinc oxide composite. These composites were prepared in three different ratios: GO:ZnO (1:1), (1:2), and (2:1) by mechanochemical method (grinding). GO was synthesized namely by Hummer's, modified Hummer's and Tour's methods. It was reported that composite of GO-ZnO (2:1) gave better results out of these three. The photocatalytic efficiencies of all three composites were evaluated selecting Victoria blue B as a model pollutant. It was found to be even better than simple zinc oxide. The composite can be reused again without any significant loss in activity for three cycles.

Leaves extract of *Citrofortunella microcarpa* (Calamondin) was used as reducing agent as well as capping agents by Rafique et al. (2020) for synthesis of copper oxide NPs (CuO NPs). The purity and spherical-like shape of CuO NPs (54–68 nm) was confirmed. As-synthesized CuO-NPs were used as photocatalyst for degradation of rhodamine B (RhB) dye in polluted water. About 98% degradation efficiency was observed under UV light exposure.

The Co and Co_3O_4 NPs were synthesized by Adekunle et al. (2020) using microwave and reductive chemical heating methods and then their morphology and photocatalytic properties were studied. They investigated photocatalytic degradation of murexide and erichromeblak-T (EBT) dye in wastewater in the presence of sunlight using these synthesized NPs. It was reported that chemically synthesized conanoparticles gave highest degradation (43.6%) of murexide at 25 mg loading and 40 min exposure, while Co_3O_4 NPs (microwave synthesized) gave the highest degradation (39.4%) at 10 mg loading and 40 min on sunlight exposure time.

The synthesis of Cu NPs by a native copper-resistant bacterial strain *Escherichia sp.* SINT7 was reported by Noman et al. (2020). The photocatalytic activity of the biogenic Cu NPs was evaluated for degradation of an azo dye and treatment of textile effluents. It was revealed that the biogenic Cu NPs were spherical with particle size ranging from 22.33 to 39 nm. It was reported that Cu NPs could decolorize CR (97.07%), MG (90.55%), direct blue-1 (88.42%), and reactive black-5 (83.61%) at concentration of 25 mg L^{-1} on 5 h of exposure to sunlight. However, at higher concentration of dye (100 mg L^{-1}), the degradation percentage was reduced to 83.90, 31.08, 62.32, and 76.84% for CR, MG, direct blue-1, and reactive black-5, respectively. Treatment of textile effluents with Cu NPs was found to reduce electrical conductivity, pH, turbidity, total dissolved solids, total suspended solids, hardness, sulfates, and chlorides as compared with the untreated samples.

Some more dyes were degraded/removed using nanomaterials and are reported in Table 8.1.

8.3 PESTICIDES

Pesticides are mostly used in commercial, agricultural, as well as residential applications. These are low cost and highly efficient. Their persistent nature, bioaccumulation, and chronic toxicity are useful for agriculture, but harmful for the environment. Therefore, there is an urgent need to remove them from water. Here, the application of nanomaterials has found a new pathway to solve the problem for the removal/degradation of pesticides.

TABLE 8.1 Degradation of Dyes by Nanomaterials.

S. no.	Pollutant	Nanomaterials	References
1	Violet GL2B azo dye	CaO and TiO_2	Madhusudhana et al. (2012)
2	Cationic dyes	Sn–SnO_2	Sinha et al. (2013)
3	Azo dyes	Au@Pd@Ru	Sahoo et al. (2015)
4	Methylene blue	SnO_2 NPs	Elango and Roopan (2016)
5	Methylene blue	SnO_2 Nanotube	Sadeghzadeh-Attar (2018)
6	Methyl red and methyl orange	ZnO	Comparelli et al. (2005)
7	Methyl red	SiO_2NPs	Badr et al. (2008)
8	Reactive black 5 (RB5)	Bismuth ferrite	Soltani and Entezari (2013)

The FeS NPs were synthesized by Paknikar et al. (2005) via wet chemical method and these were stabilized using a polymer from the basidiomycetous fungus, *Itajahia* sp. These NPs could degrade lindane (5 mg L^{-1}) with an efficiency of 94% in 8 h.

Yu et al. (2007) observed photocatalytic degradation of some organochlorine pesticides (cypermethrin, dicofol, and α-, β-, γ-, and δ-hexachlorobenzene) in nano-TiO_2-coated films on UV exposure. These pesticides were degraded effectively in presence of 2.24 mg cm^{-2} TiO_2 film and 400 W UV lamp (365 nm). It was reported that 20 mg α-BHC was degraded completely within 20 min.

Joo and Zhao (2008) prepared Fe–Pd bimetallic NPs. They used sodium carboxylmethylcellulose (CMC) (0.2% w/w) as stabilizer. These Fe–Pd NPs were used for the degradation of atrazine and lindane. It was observed that these NPs are strong electron donors and completely reduce lindane (1 mg L^{-1}) at an iron dose (0.5 g L^{-1}) and atrazine (1 mg L^{-1}) with iron (0.05 g L^{-1}) under anaerobic condition with a trace amount (0.05–0.8% of Fe) of Pd as a catalyst. These NPs can be also used in Fenton-like reactions leading to oxidation of lindane (65%) in aerobic condition. The degradation rate was found to be enhanced on increasing Pd content from 0.05 to 0.8% of Fe.

It was reported by Elliott et al. (2009) that lindane disappeared within 24 h in presence of nanoiron concentrations in the range 0.015–0.39 g L^{-1} in aqueous solutions. Approximately 40% of lindane dose was retained in solution even after 24 h in the presence of 0.53 g L^{-1} of larger microscale iron particles. Gamma-3,4,5,6-tetrachlorocyclohexene was identified as an intermediate during dihaloelimination of lindane. About 80% chlorine from these lindane molecules went as chloride in water.

Tian et al. (2009) synthesized Fe_3O_4 nanocrystals by a low-temperature solvothermal process. Later these were encapsulated in hexagonal mesoporous silica (HMS) through a packing approach, affording core-shell structured Fe_3O_4@HMS microspheres. It was revealed that the silica shell retain its mesoporous structure even after the surfactant template was removed. It was revealed that pore volume, thickness, and surface area of silica shell can be controlled by varying the reaction conditions. As-prepared Fe_3O_4@HMS materials showed fast adsorption rate and high adsorption capacity for DDT.

The degradation of mixed pesticides lindane, methyl parathion, and dichlorvos was studied by Senthilnathan and Philip (2010a) using UV irradiation. Different grades of TiO_2 were prepared by them through the acid route, alcohol route, and surfactant route (SR) and compared for their activity. The rate of degradation of pesticides was found to be higher for TiO_2 prepared through the SR as compared to the others.

Poursaberi et al. (2012) synthesized nanoscale zero-valent iron (NZVI) and then used for the removal of DDT from the contaminated water. The effect of different operating parameters such as pH, NZVI particle dosage, and reaction time was observed. It was reported that mixing an aqueous solution of DDT (3 mg L^{-1}) with Fe^0 (30 mg L^{-1}) resulted in 99.2% loss of DDT within 4 h. The formation of completely dechlorinated hydrocarbon skeleton of DDT was obtained as the end product diphenylethane (DPE), which indicates removal of all the five chlorine atoms of DDT.

Liu et al. (2014) reported that magnetic mesoporous silica Fe_3O_4@ $nSiO_2$@$mSiO_2$ microspheres exhibited excellent adsorptive performance for DDT removal from aqueous solutions, but they also display high catalytic activity. More than 97% of DDT could be removed from aqueous media in 60 min. It was reported that dichlorodiphenyldichloroethylene (DDE) content increases as DDT disappears completely followed by complete decomposition of DDE after thermal treatment at 450°C. It was indicated that Fe_3O_4@$nSiO_2$@$mSiO_2$ microspheres had strong superparamagnetism with high specific surface area (577 m^2 g^{-1}).

A novel was synthesized by Yola et al. (2014) prepared a photocatalyst based on TiO_2 NPs. They used only a waste material (BEW) not any reducing agent. They degraded atrazine using this as-prepared photocatalyst (TiO_2-BEW). It was observed that BET surface area increased, when TiO_2 NPs were intercalated into BEW. The effect of different operating variables was investigated such as catalyst dosage, initial atrazine concentration, and contact time. This photocatalyst was synthesized by a clean, simple, highly efficient, fast, and ecofriendly pathway. The TiO_2-BEW is advantageous because of stability, efficiency, reusability, and relatively lower cost.

Tian et al. (2015) prepared triple-functional nanocomposite by integrating superparamagnetic Fe_3O_4 and palladium (Pd) NPs onto mesoporous Fe_3O_4@ $nSiO_2$@$mSiO_2$ nanospheres. As-prepared materials exhibited potential capabilities of capturing and then degrading DDT in water catalytically. After use, these nanocomposites can be magnetically separated from the dispersion and then it can be easily regenerated.

Shrivas et al. (2017) reported removal of endrin and dieldrin pesticides in a stereoelectiuve manner using nanocomposite of GO and iron oxide (Fe_3O_4) magnetic nanoparticles (MNPs). It was found that removal efficiency of endrin and dieldrin was higher with GO-MNPs than using GO and MNPs alone. It was reported that removal efficiency of both these pesticides was found quite favorable, when the GO-MNPs (30 mg) was used for 30-min (contact time) keeping pH 4.0 at room temperature. A higher removal percentage (86.0%) was achieved for endrin as compared to dieldrin (74.0%). The higher adsorption percentage of endrin may be due to the endo-position of oxygen atom in molecule, which favored more interaction of molecules with GO-MNPs as compared with the exo-position of oxygen present in the case of dieldrin. This method is advantageous because simple UV–vis spectrophotometry can be used to monitor the removal and low cost of composites nanomaterial.

The sorption properties of cyclodextrin polymers (nanosponges; NS) have been studied by Salazar et al. (2018). They selected pesticides such as 4-chlorophenoxyacetic acid (4-CPA) and 2,3,4,6-tetrachlorophenol (TCF), as model pollutants and results were compared with other materials, like granulated activated carbon. It was reported that there are interactions of these pesticides with nanosponges and polymers have desired sorption capacities for chlorinated aromatic pesticides. It was revealed that an inclusion complex was more favored for NS/CPA as compared to in between TCF and NS, which may be due to the steric effects and size of the adsorbate. It was demonstrated that use of NS polymers has a potential for removal of pollutant from aquatic environments, because they are efficient as well as reusable materials. The usefulness of the polymer is improving, as the nanosponges could be recovered from aqueous solution using a neodymium magnet without any significant loss in efficiency.

Wanjeri et al. (2018) functionalized GO-based silica-coated magnetic NPs (Fe_3O_4@SiO_2@GO) with 2-phenylethylamine (PEA). It was used as an adsorbent for some organophosphorus pesticides (OPPs) (malathion, parathion, and chlorpyrifos) from aqueous solutions. It was reported that maximum adsorption was found in 15 min using 15 mg adsorbent dosage and pesticide concentration (1 μg mL^{-1}). It was also revealed that no significant difference was found in the pH conditions, which suggests that it can be used for a variety of samples. Its reusability was also tested and it was observed that there was loss in the recovery of OPPs after repeated use for ten cycles. The adsorbent was tested with real-world samples collected from the Vaal river and Dam (South Africa), where it showed 86.9% recovery.

Rani and Shanker (2018) observed the removal of some hazardous pesticides, (thiamethoxam, TH; tebuconazole, TEB and chlorpyrifos, CP). They used highly crystalline metal hexacyanoferrates (MHCFs) of different metals such as Cu, Ni-Co, and Zn. A green route for their synthesis was used using *Sapindusmukorossi* (raw ritha). Optimum conditions for degradation of pesticides were achieved as Pesticide (50 mg L^{-1}), MHCF photocatalyst (15 mg), neutral pH, and sunlight irradiation. These agrochemicals were degraded in the order:

ZnHCF (91–98%)>CuHCF (85–91%)>NiHCF (73–85%) >CoHCF (70–83%)

The order of adsorption of pesticides was

CP (83–98%) >TH (76–95%) >TEB (70–91%)

Target molecules underwent mineralization forming some minor and nontoxic by-products like 3-aminopropanoic acid, (Z) but-2-enal, and pyridin-3-ol. Higher efficiency of ZnHCF may be attributed to distorted nanocubes of CuHCF (~100 nm), sharp nanocubes of ZnHCF (~100 nm), and nanospheres of COHCF and NiHCF (~10 nm).

Jung et al. (2018) prepared metal–semiconductor composite NPs. Zn@ZnO core–shell nanocomposites, calcined ZnO, and Ag-doped ZnO (ZnO/Ag) nanostructures. The photocatalytic degradation of lindane under UV–vis irradiation showed that Zn@ZnO nanocomposites exhibited higher photocatalytic activity than rest of the samples. The optimal rate of degrading lindane was found to be 99.5% in 40 min using Zn@ZnO CS nanocomposites. It was confirmed that hydroxyl radical was the main active species, which was responsible to degrade lindane.

Aghaei et al. (2020) evaluated the potential of ZnO/CuO for degradation of methyl parathion. ZnO/CuO nanocomposites were prepared by sono-coprecipitation method. The optimal conditions were achieved as ZnO/CuO photocatalyst (90:10 M), initial concentration of parathion (20 mg L^{-1}), photocatalyst (1 g L^{-1}), no compressed air sparging, pH (8), and ultrasonic power (60 W and 80 kHz). It was reported that parathion was completely degraded in 1 h under optimal conditions on sono-photoirradiation. The ability of ZnO/CuO photocatalyst to remove the parathion was also examined with a real water sample.

Ismael et al. (2020) fabricated a TiO_2/GO/$CuFe_2O_4$ heterostructure photocatalyst by a simple and low-cost ball-milling pathway. Some chlorinated pesticides were degraded under UV light irradiation in presence of this photocatalyst. It was revealed that the nanocomposite exhibited magnetic properties, vis light absorption, and adsorption capacity due to properties of GO, TiO_2, and $CuFe_2O_4$. Nanocomposite had a well-defined crystalline phase, sizes in the range of 10–15 nm with bandgap of 2.4 eV. Seventeen different chlorinated pesticides (persistent organic pollutants, POPs) were photocatalytic degraded using as-prepared photocatalyst. The photocatalytic activity of this nanocomposite showed almost 96.5% removal efficiency of DDE from water in presence of UV irradiation. It was reported that the enhanced photocatalytic performance of TiO_2/GO/$CuFeO_4$ catalyst was due to its high separation efficiency of photo-generated carriers and adsorption performance. Photocatalyst can be reused upto five repetitive cycles for the treatment of pesticides (particularly colorless) and catalyst can be separated easily using an external magnetic field.

The bimetallic Fe/Pd NP was prepared by Ulucan-Altuntas and Debik (2020). The effect of Fe/Pd concentration, DDT concentration, and contact time was evaluated. The highest DDT concentration (0.23 μg L^{-1}) was selected in this work, which is below its carcinogenesis limit. It was observed that the highest concentration of DDT (109.95 mg L^{-1}) could be treated with these Fe/Pd bimetallic NPs. It was reported that 44.3 min contact time and 550 mg L^{-1} Fe/Pd concentration was required to carry out this treatment.

Table 8.2 includes degradation/removal of some pesticides by different nanomaterials.

8.4 PHENOLS

Phenols are widely used as disinfectant in household cleaners, mouthwash, and also as intermediates in chemical and pharmaceutical industries. Phenols show antioxidant, antimicrobial, antiinflammatory, and antiproliferative activities. These are discharged into water bodies from domestic, agricultural, and industrial activities: thus, creating water pollution. They act as carcinogens and may also damage red blood cells and liver, even if present in low concentrations. The treatment of waste water containing phenols becomes necessary before use.

TABLE 8.2 Degradation of Pesticides by Nanomaterials.

S. No.	Pollutant	Nanomaterials	References
1.	DDT	Pd, Au-on-Pd and Cu-on-Pd nanoparticle	Mendes et al. (2021)
2.	DDT	Fe-oxide nanoparticle	Pushkar and Sevak (2019)
3.	DDT and DDE	Fe_3O_4@nSiO_2@mSiO_2 nanospheres	Tian, et al. (2015)
4.	Lindane	Graphene	Sen Gupta et al. (2018)
5.	Lindane	Aluminum hydroxidenanoparticles	Nguyen et al. (2020)
6.	Lindane	N-doped TiO_2	Senthilnathan and Philip (2010b)
7.	Lindane	Fe–Pd bimetallic nanoparticles	Nagpal et al. (2010)

Zhang et al. (2009a) synthesized Fe_3O_4 MNPs (dia = 10 nm) and these were used to remove phenol and aniline from aqueous solution. It was observed that phenol and aniline could be removed completely, when the concentrations of Fe_3O_4 MNPs and H_2O_2 were kept 5 g L^{-1} and 1.2 M, respectively, in 6 h.

The dechlorination of 2,4-dichlorophenol (2,4-DCP) in the presence of Ni–Fe NPs and humic acid (HA) was carried out by Zhang et al. (2009b). It was reported that 2,4-DCP was first adsorbed by Ni–Fe NPs and then, it is reduced to o-chloro- and p-chlorophenols and finally it gives phenol. The introduction of HA was found to reduce the removal percentage of 2,4-DCP and the phenol production rates were decreased from 86 to 29% within 2 h in the absence of HA. The removal percentage of 2,4-DCP was found to be 100, 99, 95, 84, and 69%, for HA concentrations of 0, 5, 10, 30, and 40 mg L^{-1}, respectively.

Morales et al. (2011) prepared nanostructured Pt–ZnO composite thick films. It was observed that only a fraction of added Pt was incorporated into the ZnO lattice while rest one is segregated on ZnO NP surface and formed metallic nanoclusters. It was found that phenol degradation rate was higher with Pt-ZnO as compared to pure ZnO nanostructures.

Shih et al. (2011) used bimetallic NPs for reduction of chlorinated compounds. Degradation of pentachlorophenol (PCP) by Pd/Fe NPs was carried out in aqueous solutions in presence of different cations, Na^+, Cu^{2+}, Ni^{2+}, and Fe^{3+}. An inhibitory effect was observed, when Na_2SO_4 was used. It was attributed to the presence of sulfate ion.

Su et al. (2011) synthesized Fe/Pd NPs (B/Fe/Pd) and used for the dechlorination of p-chlorophenol (p-CP) from aqueous solution. Aggregation of Fe/Pd NPs to 29% was observed within 2 h in the presence of bentonite; thus, enhancing their reactivity. Almost 100% p-CP dechlorination could be achieved with B/Fe/Pd as compared to only 25.0 and 15.0% using B/Fe and bentonite. The dechlorination rate was found to increase with increasing temperature and Pd loading. Here, Fe acts as the reducing agent, Pd as a catalyst, and bentonite as a support for generation of hydrogen from water.

Fenton-like degradation of 2,4-dichlorophenol (2,4-DCP) was carried out by Xu and Wang (2012) over Fe_3O_4 MNPs. It was reported that 2,4-DCP was completely decomposed and 51% of total organic compound (TOC) was removed in 180 min at H_2O_2 (12 mM), Fe_3O_4 MNPs (1.0 g L^{-1}), 2,4-DCP (100 mg L^{-1}), and pH (3.0) at T (30 °C). A high utilization efficiency of H_2O_2 (73%) was also observed. Apart from it, reusability and stability of Fe_3O_4 MNPs were also tested.

Photocatalytic activity of TiO_2 NPs was tested by Ba-Abbad et al. (2012) for the degradation of these chlorophenols (2-chlorophenol, 2-CP), (2, 4-dichlorophenol 2,4- DCP), and (2,4,6-trichlorophenol 2,4,6T-CP) in range (50–150 mg L^{-1}) under direct solar exposure. About, 99, 98, and 92% photodegradation were obtained for 50 mg L^{-1} of 2-CP, 2,4-DCP and 2,4,6-TCP, respectively, after 90 min irradiation time at pH 6. The efficiency was found to decrease, if the initial concentration was taken more than 50 mg L^{-1}.

The Pd/Al NPs were prepared by Yang et al. (2013) and used for hydrodechlorination (HDC) of 4-chlorophenol (4-CP) and 2-chlorobiphenyl (2-PCB) in aqueous and surfactant solutions, respectively. Hydrochloric acid was found to be the proper pH adjustor out of different acids used, i.e., formic acid, acetic acid, HCl, H_2SO_4, and H_3PO_4, to initiate HDC of 4-CP with Pd/Al NPs at pH 3.0. Completely dechlorination of 4-CP (0.389 mM) into phenol could be achieved in 30 min using 2 g L^{-1} of Pd/Al NPs with 1.2 wt.% of Pd loading. An adverse effect was observed on HDC activity of Pd/Al NPs, when surfactants were added such as SDS, Brij 30, cetyltrimethylammonium bromide (CTAB), and Tween 80, or because of their inhibiting effect on Al corrosion.

PCP was treated photocatalytically with Ni/Fe NPs by Lin et al. (2015). The 0.5% nickel percentage was found to be most effective. Different surfactants improved the removal of PCP by Ni/Fe, like Triton X-100, carboxymethyl cellulose, and CTAB. Small particle size of Ni/Fe-CTAB was considered responsible for this activity supported by electrostatic interaction between electropositive Ni/Fe-CTAB and electronegative phenolate group of PCP. It was revealed that of chloro group present at ortho position is more easily reduced as compared to chloro group present at other positions.

Various Fe materials such as Fe^{2+}, nanoscale zero-valent iron (nano-ZVI), and nano-Fe_3O_4 (nFe_3O_4) were used by Li et al. (2015) as heterogeneous catalysts for oxidation of 2,4-dichlorophenol (DCP). The removal of DCP by Fe^{2+}, nZVI, and nFe_3O_4 were found to be 11.9, 9.0, and 5.5%, respectively, which increases in the presence of persulfate to 34.4, 37.8, and 5.8%, respectively. It was observed that DCP degradation increased on increasing dosage of nZVI and persulfate concentration but it decreased on increasing initial pH and DCP concentration. It was claimed that DCP degradation (>98%) could be achieved within 180 min at optimum conditions, but COD was removed lower than 40%.

Boruah et al. (2017) decorated ammonia-modified graphene (AG) sheets with Fe_3O_4 NPs (AG/Fe_3O_4) via in situ solution chemistry approach. It was used as a magnetically recoverable photocatalyst. They selected amide functional group to functionalize GO sheets. Then Fe_3O_4 NPs were doped onto this functionalized GO surface. It was reported that AG/Fe_3O_4 nanocomposite exhibited good photocatalytic activity for degradation of some phenols: Phenol (92.43%), 2-nitrophenol (2-NP) (98%), and 2-chlorophenol (2-CP) (97.15%) within 70–120 min. Slightly higher degradation was observed in case of photo-Fenton reagent, Phenol (93.56%), 2-NP (98.76%), and of 2-CP (98.06%) degradation within 50–80 using AG/Fe_3O_4 nanocomposite under sunlight irradiation. A synergistic effect between amide functionalized graphene and Fe_3O_4 NPs was considered responsible for preventing the recombination rate of electron–hole pair and as a result, photocatalytic activity Fe_3O_4 NPs was enhanced. A significant reusability of the AG/Fe_3O_4 nanocomposite was observed for consecutive ten cycles in photocatalytic degradation of these phenolic compounds without any major loss in its activity.

It has been observed that combination of TiO_2 photocatalyst and magnetic oxide NPs enhanced the separation and recoverable properties. Wysocka et al. (2018) prepared metal-modified (where Me = Cu, Pt, Pd, Au,) $TiO_2/SiO_2@Fe_3O_4$ nanocomposites via ultrasonic-assisted sol–gel method. Oxidation of phenol by metal-modified magnetic photocatalysts involved reactive oxygen species. Phenol was oxidized to catechol, benzoquinone, and hydroquinone. Maleic acid and benzoquinone were detected as final products. The highest mineralization rate was obtained for Pd-$TiO_2/SiO_2@Fe_3O_4$ and Cu-$TiO_2/SiO_2@Fe_3O_4$ photocatalysts, where highest concentration of catechol was obtained. It was suggested that this increased photocatalytic activity of these two photocatalysts (palladium- and copper-modified samples) may be attributed to effective charge carrier separation and increased adsorption sites, but keto-enol tautomeric equilibrium affects the rate of phenol photomineralization adversely on Au-TiO_2/SiO_2@Fe_3O_4.

The degradation of 2,4-DNP by Fe_3O_4@SiO_2@TiO_2/rGO photocatalyst has been reported by Hedayat et al. (2018). The effect of different variables such as pH, initial concentrations of 2,4-DNP, and photocatalyst dosages were studied on degradation of 2,4-DNP. The reusability of magnetic photocatalyst was also determined. The optimum conditions were obtained as pH = 7, 2,4-DNP = 40 mg L^{-1} and photocatalyst dosage = 0.2 g L^{-1}, where maximum removal of 2,4-DNP was obtained (88.84%) within 30 min with 50.04% TOC removal. The presence of interfering ions also affect degradation of 2,4-DNP in order:

$$CO_3^{2-} < Cl^- < SO_4^{2-}$$

It was revealed that photocatalyst can be recovered and used four times and removal efficiency was only reduced within 16.58%.

Three bimetallic oxide (BMO) NPs were synthesized by Rani and Shanker (2018); Ni-Cu oxide (NiCuO), Cu-Cr oxide ($CuCr_2O_4$), and Ni-Cr oxide ($NiCrO_3$). They used *Aegle marmelos* leaf extract for this purpose. It was confirmed that highly crystalline boimetallic oxides were formed with size <50 nm with variable shapes, such Ni-Cu oxide (nanorods), Ni-Cr oxide (nanospheres), and Cu-Cr oxide (NFs). Later, photocatalytic potential of BMO NPs was evaluated for removal of some phenols (phenol, 3-aminophenol, and 2, 4-dinitrophenol) from simulated water. The conditions were optimized as Phenol concentration, (1×10^{-4} M); catalyst dose (15 mg), and pH: (~7.0). Out of these three phenols, 3-aminophenol was better removed (97% Ni-Cu oxide > 95% Ni-Cr oxide > 92% Cu-Cr oxide) followed by phenol (95% Ni-Cu oxide > 92% Ni-Cr oxide > 89% Cu-Cr oxide) and 2,4-DNP (91% Ni-Cu oxide > 89% Ni-Cr oxide > 87% Cu-Cr oxide). Maximum catalytic degradation of 3-aminophenol may be due to its highest basicity. These green synthesized BMO nanocomposites are relatively low cost, and can be reused up to ten times retaining properties such as semiconducting nature, high surface activity, greater active sites, low band gap, and charge separation.

Tripathi and Narayanan (2018) prepared g-C_3N_4 by melamine pyrolysis. This was then added to TiO_2 at different weight percentage to form TiO_2/g-C_3N_4 nanocomposites. The 80% TiO_2/g-C_3N_4 nanocomposite had a band gap of 2.88 eV and it was selected for the photocatalytic experiment. The photocatalytic performance of TiO_2 and TiO_2/g-C_3N_4 nanocomposite was compared. It was reported that optimum conditions could be achieved at UV-irradiance time (180 min), pH (5), H_2O_2 concentration (0.8 gL^{-1}), and flow rate (3 L h^{-1}). It was also revealed that TiO_2/g-C_3N_4 composite exhibited higher photocatalytic efficiency as compared to pure TiO_2. About 93.06% phenol, 85.62% COD, and 80.23% color reduction was there using TiO_2/g-C_3N_4 nanocomposite while it was only 72.39, 68.95, and 65.80%, respectively, when only TiO_2 was used.

A glycerol-mediated method was developed by Nawaz et al. (2019) to synthesize colored titania NPs via solution route, which is a safe and facile method. The $TiCl_4$ precursor was hydrolyzed in aqueous solution in presence of different concentrations of glycerol (0.0, 1.163, 3.834, and 5.815 mol L^{-1}) followed by calcination at 300°C for 1 h to obtain these colored titania NPs. This synthesis did not affect the anatase crystalline structure of TiO_2; sidewise coloration, band gap narrowing, and remarkable bathochromic red shift were observed in the optical response of titania. The major factor responsible for coloration of TiO_2 was the presence of Ti^{3+}. These Ti^{3+} species induced some mid-gap states. It was reported that 48.17 and 62.18% of phenolic compounds and color were removed, respectively, from treated palm oil mill effluent within 3 h in presence of vis light and TiO_2 NPs prepared in 1.163 mol L^{-1} aqueous glycerol solution.

Shen et al. (2019) prepared Fe_3O_4 magnetic NPs with carboxymethyl cellulose-g-poly(acrylic acid-co-acrylic amide) hydrogel coating (Fe_3O_4@hydrogel). They used coprecipitation method, which was followed by an in situ graft copolymerization. It was indicated that this hydrogel material was coated on the Fe_3O_4 surface and the weight percent of 17.5%. The average size of Fe_3O_4@hydrogel was found to be 10 nm. This could be easily collected with the help of an external magnet after use. Fe_3O_4@hydrogel was used as heterogeneous catalyst in degradation of phenol. It was revealed that about 98.2 and 80.4% phenol and COD removal could be achieved, respectively, in 3 h. An effective phenol degradation was observed in wide pH range (3.0–5.0). Almost 50% COD could be removed at pH 5.0. It was found that an improved catalytic performance of this synthesized catalyst may be due to the excellent dispersion, pH-sensitive swelling, permeability, and hydrophilicity of the polymer hydrogel coating. This Fe_3O_4@hydrogel showed reusability and stability with more than 55% COD removal efficiency even after three cycles.

Ahmad et al. (2020) reported the formation of CeO_2 NPs in presence of *Elaeis guineensis* leaves. The formation of uniform agglomerated cubic structured CeO_2 NPs was reported having particle size in range 13–16 nm. An excellent photocatalytic degradation of phenol was exhibited by biofabricated CeO_2 NPs with 92.24% removal in 360 min under vis light.

Au-ZnO nanomaterials were used by Mesa et al. (2020) in photodegradation of phenols under UV–vis light. They selected phenol, catechol, and hydroquinone selected as target pollutants. It was found that hydroquinone is the most sensitive molecule for degradation under illumination. The presence of Au significantly increased the photocatalytic activity of ZnO in the degradation of the phenolic compounds. They could achieve highest effectiveness in the removal of phenolic compounds using ZnO modified by the addition of 2 wt.% of gold.

Tetteh et al. (2020) reported photocatalytic degradation of a local South Africa oil refinery wastewater under UV radiation in presence of titanium dioxide and Degussa P25 (20% rutile, 80% anatase). The effects of three different operational variables were examined such as TiO_2 dosage (2–8 g L^{-1}), run time (30–90 min.), and airflow rate (0.768–1.48 L min^{-1}) for the removal of phenol and soap oil and grease (SOG). The optimum conditions for pollutant removal were found as 90 min, TiO_2 dosage (8 $g^{-1}L^{-1}$), and an aeration flow rate (1.225 L min^{-1}). About 76 and 88% phenol (300 ± 7) and SOG (4000 ± 23), were removed under these conditions, respectively.

Saputra et al. (2013) prepared spinel structured Fe_3O_4, Mn_3O_4, and Co_3O_4 NPs and used for degrading aqueous phenol in the presence of peroxymonosulfate. Their activity was found to be in the order:

$$Mn_3O_4 > Co_3O_4 > Fe_3O_4.$$

Out of these, Mn_3O_4 exhibited excellent catalytic stability. Mn_3O_4 could completely remove phenol in 20 min at optimum conditions: Phenol (25 ppm), catalyst (0.4 g L^{-1}), oxone® (2 g L^{-1}), and 25°C.

A magnetic photocatalyst composed of Fe_3O_4-ZnO hybrid NPs was prepared by Feng et al. (2014). Phenol degradation efficiency of 82.3% could be achieved with these hybrid NPs as compared to only 52% with pure ZnO NPs. It was suggested that presence of free iron ions can act as an electron-trapping site, which suppressed the fast recombination of photogenerated charge carriers, and as a result, photocatalytic properties are improved. It was revealed that photocatalytic activity was well maintained even after three cyclic treatments, which indicates both; good recoverability and high performance of such photocatalyst.

Iodine-doped tin oxide (SnO_2:I) NPs were prepared by Al-Hamdi et al. (2015). Phenol degrades very rapidly within 30 min upon UV irradiation in the presence of SnO_2:I NPs.

Magnetic core–shell–shell $Fe_3O_4/SiO_2/TiO_2$ NPs were synthesized by Rashid et al. (2015). Complete degradation of 25 mg L^{-1} 2-chlorophenol (2-CP) could be achieved with catalyst (0.5 g L^{-1}) within 130 min of 100-W UV irradiation.

Some more phenols are listed in Table 8.3, which are removed from water using various nanomaterials.

TABLE 8.3 Degradation of Phenols by Nanomaterials.

S. No.	Pollutant	Nanomaterials	References
1	Phenol	Iodine-doped tin oxide	Al-Hamdi et al. (2015)
2	2-Chlorophenol	$Fe_3O_4/SiO_2/TiO_2$	Rashid et al. (2015)
3	p-Nitrophenol	TiO_2	Shaoqing et al. (2010)
4	Pentachlorophenol	Pd/Fe bimetallic	Shih et al. (2011)
5	p-Chlorophenol	B/Fe/Pd	Su et al. (2011)

8.5 CHLORO COMPOUNDS

Chloro compounds are reported to be carcinognic, teratogenic, and mutagenic for human health. Chlorinated compounds (aliphatic and aromatic) are major contaminants commonly found in groundwater.

A large number of chlorinated organic compounds are manufactured for industrial, agricultural, pharmaceutical, and household purposes. These chloro compounds are having many toxic effects on health of humans such as heart rate, blood pressure, endocrine-disrupting effect, gene interaction, damaging DNA, and infertility. Therefore, it is important to remove such pollutants from water to make it safe.

The potential of nanoscale Pd/Fe bimetallic particles was evaluated by Lien and Zhang (2001) to reduce chlorinated ethenes. Complete dechlorination could be achieved for six chlorinated ethenes such as tetrachloroethene (PCE), trichloroethene (TCE), 1,1-dichloroethene (1,1-DCE), cis- and trans-1,2-dichloroethene (c-DCE, t-DCE), and vinyl chloride (VC). It was found that these chlorinated ethenes (20 mg L^{-1}) were completely reduced within 90 min, when the metal loading was kept 5 g L^{-1}. Ethane was the primary product (60–90% of the total carbon) and ethene (3–20%) was produced during the transformation of VC, TCE, and DCEs. Such a significant performance can be attributed to (1) high-specific surface area of the nanoscale metal particles (35 m^2 g^{-1}), and (2) increased reactivity per unit metal surface area, mainly because of the presence of the noble metal (Pd) on the surface.

Schrick et al. (2002) used high surface-area nickel–iron NPs (1:3 Ni:Fe) for dehalogenation of TCE. Ni–Fe (0.1 g) NPs reduced it from a 40-mL saturated aqueous solution of 24 ppm to <6 ppb within 120 min. It was reported that dehalogenation reaction was almost 50–80 times slower, when nanoiron or iron filings were used. It was found that hydrogenation was formed by both; linear and branched saturated and unsaturated hydrocarbons of the C–Cl bond, such as Butane, hexane, and octane (even-numbered saturated hydrocarbons) were the major products in TCE degradation, while toxic dehalogenation products such as 1,1-dichloroethylene, VC, cis-, and trans-dichloroethylenes were formed only in trace amounts and these do not persist. VC and DCE are formed at zerovalent iron surfaces by electron transfer in the rate-determining step.

The potential of nanoscale bimetallic (Fe⁄Pd)(Fe⁄Pd) particles was observed by Lien and Zhang (2005) for the HDC of some chlorinated ethanes such as 1,1,2,2-tetrachloroethane, pentachloroethane, and hexachloroethane. It was observed that these were rapidly hydrodechlorinated (9–28 min half-lives) at a NP loading of 5 g L^{-1}. End products were found to be ethane (61–87%) and ethylene (6–16%). It was revealed that Pd/Fe bimetallic NPs exhibited much higher reactivity as compared with conventional micrometer and millimeter scale iron powders.

Tee et al. (2005) synthesized bimetallic Ni/Fe NPs with different compositions. These were used for degradation of trichloroethylene. Ethane was formed as the main product. The reaction rate constant for the Ni/Fe particles was observed to be two orders of magnitude higher than the bulk and nanoiron. A higher degradation rate was observed on increasing the nickel weight percentage from 2 to 25 wt.%.

He and Zhao (2005) developed a green approach to synthesize palladized iron (Fe–Pd) NPs via modified conventional methods. They used water-soluble starch as a stabilizer. It was reported that these starched NPs showed much less agglomeration but enhanced dechlorination power as compared to those prepared without a stabilizer. It was revealed that starched NPs were discrete particles, while dendritic flocs were obtained for nonstarched particles as evident by TEM analysis. The average particle size was found to be 14.1 nm with the surface area of ~55 m^2 g^{-1}. It was interesting to know that starched NPs remained suspended in water for some days, while nonstarched particles agglomerated and precipitated within few minutes. It was reported that starched NPs exhibited much higher reactivity for dechlorination of PCBs or TCE in water. The starched particles were able to degrade about 98% of TCE (25 mg L^{-1}) at a dose of 0.1 g L^{-1} within 1 h. Trace amounts (< 25 μg L^{-1}) of 1,1-DCE were detected in the first twenty min of degradation, and no other intermediate byproducts were detected like VC, cis-, or trans-dichloroethene. The starched NPs at ~1 g L^{-1} were able to convert over 80%

of PCBs (2.5 mg L^{-1}) in <100 h, whereas nonstarched Fe–Pd NPs degrade only 24%.

Wu and Ritchie (2008) prepared cellulose acetate (CA) supported Ni/Fe NPs and used for removal of trichloroethylene from water. It was found that rate constant reached a maximum between 14.3 and 21.4 wt.%, but it decreased on further increasing Ni content.

The effects of CMC (stabilizer) were observed by He and Zhao (2008) on the reactivity of CMC-stabilized Fe-Pd bimetallic NPs for degradation of TCE. Here, the particle agglomeration was prevented by particle stabilization, which is responsible for its higher reactivity. It was reported that TCE degradation rate was increased from 0.86 to 6.8 h^{-1} as the CMC-to-Fe molar ratio was increased from 0 to 0.0124, but the higher CMC-to-Fe ratio inhibited the degradation of TCE. It was reported that hydrogen (hydrogen evolved from ZVI corrosion or residual hydrogen from zerovalent iron (ZVI) NP synthesis) can serve as effective electron donors for dechlorination of TCE in the presence of Pd (separate NPs or coated on ZVI). A decrease in rate by 11.5 times was observed, when reaction pH was decreased from 9.0 to 6.0.

Wang et al. (2009) evaluated the potential of Pd/Fe bimetallic NPs to dechlorinate chlorinated methanes such as dichloromethane, chloroform, and carbon tetrachloride. Chemical precipitation method was used to prepare these bimetallic NPs with diameters 30–50 nm. The effect of initial pH value, Pd loading (weight ratio of Pd to Fe), and Pd/Fe addition (Pd/Fe bimetallic NPs to solution ratio) was monitored on dechlorination efficiency of these compounds. Maximum dechlorination efficiency was obtained with 0.2 wt.% Pd loading. The pH 7 was found to be optimum for dechlorination reaction of chlorinated methanes.

The Pd/Fe NPs (Pd/Fe-1 and Pd/Fe-2) were prepared by Wang et al. (2010). A comparison was made in catalytic activities of Pd/Fe-1 and Pd/Fe-2 NPs using dechlorination reaction of some chloroacetic acids (monochloroacetic acid, MCAA; dichloroacetic acid, DCAA; and trichloroacetic acid TCAA). The removal efficiency of chloroacetic acids with Pd/Fe-2 was about 58.5–22.7%, which was greater as compared with Pd/Fe-1. The removal efficiencies of chloroacetic acids followed the order:

$$\text{TCAA} > \text{DCAA} > \text{MCAA}$$

Monodispersed CMC-stabilized Fe–Cu bimetal NPs were synthesized by Cao et al. (2011) (average diameter < 20 nm). As-prepared particles had a core–shell structure, which are uniform in shape and size. It was reported that these NPs could dechlorinate 1,2,4-trichlorobenzene (1,2,4-TCB) effectively, and about 90% reduction efficiency was achieved within 24 h. It was found that 1,2,4-TCB degradation is in the sequence: First, hydrogenolysis to 1,2-dichlorobenzene, then chlorobenzene, and finally benzene.

Lugaresi et al. (2015) reported the synthesis of three different sized Ag NPs. They tested their electrocatalytic performance in the degradation of a model pollutant (trichloromethane) in aqueous media. It was suggested that proper size range and the presence of abundant superficial defective sites, such as steps or kinks are the main reasons for Ag NPs exhibiting the highest catalytic performance in electrochemical reduction of trichloromethane.

Magnetic copper ferrite NPs ($CuFe_2O_4$, MCFNs) were used by Jaafarzadeh et al. (2017) in ozone/PMS system to degrade 2,4-dichlorophenoxyacetic acid (2,4-D). It was reported that 2,4-D (20.0 mg L^{-1}) was completely degraded at following reaction conditions: pH = 6.0, PMS = 2.0 mM, ozone = 16.0 mg L^{-1}, MCFNs = 0.20 g L^{-1} and 40 min. Stability and reusability experiments were also performed, which indicated that MCFNs were effective for several consecutive runs. It was revealed that MCFNs achieved higher degradation as compared to Fe_2O_3 and CuO in ozone/PMS system. Bicarbonate, phosphate, and nitrite showed an adverse effect on the performance of ozone/MCFNs/PMS system. Mineralization of 2,4-D was evaluated by total organic carbon and chloride release, which were found to be 67.3 and 42.7%, respectively. It was also reported that there was a considerable decrease in toxicity of 2,4-D and a significant increase in biodegradability after treatment by PMS/MCFNs/ozone system.

Several adsorbents have been tried to remove trichloroacetic acid, TCAA. Long et al. (2017) applied crayfish shell (CFS) biochar as an adsorbent to remove TCAA. The removal efficiency of synthesized ZnO-CFS600 was found to be 17.8 mg g^{-1}. It was found that positive charge condition was provided by low pH, which increases adsorption capacity, while the removal process was endothermic in nature. The ZnO-CFS600 had a better pore structure with numerous apparent ZnO NPs as evident from SEM images. The ZnO NPs were of size 37.8 nm. It was proposed that the removal of TCAA was mainly due to surface adsorption, which was supported by electrostatic attraction.

Khammar et al. (2020) used surface-modified magnetic NPs for degrading photocatalytically PCBs from transformer oil. They prepare CMCD-Fe_3O_4@TiO_2 for this purpose via grafting of carboxymethyl-β-cyclodextrin (CM-β-CD) onto the core-shell titania magnetic NPs surface. The photocatalytic efficiency of as-prepared catalyst was evaluated for degradation of PCBs. Optimal degradation yields of PCBs (83%) could be achieved at a temperature of 25°C with 8.35 mg catalyst of and oil: ethanol ratio of 1:5 in 16 min. There is additional benefit that this catalyst can be separated by magnet and reused effectively.

The bimetallic Fe/Pd NP was synthesized by Ulucan-Altuntas and Debik (2020) nano zero-valent iron (nZVI). The effect of Fe/Pd concentration, DDT concentration, and contact time was evaluated. The highest possible initial DDT concentration (0.23 μg L^{-1}) was selected, which is below its carcinogenesis limit. It was observed that the highest concentration that could be treated was 109.95 mg L^{-1} with Fe/Pd bimetallic NPs. It was reported that 550 mg L^{-1} Fe/Pd concentration and 44.3 min contact time was required for this treatment.

NZVI was prepared by Poursaberi et al. (2012) and then used for the removal of DDT from the contaminated water. The effect of different operating parameters such as reaction time, NZVI particle dosage, and pH was observed. It was reported that when an aqueous solution of DDT (3 mg L^{-1}) was mixed with Fe^0 (30 mg L^{-1}), it resulted in 99.2% loss of DDT within 4 h. The formation of completely dechlorinated hydrocarbon skeleton of DDT was obtained as the end product DPE, which indicates removal of all the five chlorine atoms of DDT.

Tian et al. (2009) synthesized Fe_3O_4nanocrystals by a low-temperature solvothermal process, and then encapsulated these in HMS through a packing approach, affording core-shell structured Fe_3O_4@HMS microspheres. It was indicated that the silica shell conserves mesoporous structure even after the removal of surfactant templates. It was revealed that thickness, pore volume, and surface area of silica shell can be controlled by adjusting the reaction conditions. As-prepared Fe_3O_4@HMS materials showed high adsorption capacity and fast adsorption rate for DDT.

8.6 NITRO COMPOUNDS

Nitro compounds are used to produce fungicides, insecticides, dyes, and also to darken leather products. Inhalation of these compounds or ingestion cause headaches, drowsiness, nausea, and lividness (in lips, ears, and finger nails) in humans. Their presence in waste water pose a considerable environmental risk. Therefore, their removal is necessary from wastewater.

Adsorption and photocatalytic degradation of nitrobenzene (NB) was investigated by Cropek et al. (2008) in the presence and absence of phenol over UV-illuminated arginine-modified TiO_2 colloids. It was found that photodegradation of NB strongly depends on the nature of TiO_2 surface. A threefold increase was observed in decomposition rate of NB using arginine-modified TiO_2 NPs.

Size tunability of thiophenol capped CdS NPs was achieved by Datta et al. (2008). It was reported that stable particles of smallest size having narrow size distribution and high photoluminescence quantum yield could be obtained at 5°C. As-prepared CdS NPs were found to display size-dependent photocatalytic activity for degradation of some nitroaromatics. The catalytic efficiency of CdS quantum particles was increased five times on decreasing particle size from 5.8 to 3.8 nm.

A greener synthesis of nanogold-biocomposite was reported by Narayan and Sakthivel (2011) using fungus, *Cylindrocladium floridanum*. It was revealed that the fungus accumulated gold NPs on the surface of the mycelia, when it is cultured in static condition for a period of 7 days. It was confirmed that gold NPs are present on the surface of the fungal mycelia as evident from SPR band. It was found that synthesized particles were spherical in shape. The reduction of 4-nitrophenol (4-NP) to 4-aminophenol (4-AP) with reducing agent sodium borohydride can be carried out using biocomposites with Au NPs (~25 nm). The reaction rate was increased with an increase in concentration of gold NPs (2.54 to 12.67×10^{-6} mol dm^{-3}) and with reduced size (53.2–18.9 nm). These AuNPs s can be recovered and reused.

Pure and Ag-NPs were synthesized by Ilyas et al. (2011) with metallic doping using the liquid impregnation (LI) method. The crystallite size was in the range 27–42 nm. The photocatalytic activity of pure and silver doped titanium dioxide NPs was evaluated in photocatalytic degradation of p-nitrophenol. It was reported that Ag-doped NPs exhibited better degradation of p-nitrophenol (98%) as compared to pure NPs (83%) in 1 h. Photocatalytic degradation of some chlorophenol (2,4-dichlorphenol, 2,5-dichloropheno-land 2,4,6-trichlorophenol) was also investigated. It was observed that their degree of mineralization in 2 h was in the following order:

$$\text{2,4-DCP (100\%)} > \text{2,5-DCP (95\%)} > \text{2,4,6-TCP (86\%)}.$$

Perovskite-type $LaFeO_3$ NPs were synthesized by Farhadi and Siadatnasab (2011) via thermal decomposition of $La[Fe(CN)_6]\cdot 5H_2O$ complex. This nanosized perovskite-type oxide was having an average particle size of 35 nm with a specific surface area 38.5 m^2g^{-1}. It was used as highly efficient heterogeneous catalyst for selective reduction of aromatic nitro compounds. Here, propan-2-ol was used as the hydrogen donor on microwave exposure. This method is clean, compatible, inexpensive, and regio- and chemoselective, even with the substrates having reducible functional groups. It was also reported that this catalyst can be reused again without any significant loss of its activity.

The synthesis of a composite hydrogel based on acrylamide-chicken biochar—(AAm)-CB was reported by Sahiner et al. (2013) and it was used as a template in the preparation of silver metal NPs. An efficient utilization of p(AAm)-CB composite hydrogel was made in the reduction of 4-nitrophenol to 4-aminophenol. As-prepared p(AAm)-CB-Ag composites retained 99.45% of its activity even after five consecutive runs.

Kumar and Deka (2014) developed $Ag_{1-x}Ni_x$ bimetallic alloy NPs. Alloying between Ag and Ni atoms on a nanoscale was confirmed. As-synthesized NPs showed higher catalytic activity in electrocatalytic reduction of 4-nitrophenol. The highest activity was reported for $Ag_{0.6}Ni_{0.4}$ composition to be 156 $s^{-1}g^{-1}$ with 0.2 mg catalyst and there was complete conversion of 4-nitrophenol to 4-aminophenol in 60 s.

A simple, one step, and economical procedure for the synthesis of gold NPs (Au NPs) was developed by Bhosale et al. (2017) via ultrasound probe sonicator. They used dimethyl sulfoxide as reducing agent as well as capping agent. No surfactant or stabilizer was required in this synthesis of Au NPs. It was revealed that spherical NPs were obtained with uniform particle size ranging from 15 to 40 nm within quite a short reaction time. The catalytic activity of as-prepared gold catalyst was evaluated spectrophotometrically for the degradation of aromatic nitro compounds at room temperature.

Ag NPs fabricated on MWCNTs (Ag/MWCNTs nanocomposites) were prepared by Song and Shi (2017) via in situ assisted by *Shewanellaoneidensis* MR-1 (S. oneidensis MR-1). It was indicated that Ag NPs (< 20 nm in diameter) were obtained on the MWCNTs without any aggregation. The catalytic activities of the Ag/MWCNTs nanocomposites were observed in the reduction of 4-nitrophenol (4-NP) by sodium borohydride. It was also indicated that catalytic degradation ratio of 4-NP (94.0%) could be achieved in 10 min in the presence of Ag/MWCNTs nanocomposites.

A convenient and low-cost method was adopted by Khan et al. (2017a) to synthesize efficient silver, copper, and copper–silver mixed NPs on the surface of pure CA and CA-copper oxide nanocomposite (CA-CuO). These loaded NPs exhibited excellent catalytic ability and out of these, Cu-Ag NPs loaded onto CA-CuO (Cu0-Ag0/CA-CuO) exhibited much higher catalytic efficiency for conversion of 4-nitrophenol into 4-aminophenol in the presence of $NaBH_4$. It was also reported that Cu0-Ag0/CA-CuO can be easily recovered by taking out the sheet from the reaction media, and it can be recycled without significant loss of catalytic ability even up to four cycles.

CTAB-capped nanosized α-cobalt hydroxides (Co NPs) were prepared by Khan et al. (2017b) and used for the removal of CR at room temperature in both; absence and presence of sunlight. It was revealed that Co NPs and $NaBH_4$ had no catalytic effects toward degradation of 4-nitrophenol and CR, but Co NPs act as an efficient catalyst even with a small amount of $NaBH_4$. It was revealed that catalytic reaction has an initial induction period, which was followed by an auto-acceleration, depending on concentration of $NaBH_4$.

A bacterial strain *Staphylococcus warneri*, isolated from the estuarine mangroves of Sundarbans region, was used to produce highly stable gold NPs Au NPs by Nag et al. (2018) via reducation of hydrogen auric chloride salt. Highly dispersed and spherically shaped Au NPs were obtained with the size 15–25 nm. Complete degradation of nitroaromatic pollutants (2-nitroaniline, 4-nitroaniline, 2-nitrophenol, and 4-nitrophenol) were observed on as-synthesized Au NPs.The Au NPs showed excellent and efficient catalytic activity with high rate constant with high turnover frequency and it can be reused up to three times.

Islam et al. (2018) developed a simple method for the preparation of gold NPs (AuNPs) supported on TiO_2. The average size of as-prepared TiO_2 supported AuNPs was found to be 8 nm as evident from TEM. The TiO_2 supported AuNPs exhibited significant photocatalytic activity in the degradation of MO as well as catalytic reduction of 4-nitrophenol (4-NP) in water. It was observed that catalytic reduction of 4-NP was influenced by the size and the percent AuNPs loading. The nanocomposite having Au NPs (4.76 wt%) showed about 132 times more activity in the reduction of 4-NP as compared to the pristine TiO_2. The stability of this nanocomposite was also examined for ten cycles and it was found active.

A simple green synthesis of 2D CuO nanoleaves (NLs) was suggested by Bhattacharjee and Ahmaruzzaman (2018) using glutamic acid with microwave heating. Here, amino acid acts as a complexing/capping agent. It was reported that self-assembled 2D CuO NLs were obtained having an average length and width of ~720–800 nm and ~136–160 nm, respectively. It was revealed that as-synthesized CuO NLs can be used as an efficient catalyst for the reduction of some toxic nitro aromatic compounds (4-nitroaniline, 4-nitrophenol, and 2,4,6-trinitrophenol).

A composite of TiO_2 nanospheres decorated with SnO_2 quantum dots (QDs) was prepared by Bhatt et al. (2019). They used this composite as photocatalyst for photodegradation of 4-nitrophenol. It was revealed that the SnO_2 QDs were uniformly dispersed on the surface of TiO_2 nanospheres in TiO_2/SnO_2 QD composites. The effect of various parameters such as pH, amount of semiconductor, light intensity, and concentration of dye was observed. The photocatalytic activity of as-prepared samples was found to be better than pure TiO_2 nanospheres.

Ramya et al. (2019) investigated catalytic degradation of 4-nitrophenol over $CuO@SiO_2$ NPs. The CuO NPs were doped onto SiO_2 NPs, which was synthesized from rice husk. The reduction of 4-nitrophenol to 4-aminophenol was investigated by $NaBH_4$ with synthesized and doped NPs as catalyst. It was found that the catalytic efficiency of $CuO@SiO_2$ was higher than the CuO and SiO_2 NPs alone and it was also observed that doped NPs show significant antimicrobial activity.

Superporouspoly(vinyl imidazole) (p(VI)) cryogel was used by Tercan et al. (2020) to synthesize Cu, Ni, and Co NPs in situ. The catalytic performance of as-prepared composites in the degradation of MB & eosin yellow (EY) and reduction of 4-nitrophenol in aqueous media in the presence of $NaBH_4$ was determined, which followed the order

$$\text{Co} > \text{Cu} > \text{Ni for MB, Co} > \text{Ni} > \text{Cu for eosin Y, and}$$
$$\text{Ni} \approx \text{Co} > \text{Cu for 4-nitrophenol}$$

Binary and tertiary mixtures of these (dyes and 4-NP) were also used with their p(VI)–M composites, and the highest catalytic efficiency (95% conversion in 5 min) was observed, whether single or binary and tertiary substrate mixtures. It was further indicated that over 99% conversion was achieved for individual substrate in the triple mixture (MB + EY + 4-NP) within ~30 min. It was claimed that p(VI)–Cocomposite catalyst has a great potential in applications to real samples as it is capable of simultaneous removal of multiple organic compounds (toxic) from aqueous environments.

Yin et al. (2020) prepared a novel polyvinyl alcohol (PVA)/polyacrylic acid (PAA)/MXene fiber membrane. They used electrospinning method for this purpose. The PVA/PAA/MXene@PdNPs composite nanofiber membrane was obtained on heat treatment of film and then it was modified with Pd NPs. It had a high specific surface area and excellent catalytic performance. A uniform distribution of MXene sheets in this composite fiber membrane solved the problem otherwise MXene sheet cannot be easily monolayerized; sidewise, it can grow the self-reduced Pd NPs on the MXene sheets. This composite nanofiber membrane exhibited significant stability and catalytic ability for degradation of 4-nitrophenol and 2-nitrophenol.

Sargin et al. (2020) prepared chitosan-carbon nanotube supported palladium catalyst. This catalyst was found effective in catalytic reduction of 2-nitroaniline, 4-nitrophenol, 4-nitro-o-phenylenediamine, and 2,4-dinitrophenol.

8.7 METAL IONS

Heavy metals are used in cars, antiseptics, self-cleaning ovens, plastics, and so on. The heavy metals include lead (Pb), cadmium (Cd), chromium (Cr), mercury (Hg), arsenic (As), thallium (Tl), and so on. Presence of heavy metals adversely affects the environment and living organisms. Their toxicity includes the generation of free radicals, which causes oxidative stress, damage of enzymes, proteins, lipids, DNA, and so on. They also affect function of vital organs and glands adversely such as the heart, brain, kidneys, bone, liver. Therefore, their removal from waste water is necessary to make this water useful for living beings.

Khaydarov et al. (2010) prepared nanocarbon-conjugated polymer nanocomposites (NCPC) using polyethylenimine (PEI) and nanocarbon colloids (NCC). It was reported that NCPC has a bonding capacity of 4.0–5.7 mmol g^{-1} at pH 6 for most of the divalent metal ions. They claimed that the percent of sorption of Cu^{2+}, Ni^{2+}, Zn^{2+}, Cd^{2+}, Cr^{2+}, and Hg^{2+} ions was more than 99%. Before coagulation, lifetime of NCPC in the treated water was found to be ranging between 1 s and 1000 min and it depends on the ratio of carbon NP concentrations and polymeric molecules.

A facile thermodecomposition process has been reported by Zhu et al. (2012) to prepare magnetic graphene nanocomposites (MGNCs). The MGNCs have a potential for Cr(VI) removal from the wastewater. It can remove almost remove Cr(VI) completely within 5 min. It was reported that recycling processes of both; MGNC and adsorbed Cr(VI) are energetically and economically sustainable.

Badruddoza et al. (2013) synthesized carboxymethyl-β-cyclodextrin (CM-β-CD) polymer-modified Fe_3O_4 NPs (CDpoly-MNPs) and used for removal of Pb^{2+}, Cd^{2+}, and Ni^{2+} ions from water. It was reported that maximum uptake for Pb^{2+}, Cd^{2+}, and Ni^{2+} on CDpoly-MNPs is in the order:

$$Pb^{2+} >> Cd^{2+} > Ni^{2+}$$

Hard and soft acids and bases theory can explain this order. The recyclability of CDpoly-MNPs was also evaluated.

Jiang et al. (2013) prepared magnetically separable millimeter-sized chitosan beads containing nanosized γ-Fe_2O_3 and used for the removal of Cr(VI). It was reported that the sorption of Cr(VI) onto γ-Fe_2O_3–chitosan beads was exothermic and spontaneous. This sorbed Cr(VI) can be effectively removed from the sorbent with 0.1 M NaOH. The sorbents can be regenerated even after six sequential cycles of sorption–desorption without any significant loss in sorption capacity.

GO/Fe_3O_4 composites were fabricated by Zhang et al. (2013) via copper-catalyzed azide-alkyne cycloaddition reaction. Then these were modified with polyacrylic acid (PAA) to afford water-soluble magnetic graphene nanocomposites. An extraordinary removal capacity for Cu^{2+}, Cd^{2+}, and Pb^{2+} ions was there in PAA/GO/Fe_3O_4 nanocomposites. These nanocomposites are very easy to recycle and separate due to the superparamagnetic behavior of Fe_3O_4. The removal efficiency of the nanoadsorbents was over 85% even after five cycles.

Mahapatra et al. (2013) prepared iron oxide–alumina mixed nanocomposite fiber by electrospinning method. They carried out the experiments to study the sorption behavior of Cu^{2+}, Pb^{2+}, Ni^{2+}, and Hg^{2+} ions as a function of initial concentration, contact time, and pH. The removal percentage was found to be in the order:

$$Cu^{2+} < Pb^{2+} < Ni^{2+} < Hg^{2+}$$

The maximum sorption capacities were found to be 4.98, 32.36, 23.75, and 63.69 mg g^{-1} for Cu^{2+}, Ni^{2+}, for Pb^{2+} and Hg^{2+} ions, respectively.

Polyvinylchloride-blend-CA /iron oxide NPs (nanocomposite membranes) were synthesized by Gholami et al. (2014) via casting technique and used for removal of lead from wastewaters. It was reported that modified membrane containing 10 wt.% CA and 0.1 wt.% Fe_3O_4 NPs exhibited better performance in removal of lead as compared to pristine as well as modified membranes also. Various concentrations (2.5, 5, 7.5, 10, 15, 25, 50, and 75 wt.%) of CA were used as an additive to improve hydrophilicity of membranes, but finally, 10% of CA was found to be the best concentration.

A facile template synthesis combined with chemical coprecipitation was used by Wan and Li (2015) to prepare hybrid γ-Fe_2O_3@cellulose aerogels (γ-Fe_2O_3@CA). It was observed that γ-Fe_2O_3 NPs were well dispersed and immobilized in the micro/nanoscale pore structure of the aerogels and exhibited superparamagnetic behavior. The particle size, magnetic property, pore characteristic parameters, and mechanical strength of γ-Fe_2O_3@CA could be controlled by adjusting the concentrations of the reactants. It exhibited rapid adsorption rate and excellent adsorption ability to remove Cr(VI) ions.

Starch-based nanocomposite (starch/SnO_2) was synthesized by Naushad et al. (2016). Then it was used as an adsorbent for the removal of Hg^{2+} from aqueous medium. It was reported that starch/SnO_2 nanocomposite showed excellent ability to remove Hg^{2+} ion from aqueous solutions pH 6. It was reported that equilibrium was achieved within 1 h. The optimum adsorption capacity 192 mg g^{-1} at 25°C and 94% of adsorption capacity of regenerated adsorbents was there even after the fourth cycle (adsorption–desorption).

Reduced GO was modified by magnetic iron oxide (Fe_3O_4/rGO) by Vuong Hoan et al. (2016) and it was applied for the removal of heavy metals such as Pb(II), Ni(II), and As(V). It was reported that Fe_3O_4/rGO was spherical iron nanooxide and very fine particles.

Titanate nanotubes (T), cobalt ferrite nanoparticles (CF), alginate (G), and their nanocomposites (CF/G and T/G) were fabricated by Esmat et al (2017) and used for efficient removal of some metal ions such as Cu^{2+}, Fe^{3+}, and As^{3+} ions from water. The effects of different parameters on the removal efficiency were also investigated such as contact time, pH, heavy metal ion concentration, and adsorbent weight. The optimum conditions were found to be pH (6.5), contact time (2 h), and adsorbent weight (0.15 g). It was reported that removal efficiencies for Cu^{2+} using T, G, CF, T/G, and CF/G were found to be 91, 100, 99.9, 95, and 98%, respectively, while it was 60, 100, 100, 60, and 82%, respectively, for removal of Fe^{3+}. Efficient removal of As^{3+} ions (98 %) was also attained upon titanate nanotubes.

Zeolite/zinc oxide nanocomposite (Zeolite/ZnO NCs) was prepared by Alswata et al. (2017) via coprecipitation method. Then, they used nanocomposite for adsorption of arsenic and lead from aqueous solutions. It was revealed that ZnO NPs with average diameter 4.5 nm were loaded on zeolite. It was indicated that 93 and 89% of Pb(II) and As(V), respectively, were removed from 100 mg L^{-1} solutions at pH 4.0.

Vinod et al. (2017) stabilized nanoscale zerovalent iron (NZVI) suspensions using a "green" biopolymer, Gum Karaya (GK). They investigated stability, sedimentation, aggregation behavior, and reactivity of NZVI–GK (GK stabilized NZVI) and bare NZVI toward removal of Cr(VI) and volatile organic compounds (*cis*-1,2-dichloroethene, perchloroethene, and TCE). It was reported that NZVI–GK NP suspension remained stable for at least three months, suggesting that GK is stable forming a scaffold to prevent NZVI from aggregation. The Cr(VI) reduction and degradation of volatile organic compounds was carried out using that NZVI–GK and it was found more reactive than bare NZVI. It was revealed that Cr(VI) was reduced to Cr(III) in this process and the remaining Cr(III) in solution was adsorbed onto GK. It has properties such as nontoxicity, biodegradability, and cost-effectiveness.

Ge et al. (2018) prepared magnetic Fe@MgO nanocomposites via a facile precipitation-calcination method. They used this composite for removal of heavy metal ion from water. Fe@MgO nanocomposites exhibited excellent adsorption properties and the maximum adsorption capacities for Pb(II) was 1476.4 mg·g^{-1}.

Mahmoud et al. (2019) converted iron oxide fine waste by-product (obtained from steel industry) into nanoparticulates (Fe_2O_3 NPs) and then it was crosslinked with starch (biodegradable polymer as well as stabilizer) using formaldehyde to form Fe_2O_3 NPs-Starch nanocomposite. The average particle size was found to be in the range of 40–70 nm. The sorption behavior of this nanocomposite was evaluated using Pb(II), Cd(II), and Hg(II). Different factors were monitored to determine the adsorptive capacity such as pH, contact time, nanocomposite dosage, and metal ion concentration. The maximum sorption capacities were found to be 2000, 322.58, and 133.3 mg g^{-1} for Pb(II), Cd(II), and Hg(II), respectively. It was revealed that extraction from tap water, industrial wastewater, and marine water with percentage recovery was 93–97, 76–93, and 70–94%, respectively.

A layer-by-layer (LbL) cation exchange membrane was fabricated by Hosseini et al. (2020). LBL membranes prepared by the coating of chitosan-coactivated carbon NPs layer on polyvinyl chloride-based heterogeneous cation exchange membrane. Such LbL membrane had smoother and more hydrophilic surface as compared to pristine membrane. Electrodialysis experiment showed a good ability in removal of heavy metal ions with LBL membrane in the order: $Cu^{2+} > Ni^{2+} > Pb^{2+}$ while the order was $Pb^{2+} > Cu^{2+} \geq Ni^{2+}$ based on EDX analysis.

Some metal ions have been removed/reduced by the use of nanomaterials and these are tabulated in Table 8.4.

TABLE 8.4 Reduction of Metal Ions by Nanomaterials.

S. No.	Pollutant	Nanomaterials	References
1	Cu (II)	Polyethersulfone nanocomposite membrane prepared by PANI/Fe_3O_4	Daraei et al. (2012)
2	Hospital wastewater	ZnO-NPs on PVC membrane	Alsalhy et al. (2018)
3	Chromium	Chemically modified MWCNTs	Dokmaji et al. (2020)
4	Arsenic	Iron oxide	Kamath et al. (2020)
5	Chromium	Cobalt ferriete	Hosseini et al. (2019)

8.8 OTHERS

Deng et al. (2013) prepared a magnetic nanoscaled catalyst cobalt ferrite ($CoFe_2O_4$) and then it was used for activating oxone to generate sulfate radicals, which were responsible for degradation of diclofenac. The effects of different parameters such as calcination temperature, pH, oxone, and catalyst dosage on the degradation efficiency were monitored. It was reported that $CoFe_2O_4$-300 exhibited the best performance and removed diclofenac completely in 15 min. It was also found that $CoFe_2O_4$ could retain better degradation efficiency after consecutive five cycles also.

Polyvinyl chloride/polycarbonate hollow fiber membranes were prepared by Behboudi et al. (2018) by incorporating modified silver NPs via wet spinning method. It was observed that structure of hollow fiber membranes was changed on the addition of modified silver NPs and the size of macrovoids in the middle layer was decreased, while thickness of sponge-like outer layer was increased. It was revealed that number of pores on the membrane surface increased because of the presence of NPs but their mean pore radii decreased. An improvement in tensile strength, hydrophilicity, and elongation of hollow fiber membranes was reported in addition to modified NPs membranes and these were used in membrane bioreactor system for the treatment of pharmaceutical wastewater. It was found that antifouling properties of hollow fiber membranes were improved on incorporation of modified silver NPs.

Şimşek et al. (2019) compared the performance of GO-based filtration with traditional method. It was observed that there was 40, 50, 86.8, and 45.5% improvement in TOC, TIC, hardness, and LSI, but there was only 39.8, 10.3, 69.8, and 28.6 % improvement using GO-Ag-based filtration. GO-Ag affects adversely microorganisms such as *Escherichia coli* fecal (gaita-infected), *Salmonella typhi*, *Enterococcus faecalis*, *Pseudomona aeruginosa*, and *Staphylococcus aureus*.

The TiO_2 nanotube arrays (TNAs), and TiO_2 nanowires on nanotube arrays (TNWs/TNAs) were prepared by Do et al. (2019). These TNWs/TNAs and Au NP (NP)-decorated-TNAs were applied for the degradation of eight antibiotics, simultaneously. The photocatalytic performance of TNAs, Au-TNAs TNWs/TNAs, and Au-TNWs/TNAs was evaluated by monitoring the degradation of doxycycline, amoxicillin, ampicillin, oxytetracycline, vancomycin, lincomycin, sulfamethoxazole, and sulfamethazine in presence of vis or UV–vis radiation. It was reported that these nanomaterials degraded the antibiotics rapidly and effectively, where most of the antibiotics were removed completely within 20 min treatment. Highest photocatalytic activity was exhibited by Au-TNWs/TNAs in degrading all these antibiotics. The rate constants for degradation of lincomycin with Au-TNWs/TNAs were 0.26 and 0.096 min^{-1} in presence of UV–vis and vis irradiation, respectively. The enhanced photocatalytic activity of Au-TNWs/TNAs may be attributed to the synergistic effects. These are

- Larger surface area of TNWs/TNAs in comparison to TNAs, and
- Surface plasmonic effect in Au NPs, which increases the vis light harvesting.

The use of magnetically separable Fe_3O_4-red mud NPs (Fe_3O_4-RM-NPs) was investigated by Aydin et al. (2019) for the removal of antibiotics from wastewater. There are disadvantages of red mud and Fe_3O_4 as these are difficult to separate from aqueous media, undergo iron leaching and agglomeration, which can be overcome by combining these two materials. The magnetic Fe_3O_4-RM-NPs were used for the removal of ciprofloxacin (CIPRO), a model pollutant. The effect of red mud content of synthesized Fe_3O_4-RM-NPs, pH, reaction time and temperature was monitored. Optimum conditions were achieved for the efficient removal of CIPRO from aqueous media as: 1.5 g red mud for Fe_3O_4-RM-NPs, pH 6.0, reaction time 60 min, 3 gL^{-1} Fe_3O_4-RM-NPs dosage at 25°C. They proposed magnetically separable Fe_3O_4-RM-NPs as efficient adsorbent to remove antibiotics from wastewater samples. As red mud is a process waste, this nanomaterial will prove to be a promising alternative to commercial adsorbents.

Al-Anazi et al. (2020) developed a novel redox technology, which was based on synthetic franklinite-like zinc-ferrite nanomaterial. The redox activity of these NPs was tested for degradation of diclofenac (DCF) in water, in absence of any external oxidants and that too in dark experimental conditions. It was revealed ~90% degradation of DCF (10 μM) was there within 2 min of reaction, when 0.17 g L^{-1} $Zn_{1.0}Fe_{2.0}O_4$ was used. It was revealed that degradation of DCF was mainly due to oxidation by oxygen-based radicals and reduction by surface electrons on zinc ferrite.

Water pollution has resulted in scarcity of potable and useful water and it is increasing rapidly due to exponential population growth, rapid industrialization, transportation, and construction activity as well as demand for more active pharmaceutical drugs, wide range of colors for fabrics, paints and varnishes, pesticides for better output in agriculture, use of detergents in place of soap, etc. It is, therefore, become necessary to turn toward green chemical technology (processes, reagents, products, and treatment methods) to combat against this burning problem of water pollution. Nanomaterials may play an interesting and important role here. Although, nanomaterials have many advantages but it is also important to point out that these nanomaterials may have certain adverse effects and therefore, they may be used with cautions. In this chapter, major water pollutants have been discussed but the list is still incomplete as there are many more water pollutants and few more are emerging ones.

KEYWORDS

- **nanoparticles**
- **nanomaterials**
- **water purification**
- **waste water treatment**
- **dyes**
- **pesticides**

REFERENCES

Abbasi, M.; Asl, N. R. Sonochemical Degradation of Basic Blue 41 Dye Assisted by Nano TiO_2 and H_2O_2. *J. Hazard. Mater.* **2008,** *153* (3), 942–947.

Adekunle, A. S.; Oyekunle, J. A.; Durosinmi, L. M.; Oluwafemi, O. S.; Olayanju, D. S.; Akinola, A. S. et al. Potential of Cobalt and Cobalt Oxide Nanoparticles as Nanocatalyst Towards Dyes Degradation in Wastewater. *Nano-Struct. Nano-Objects*, **2020,** *21*. DOI: 10.1016/j.nanoso.2019.100405.

Aghaei, M.; Sajjadi, S.; Keihan, A. H. Sono-Coprecipitation Synthesis of ZnO/CuO Nanophotocatalyst for Removal of Parathion from Wastewater. *Environ. Sci. Poll. Res.* **2020,** *27*. DOI: 10.1007/s11356-020-07680-0.

Ahmad, T.; Iqbal, J.; Bustam, M. A.; Zulfiqar, M.; Muhammad, N.; Al Hajeri, B. M. et al. Phytosynthesis of Cerium Oxide Nanoparticles and Investigation of Their Photocatalytic Potential for Degradation of Phenol under Visible Light. *J. Mol. Struct.* **2020,** *1217*. DOI: 10.1016/j.molstruc.2020.128292.

Ahmed, M. A.; El-Katori, E. E.; Gharni, Z. H. Photocatalytic Degradation of Methylene Blue Dye Using Fe_2O_3/TiO_2 Nanoparticles Prepared by Sol–Gel Method. *J. Alloys Compd.* **2013,** *553*, 19–29.

Al-Anazi, A.; Abdelraheem, W. H.; Scheckel, K.; Nadagouda, M. N.; O'Shea, K.; Dionysiou, D. D. Novel Franklinite-Like Synthetic Zinc-Ferrite Redox Nanomaterial: Synthesis, and Evaluation for Degradation of Diclofenac in Water. *Appl. Catal. B: Environ.* **2020,** *275*, 119098. DOI: org/10.1016/j.apcatb.2020.119098

Al-Hamdi, A. M.; Sillanpää, M.; Dutta, J. Photocatalytic Degradation of Phenol by Iodine Doped Tin Oxide Nanoparticles under UV and Sunlight Irradiation. *J. Alloys Compd.* **2015,** *618*, 366–371.

Alsalhy, Q. F.; Al-Ani, F. H.; Al-Najar, A. E.; Jabuk, S. I. A Study of the Effect of Embedding ZnO-NPs on PVC Membrane performance Use in Actual Hospital Wastewater Treatment by Membrane Bioreactor. *Chem. Eng. Process.-Process Intens.* **2018,** *130*, 262–274.

Alswata, A. A.; Ahmad, M. B.; Al-Hada, N. M.; Kamari, H. M.; Hussein, M. Z. B.; Ibrahim, N. A. Preparation of Zeolite/Zinc Oxide Nanocomposites for Toxic Metals Removal from Water. *Results Phys.* **2017,** *7*, 723–731.

Ameen, S.; Akhtar, M. S.; Nazim, M.; Shin, H. S.; Rapid Photocatalytic Degradation of Crystal Violet Dye Over ZnO Flower Nanomaterials. *Mater. Lett.* **2013,** *96*, 228–232.

Anandan, S.; Kumar, P. S.; Pugazhenthiran, N.; Madhavan, J.; Maruthamuthu, P. Effect of Loaded Silver Nanoparticles on TiO_2 for Photocatalytic Degradation of Acid Red 88. *Sol. Energy Mater. Solar Cells*, **2008,** *92* (8), 929–937.

Aydin, S.; Aydin, M. E.; Beduk, F.; Ulvi, A. Removal of Antibiotics from Aqueous Solution by Using Magnetic Fe_3O_4/Red Mud-Nanoparticles. *Sci. Total Environ.* **2019,** *670*, 539–546.

Ba-Abbad, M. M.; Kadhum, A. A. H.; Mohamad, A. B.; Takriff, M. S.; Sopian, K. Synthesis and Catalytic Activity of TiO_2 Nanoparticles for Photochemical Oxidation of Concentrated Chlorophenols under Direct Solar Radiation. *Int. J. Electrochem. Sci.* **2012,** *7* (6), 4871–4888.

Badr, Y.; Abd El-Wahed, M. G.; Mahmoud, M. A. Photocatalytic Degradation of Methyl Red Dye by Silica Nanoparticles. *J. Hazard. Mater.* **2008,** *154* (1–3), 245–253.

Badruddoza, A. Z. M.; Shawon, Z. B. Z.; Tay, W. J. D.; Hidajat, K.; Uddin, M. S. Fe_3O_4/Cyclodextrin Polymer Nanocomposites for Selective Heavy Metals Removal from Industrial Wastewater. *Carbohyd. Polym.*; **2013,** *91* (1), 322–332.

Behboudi, A.; Jafarzadeh, Y.; Yegani, R. Enhancement of Antifouling and Antibacterial Properties of PVC Hollow Fiber Ultrafiltration Membranes Using Pristine and Modified Silver Nanoparticles. *J. Environ. Chem. Eng.* **2018,** *6* (2), 1764–1773.

Behboudi, A.; Jafarzadeh, Y.; Yegani, R. Incorporation of Silica Grafted Silver Nanoparticles into Polyvinyl Chloride/Polycarbonate Hollow Fiber Membranes for Pharmaceutical Wastewater Treatment. *Chem. Eng. Res. Design*, **2018,** *135*, 153–165.

Bhatt, J.; Benjamin, S.; Ameta, R.; Ameta, S. C. Enhanced Visible Light Photocatalytic Activity of TiO_2 Nanospheres/SnO_2 Quantum Dots Composite for Degradation of 4-Nitrophenol. *J. Nanosci. Technol.*; **2019,** *5* (4), 748–750.

Bhattacharjee, A.; Ahmaruzzaman, M. Microwave Assisted Facile and Green Route for Synthesis of CuO Nanoleaves and Their Efficacy as a Catalyst for Reduction and Degradation of Hazardous Organic Compounds. *J. Photochem. Photobiol. A: Chem.* **2018,** *353*, 215–228.

Bhosale, M. A.; Chenna, D. R.; Bhanage, B. M. Ultrasound Assisted Synthesis of Gold Nanoparticles as an Efficient Catalyst for Reduction of Various Nitro Compounds. *Chem. Select*, **2017,** *2* (3), 1225–1231.

Bokare, A. D.; Chikate, R. C.; Rode, C. V.; Paknikar, K. M. Iron-Nickel Bimetallic Nanoparticles for Reductive Degradation of Azo Dye Orange G in Aqueous Solution. *Appl. Catal. B: Environ.* **2008,** *79* (3), 270–278.

Boruah, P. K.; Sharma, B.; Karbhal, I.; Shelke, M. V.; Das, M. R.; Ammonia-Modified Graphene Sheets Decorated with Magnetic Fe_3O_4 Nanoparticles for the Photocatalytic and Photo-Fenton Degradation of Phenolic Compounds under Sunlight Irradiation. *J. Hazard. Mater.* **2017,** *325*, 90–100.

Cao, J.; Xu, R.; Tang, H.; Tang, S.; Cao, M. Synthesis of Monodispersed CMC-Stabilized Fe–Cu Bimetal Nanoparticles for in Situ Reductive Dechlorination of 1, 2, 4-Trichlorobenzene. *Sci. Total Environ.* **2011,** *409* (11), 2336–2341.

Chauhan, P. S.; Kant, R.; Rai, A.; Gupta, A.; Bhattacharya, S.; Facile Synthesis of ZnO/GO Nanoflowers over Si Substrate for Improved Photocatalytic Decolorization of MB Dye and Industrial Wastewater under Solar Irradiation. *Mater. Sci. Semicond. Process.* **2019,** *89*, 6–17.

Comparelli, R.; Fanizza, E.; Curri, M. L.; Cozzoli, P. D.; Mascolo, G.; Agostiano, A. UV Induced Photocatalytic Degradation of Azo Dyes by Organic-Capped ZnO Nanocrystals Immobilized Onto Substrates. *Appl. Catal. B: Environ.* **2005,** *60* (1–2), 1–11.

Cropek, D.; Kemme, P. A.; Makarova, O. V.; Chen, L. X.; Rajh, T. Selective Photocatalytic Decomposition of Nitrobenzene Using Surface Modified TiO_2 Nanoparticles. *J. Phys. Chem. C*, **2008,** *112* (22), 8311–8318.

Cruz, J. C.; Nascimento, M. A.; Amaral, H. A.; Lima, D. S.; Teixeira, A. P. C.; Lopes, R. P. Synthesis and Characterization of Cobalt Nanoparticles for Application in the Removal of Textile Dye. *J. Environ. Manage.* **2019,** *242*, 220–228.

Dai, K.; Chen, H.; Peng, T.; Ke, D.; Yi, H. Photocatalytic Degradation of Methyl Orange in Aqueous Suspension of Mesoporous Titania Nanoparticles. *Chemosphere*, **2007,** *69* (9), 1361–1367.

Daraei, P.; Madaeni, S. S.; Ghaemi, N.; Salehi, E.; Khadivi, M. A.; Moradian, R.; Astinchap, B. Novel Polyethersulfone Nanocomposite Membrane Prepared by PANI/Fe_3O_4 Nanoparticles with Enhanced Performance for Cu(II) Removal from Water. *J. Membr. Sci.* **2012,** *415*, 250–259.

Datta, A.; Priyam, A.; Bhattacharyya, S. N.; Mukherjea, K. K.; Saha, A. Temperature Tunability of Size in CdS Nanoparticles and Size Dependent Photocatalytic Degradation of Nitroaromatics. *J. Colloid Interface Sci.* **2008,** *322* (1), 128–135.

Deng, J.; Shao, Y.; Gao, N.; Tan, C.; Zhou, S.; Hu, X. $CoFe_2O_4$ Magnetic Nanoparticles as a Highly Active Heterogeneous Catalyst of Oxone for the Degradation of Diclofenac in Water. *J. Hazard. Mater.* **2013,** *262*, 836–844.

Do, T. C. M. V.; Nguyen, D. Q.; Nguyen, K. T.; Le, P. H. TiO_2 and Au-TiO_2 Nanomaterials for Rapid Photocatalytic Degradation of Antibiotic Residues in Aquaculture Wastewater. *Materials* **2019,** *12* (15). DOI: 10.3390/ma12152434.

Dokmaji, T.; Ibrahim, T.; Khamis, M.; Abouleish, M.; Alam, I. Chemically Modified Nanoparticles Usage for Removal of Chromium from Sewer Water. *Environ. Nanotech. Monitoring Manage.* **2020,** *14*. DOI: org/10.1016/j.enmm.2020.100319

Elliott, D. W.; Lien, H.-L.; Zhang, W.-X. Degradation of Lindane By Zero-Valent Iron Nanoparticles. *J. Environ. Eng.* **2009,** *135* (5), 317–324.

Esmat, M.; Farghali, A. A.; Khedr, M. H.; El-Sherbiny, I. M. Alginate-Based Nanocomposites for Efficient Removal of Heavy Metal Ions. *Int. J. Biol. Macromol.* **2017,** *102*, 272–283.

Farhadi, S.; Siadatnasab, F. Perovskite-Type $LaFeO_3$ Nanoparticles Prepared by Thermal Decomposition of the La $[Fe(CN)_6] \cdot 5H_2O$ Complex: A New Reusable Catalyst for Rapid and Efficient Reduction of Aromatic Nitro Compounds to Arylamines with Propan-2-ol under Microwave Irradiation. *J. Mol. Catal. A: Chem.* **2011,** *339* (1–2), 108–116.

Feng, X.; Guo, H.; Patel, K.; Zhou, H.; Lou, X. High Performance, Recoverable Fe_3O_4ZnO Nanoparticles for Enhanced Photocatalytic Degradation of Phenol. *Chem. Eng. J.* **2014,** *244*, 327–334.

Ge, L.; Wang, W.; Peng, Z.; Tan, F.; Wang, X.; Chen, J. et al. Facile Fabrication of Fe@ MgO Magnetic Nanocomposites for Efficient Removal of Heavy Metal Ion and Dye from Water. *Powder Technol.* **2018,** *326*, 393–401.

Gholami, A.; Moghadassi, A. R.; Hosseini, S. M.; Shabani, S.; Gholami, F. Preparation and Characterization of Polyvinyl Chloride Based Nanocomposite Nanofiltration-Membrane Modified by Iron Oxide Nanoparticles for Lead Removal from Water. *J. Industr. Eng. Chem.* **2014,** *20* (4), 1517–1522.

Gupta, M.; Bhatt, J.; Ruchi, Ameta,R.; Ameta, S. C. Use of Graphene Oxide-Zinc Oxide Composite in Photocatalytic Degradation of Victoria Blue B. *J. Nanosci. Technol.* **2019,** *5*(2), 702–705.

He, F.; Zhao, D. Hydrodechlorination of Trichloroethene Using Stabilized Fe-Pd Nanoparticles: Reaction Mechanism and Effects of Stabilizers, Catalysts and Reaction Conditions. *Appl. Catal. B: Environ.* **2008,** *84* (3–4), 533–540.

He, F.; Zhao, D. Preparation and Characterization of a New Class Of Starch-Stabilized Bimetallic Nanoparticles for Degradation of Chlorinated Hydrocarbons in Water. *Environ. Sci. Technol.* **2005,** *39* (9), 3314–3320.

Hedayat, B. M.; Noorisepehr, M.; Dehghanifard, E.; Esrafili, A.; Norozi, R. Evaluation of Photocatalytic Degradation of 2,4-dinitrophenol from Synthetic Wastewater Using Fe_3O_4@ SiO_2@ TiO_2/rGO Magnetic Nanoparticles. *J. Mol. Liq.* **2018,** *264*, 571–578.

Helmy, E. T.; El Nemr, A.; Mousa, M.; Arafa, E. and Eldafrawy, S.; Photocatalytic Degradation of Organic Dyes Pollutants in the Industrial Textile Wastewater by Using Synthesized TiO_2, C-doped TiO_2, S-doped TiO_2 and C, S Co-doped TiO_2 Nanoparticles. *J. Water Environ. Nanotechnol.* **2018,** *3* (2), 116–127.

Hemraj-Benny, T.; Tobar, N.; Carrero, N.; Sumner, R.; Pimentel, L.; Emeran, G. Microwave-Assisted Synthesis of Single-Walled Carbon Nanotube-Supported Ruthenium Nanoparticles for the Catalytic Degradation of Congo Red Dye. *Mater. Chem. Phys.* **2018,** *216*, 72–81.

Hosseini, S. M.; Alibakhshi, H.; Jashni, E.; Parvizian, F.; Shen, J. N.; Taheri, M. et al. A Novel Layer-By-Layer Heterogeneous Cation Exchange Membrane for Heavy Metal Ions Removal from Water. J. Hazard. Mater. 2020, 381.doi.org/10.1016/j.jhazmat.2019.120884

Hosseini, S. M.; Sohrabnejad, S.; Nabiyouni, G.; Jashni, E.; Van der Bruggen, B.; Ahmadi, A. Magnetic Cation Exchange Membrane Incorporated with Cobalt Ferrite Nanoparticles for Chromium Ions Removal via Electrodialysis. *J.Membr. Sci.* **2019,** *583*, 292–300.

Hussain, M.; Sun, H.; Karim, S.; Nisar, A.; Khan, M.; Ul Haq, A. et al. Noble Metal Nanoparticle-Functionalized ZnO Nanoflowers for Photocatalytic Degradation of RhB Dye and Electrochemical Sensing of Hydrogen Peroxide. *J. Nanoparticle Res.* **2016,** *18* (4). DOI: 10.1007/s11051-016-3397-y.

Ilyas, H.; Qazi, I. A.; Asgar, W.; Awan, M. A. Photocatalytic Degradation of Nitro and Chlorophenols Using Doped and Undoped Titanium Dioxide Nanoparticles. *J. Nanomater.* **2011,** *2011*. DOI: 10.1155/2011/589185

Islam, M. T.; Jing, H.; Yang, T.; Zubia, E.; Goos, A. G.; Bernal, R. A. et al. Fullerene Stabilized Gold Nanoparticles Supported on Titanium Dioxide for Enhanced Photocatalytic Degradation of Methyl Orange and Catalytic Reduction of 4-nitrophenol. *J. Environ. Chem. Eng.* **2018,** *6* (4), 3827–3836.

Ismael, A. M.; El-Shazly, A. N.; Gaber, S. E.; Rashad, M. M.; Kamel, A. H.; Hassan, S. S. M. Novel TiO_2/GO/$CuFe_2O_4$ Nanocomposite: A Magnetic, Reusable and Visible-Light-Driven Photocatalyst for Efficient Photocatalytic Removal of Chlorinated Pesticides from Wastewater. *RSC Adv.* **2020,** *10* (57), 34806–34814.

Jaafarzadeh, N.; Ghanbari, F.; Ahmadi, M. Efficient Degradation of 2,4-Dichlorophenoxyacetic Acid by Peroxymonosulfate/Magnetic Copper Ferrite Nanoparticles/Ozone: A Novel Combination of Advanced Oxidation Processes. *Chem. Eng. J.* **2017,** *320*, 436–447.

Jat, K. K.; Bhatt, J.; Ameta, S. C. Photodegradation of Fast Green by Using SnO_2 Quantum Dots/TiO_2 Nanoparticles Composite. *J. Applicable Chem.* **2019,** *8* (1), 139–145.

Jiang, Y. J.; Yu, X. Y.; Luo, T.; Jia, Y.; Liu, J. H.; Huang, X. J. γ-Fe_2O_3 Nanoparticles Encapsulated Millimeter-Sized Magnetic Chitosan Beads for Removal of Cr(VI) from Water: Thermodynamics, Kinetics, Regeneration, and Uptake Mechanisms. *J. Chem. Eng. Data*, **2013**, *58* (11), 3142–3149.

Jin, X.; Liu, Y.; Tan, J.; Owens, G.; Chen, Z. Removal of Cr (VI) from Aqueous Solutions via Reduction and Absorption by Green Synthesized Iron Nanoparticles. *J. Clean. Prod.* **2018,** *176*, 929–936.

Joo, S. H.; Zhao, D. Destruction of Lindane and Atrazine Using Stabilized Iron Nanoparticles under Aerobic and Anaerobic Conditions: Effects of Catalyst and Stabilizer. *Chemosphere*, **2008,** *70* (3), 418–425.

Jung, H. J.; Koutavarapu, R.; Lee, S.; Kim, J. H.; Choi, H. C.; Choi, M. Y. Enhanced Photocatalytic Degradation of Lindane Using Metal–Semiconductor Zn@ ZnO and ZnO/Ag Nanostructures. *J. Environ. Sci.* **2018,** *74*, 107–115.

Kamath, V.; Chandra, P.; Jeppu, G. P. Comparative Study of Using Five Different Leaf Extracts in the Green Synthesis of Iron Oxide Nanoparticles for Removal of Arsenic from Water. *Int. J. Phytoremedia.* **2020,** *22* (12), 1278–1294.

Kapoor, N.; Ameta, R.; Ameta, S. C. Degradation of Evans Blue Using Multi-Walled Carbon Nanotubes (MWCNTs)–Barium Chromate Composite. *J. Nanosci. Technol.* **2019,** 662–665. DOI: 10.30799/jnst.227.19050204

Katata-Seru, L.; Moremedi, T.; Aremu, O. S.; Bahadur, I. Green Synthesis of Iron Nanoparticles Using *Moringa oleifera* Extracts and Their Applications: Removal of Nitrate from Water and Antibacterial Activity against *Escherichia coli*. *J. Mol. Liq.* **2018,** *256*, 296–304.

Khammar, S.; Bahramifar, N.; Younesi, H. Preparation and Surface Engineering of CM-β-CD Functionalized Fe_3O_4@ TiO_2 Nanoparticles for Photocatalytic Degradation of Polychlorinated Biphenyls (PCBs) from Transformer Oil. *J. Hazard. Mater.* **2020,** *394*. DOI: 10.1016/j.jhazmat.2020.122422.

Khan, F. U.; Khan, S. B.; Kamal, T.; Asiri, A. M.; Khan, I. U.; Akhtar, K. Novel Combination of Zero-Valent Cu and Ag Nanoparticles@ Cellulose Acetate Nanocomposite for the Reduction of 4-nitro Phenol. *Int. J. Biol. Macromol.* **2017a,** *102*, 868–877.

Khan, M. N.; Bashir, O.; Khan, T. A.; Al-Thabaiti, S. A.; Khan, Z. Catalytic Activity of Cobalt Nanoparticles for Dye and 4-nitro Phenol Degradation: A Kinetic and Mechanistic Study. *Int. J. Chem. Kinet.* **2017b,** *49* (6), 438–454.

Khataee, A. R.; Pons, M. N.; Zahraa, O. Photocatalytic Degradation of Three Azo Dyes Using Immobilized TiO_2 Nanoparticles on Glass Plates Activated by UV Light Irradiation: Influence of Dye Molecular Structure. *J. Hazard Mater.* **2009,** *168* (1), 451–457.

Khaydarov, R. A.; Khaydarov, R. R.; Gapurova, O. Water Purification from Metal Ions Using Carbon Nanoparticle-Conjugated Polymer Nanocomposites. *Water Res.* **2010,** *44* (6), 1927–1933.

Kumar, M.; Deka, S. Multiply Twinned AgNi Alloy Nanoparticles as Highly Active Catalyst for Multiple Reduction and Degradation Reactions. *ACS Appl. Mater. Interfaces,* **2014,** *6* (18), 16071–16081.

Kumar, P.; Govindaraju, M.; Senthamilselvi, S.; Premkumar, K. Photocatalytic Degradation of Methyl Orange Dye Using Silver (Ag) Nanoparticles Synthesized from *Ulva lactuca*. *Colloids Surf. B: Biointerfaces*, **2013,** *103*, 658–661.

Kumar, R.; Umar, A.; Kumar, G.; Akhtar, M. S.; Wang, Y.; Kim, S. H. Ce-doped ZnO Nanoparticles for Efficient Photocatalytic Degradation of Direct Red-23 Dye. *Ceram. Int.* **2015,** *41* (6), 7773–7782.

Li, P.; Lin, K.; Fang, Z.; Wang, K. Enhanced Nitrate Removal by Novel Bimetallic Fe/Ni Nanoparticle Supported on Biochar. *J. Clean. Prod.* **2017,** *151*, 21–33.

Li, R.; Jin, X.; Megharaj, M.; Naidu, R.; Chen, Z. Heterogeneous Fenton Oxidation of 2, 4-dichlorophenol Using Iron-Based Nanoparticles and Persulfate System. *Chem. Eng. J.* **2015,** *264*, 587–594.

Lien, H. L.; Zhang, W. X. Hydrodechlorination of Chlorinated Ethanes by Nanoscale Pd/Fe Bimetallic Particles. *J. Environ. Eng.* **2005,** *131* (1), 4–10.

Lien, H. L.; Zhang, W. X. Nanoscale Iron Particles for Complete Reduction of Chlorinated Ethenes. *Colloids Surf. A: Physicochem. Eng. Aspects*, **2001,** *191* (1–2), 97–105.

Lin, C. H.; Shih, Y. H.; MacFarlane, J. Amphiphilic Compounds Enhance the Dechlorination of Pentachlorophenol with Ni/Fe Bimetallic Nanoparticles. *Chem. Eng. J.* **2015,** *262*, 59–67.

Liu, F.; Tian, H.; He, J. Adsorptive Performance and Catalytic Activity of Superparamagnetic Fe_3O_4@nSiO_2@mSiO_2 Core–Shell Microspheres Towards DDT. *J. Colloid Interf. Sci.* **2014,** *419*, 68–72.

Long, L.; Xue, Y.; Zeng, Y.; Yang, K.; Lin, C. Synthesis, Characterization and Mechanism Analysis of Modified Crayfish Shell Biochar Possessed ZnO Nanoparticles to Remove Trichloroacetic Acid. *J. Clean. Prod.* **2017,** *166*, 1244–1252.

Lugaresi, O.; Perales-Rondón, J. V.; Minguzzi, A.; Solla-Gullón, J.; Rondinini, S.; Feliu, J. M.; Sánchez-Sánchez, C. M. Rapid Screening of Silver Nanoparticles for the Catalytic Degradation of Chlorinated Pollutants in Water. *Appl. Catal. B: Environ.* **2015,** *163*, 554–563.

Madhusudhana, N.; Yogendra, K.; Mahadevan, K. M. A Comparative Study on Photocatalytic Degradation of Violet GL2B Azo Dye Using CaO and TiO_2 Nanoparticles. *Int. J. Eng. Res. Appl*, **2012,** *2* (5), 1300–1307.

Mahapatra, A.; Mishra, B. G.; Hota, G. Electrospun Fe_2O_3–Al_2O_3 Nanocomposite Fibers as Efficient Adsorbent for Removal of Heavy Metal Ions from Aqueous Solution. *J. Hazard. Mater.* **2013,** *258*, 116–123.

Mahmoodi, N. M. Zinc Ferrite Nanoparticle as a Magnetic Catalyst: Synthesis and Dye Degradation. *Mater. Res. Bull.* **2013,** *48* (10), 4255–4260.

Mahmoodi, N. M.; Arami, M. Bulk Phase Degradation of Acid Red 14 by Nanophotocatalysis Using Immobilized Titanium (IV) Oxide Nanoparticles. *J. Photochem. Photobiol. A: Chem.* **2006,** *182* (1), 60–66.

Mahmoud, M. E.; Nabil, G. M.; Zaki, M. M.; Saleh, M. M. Starch Functionalization of Iron Oxide by-Product from Steel Industry as a Sustainable Low Cost Nanocomposite for Removal of Divalent Toxic Metal Ions from Water. *Int. J. Biol. Macromol.* **2019,** *137*, 455–468.

Mendes, L. D.; Bernardi, G.; Elias, W. C.; de Oliveira, D. C.; Domingos, J. B.; Carasek, E. A Green Approach to DDT Degradation and Metabolite Monitoring in Water Comparing the Hydrodechlorination Efficiency of Pd, Au-on-Pd and Cu-on-Pd Nanoparticle Catalysis. *Sci. Total Environ.* **2021,** *760*. DOI: org/10.1016/j.scitotenv.2020.143403.

Mesa, J. J. M.; García Arias, J. A.; Rojas, H. A.; Cárdenas Espinosa, O. E.; Photocatalytic Degradation of Phenol, Catechol and Hydroquinone Over Au-ZnO Nanomaterials. *Rev. Fac. Ing. Univ. Antioquia* **2020,** *94*, 24–32.

Morales-Flores, N.; Pal, U.; Mora, E. S. Photocatalytic Behavior of ZnO and Pt-Incorporated ZnO Nanoparticles in Phenol Degradation. *Appl. Catal. A: Gen.* **2011,** *394* (1–2), 269–275.

Nag, S.; Pramanik, A.; Chattopadhyay, D.; Bhattacharyya, M. Green-Fabrication of Gold Nanomaterials Using *Staphylococcus warneri* from Sundarbans Estuary: An Effective Recyclable Nanocatalyst for Degrading Nitro Aromatic Pollutants. *Environ. Sci. Poll. Res.* **2018,** *25* (3), 2331–2349.

Nagpal, V.; Bokare, A. D.; Chikate, R. C.; Rode, C. V.; Paknikar, K. M. Reductive Dechlorination of γ-Hexachlorocyclohexane Using Fe–Pd Bimetallic Nanoparticles. *J. Hazard. Mater.* **2010,** *175* (1–3), 680–687.

Narayanan, K. B.; Sakthivel, N. Synthesis and Characterization of Nano-Gold Composite Using *Cylindrocladiumfloridanum* and Its Heterogeneous Catalysis in the Degradation of 4-Nitrophenol. *J. Hazard. Mater.* **2011,** *189* (1–2), 519–525.

Naushad, M.; Ahamad, T.; Sharma, G.; Ala'a, H.; Albadarin, A. B.; Alam, M. M. et al. Synthesis and Characterization of a New Starch/SnO_2 Nanocomposite for Efficient Adsorption of Toxic Hg^{2+} Metal Ion. *Chem. Eng. J.* **2016,** *300*, 306–316.

Nawaz, R.; Kait, C. F.; Chia, H. Y.; Isa, M.H.; Huei, L. W. Glycerol-Mediated Facile Synthesis of Colored Titania Nanoparticles for Visible Light Photodegradation of Phenolic Compounds. *Nanomater.* **2019,** *9* (11). DOI: 10.3390/nano9111586.

Nguyen, T. H.; Nguyen, T. T. L.; Pham, T. D.; Le, T. S. Removal of Lindane from Aqueous Solution Using Aluminum Hydroxide Nanoparticles with Surface Modification by Anionic Surfactant. *Polymers*, **2020,** *12* (4). DOI: org/10.3390/polym12040960

Noman, M.; Shahid, M.; Ahmed, T.; Niazi, M. B. K.; Hussain, S.; Song, F. et al. Use of Biogenic Copper Nanoparticles Synthesized from a Native *Escherichia* sp. as Photocatalysts for Azo Dye Degradation and Treatment of Textile Effluents. *Environ. Poll.* **2020,** *257*. DOI: 10.1016/j.envpol.2019.113514.

Paknikar, K. M.; Nagpal, V.; Pethkar, A. V.; and Rajwade, J. M. Degradation of Lindane from Aqueous Solutions Using Iron Sulfide Nanoparticles Stabilized by Biopolymers. *Sci. Technol. Adv. Mater.* **2005,** *6* (3–4), 370–374.

Pathak, B.; Ameta, S. C.; Ameta, R.; Visible Light Induced Photocatalytic Degradation of Bismarck Brown Using $CaMoO_4$-MWCNT Composite. *J. Nanosci. Technol.* **2019,** *5* (5), 832–835.

Pouretedal, H. R.; Kadkhodaie, A. Synthetic CeO_2 Nanoparticle Catalysis of Methylene Blue Photodegradation: Kinetics and Mechanism. *Chin. J. Catal.* **2010,** *31* (11–12), 1328–1334.

Pouretedal, H. R.; Norozi, A.; Keshavarz, M. H.; Semnani, A. Nanoparticles of Zinc Sulfide Doped with Manganese, Nickel and Copper as Nanophotocatalyst in the Degradation of Organic Dyes. *J. Hazard. Mater.* **2009,** *162* (2–3), 674–681.

Poursaberi, T.; Konoz, E.; Sarrafi, A. M.; Hassanisadi, M.; Hajifathli, F. Application of Nanoscale Zero-Valent Iron in the Remediation of DDT from Contaminated Water. *Chem. Sci. Trans.* **2012,** *1* (3), 658–668.

Pushkar, B.; Sevak, P. Adsorption Study of Green Synthesized Fe-Oxide Nanoparticle for DDT Removal. *Int. J. Pharm. Sci. Rev. Res.* **2019,** *55* (2), 84–90.

Rabiee, H.; Vatanpour, V.; Farahani, M. H. D. A.; Zarrabi, H. Improvement in Flux and Antifouling Properties of PVC Ultrafiltration Membranes by Incorporation of Zinc Oxide (ZnO) Nanoparticles. *Sep. Purific. Technol.* **2015,** *156*, 299–310.

Rafique, M.; Shafiq, F.; Gillani, S. S. A.; Shakil, M.; Tahir, M. B.; Sadaf, I.; Eco-Friendly Green and Biosynthesis of Copper Oxide Nanoparticles Using *Citrofortunellamicrocarpa* Leaves Extract for Efficient Photocatalytic Degradation of Rhodamin B Dye form Textile Wastewater. *Optik* **2020,** *208*. DOI: 10.1016/j.ijleo.2019.164053

Rahman, Q. I.; Ahmad, M.; Misra, S. K.; Lohani, M. Effective Photocatalytic Degradation of Rhodamine B Dye by ZnO Nanoparticles. *Mater. Lett.* **2013,** *91*, 170–174.

Raliya, R.; Avery, C.; Chakrabarti, S.; Biswas, P.; Photocatalytic Degradation of Methyl Orange Dye by Pristine Titanium Dioxide, Zinc Oxide, and Graphene Oxide Nanostructures and Their Composites under Visible Light Irradiation. *Appl. Nanosci.* **2017,** *7* (5), 253–259.

Ramya, E.; Thirumurugan, A.; Rapheal, V. S.; Anand, K. CuO@ SiO_2 Nanoparticles Assisted Photocatalytic Degradation of 4-nitrophenol and Their Antimicrobial Activity Studies. *Environ. Nanotechnol. Monito. Manage.* **2019,** *12*. DOI: 10.1016/j.enmm.2019.100240

Rani, M.; Shanker, U. Photocatalytic Degradation of Toxic Phenols from Water Using Bimetallic Metal Oxide Nanostructures. *Colloids Surf. A: Physicochem. Eng. Aspects*, **2018,** *553*, 546–561.

Rani, M.; Shanker, U. Removal of Chlorpyrifos, Thiamethoxam, and Tebuconazole from Water Using Green Synthesized Metal Hexacyanoferrate Nanoparticles. *Environ. Sci. Poll. Res.* **2018,** *25* (11), 10878–10893.

Rashid, J.; Barakat, M. A.; Ruzmanova, Y.; Chianese, A. $Fe_3O_4/SiO_2/TiO_2$ Nanoparticles for Photocatalytic Degradation of 2-chlorophenol in Simulated Wastewater. *Environ. Sci. Poll. Res.* **2015,** *22* (4), 3149–3157.

Rupa, E. J.; Kaliraj, L.; Abid, S.; Yang, D. C.; Jung, S. K. Synthesis of a Zinc Oxide Nanoflower Photocatalyst from Sea Buckthorn Fruit for Degradation of Industrial Dyes in Wastewater Treatment. *Nanomaterials* **2019,** *9* (12). DOI: 10.3390/nano9121692.

Sadeghzadeh-Attar, A. Efficient Photocatalytic Degradation of Methylene Blue Dye by SnO_2 Nanotubes Synthesized at Different Calcination Temperatures. *Sol. Energy Mater. Solar Cells*, **2018,** *183*, 16–24.

Safardoust-Hojaghan, H.; Salavati-Niasari, M. Degradation of Methylene Blue as a Pollutant with N-doped Graphene Quantum Dot/Titanium Dioxide Nanocomposite. *J. Clean. Prod.* **2017,** *148*, 31–36.

Sahiner, N.; Karakoyun, N.; Alpaslan, D.; Aktas, N. Biochar-Embedded Soft Hydrogel and Their Use in Ag Nanoparticle Preparation and Reduction of 4-nitro Phenol. *Int. J. Polym. Mater. Polym. Biomater.* **2013,** *62* (11), 590–595.

Sahoo, A.; Tripathy, S. K.; Dehury, N.; Patra, S. A Porous Trimetallic Au@ Pd@ Ru Nanoparticle System: Synthesis, Characterisation and Efficient Dye Degradation and Removal. *J. Mater. Chem. A*, **2015,** *3* (38), 19376–19383.

Saikia, L.; Bhuyan, D.; Saikia, M.; Malakar, B.; Dutta, D. K.; Sengupta, P.; Photocatalytic Performance of ZnO Nanomaterials for Self Sensitized Degradation of Malachite Green Dye under Solar Light. *Appl. Catal. A: Gen.***2015,** *490*, 42–49.

Salazar, S.; Guerra, D.; Yutronic, N.; Jara, P. Removal of Aromatic Chlorinated Pesticides from Aqueous Solution Using β-Cyclodextrin Polymers Decorated with Fe_3O_4 Nanoparticles. *Polymers* **2018,** *10* (9). DOI: 10.3390/polym10091038.

Salem, M. A.; Al-Ghonemiy, A. F.; Zaki, A. B. Photocatalytic Degradation of Allura Red and Quinoline Yellow with Polyaniline/TiO_2 Nanocomposite. *Appl. Catal. B: Environ.* **2009,** *91* (1–2), 59–66.

Saputra, E.; Muhammad, S.; Sun, H.; Ang, H. M.; Tadé, M. O.; Wang, S. A Comparative Study of Spinel Structured Mn_3O_4, Co_3O_4 and Fe_3O_4 Nanoparticles in Catalytic Oxidation of phenolic contaminants in Aqueous Solutions. *J. Colloid Interf. Sci.* **2013,** *407*, 467–473.

Sargin, I.; Baran, T.; Arslan, G. Environmental Remediation by Chitosan-Carbon Nanotube Supported Palladium Nanoparticles: Conversion of Toxic Nitroarenes into Aromatic Amines, Degradation of Dye Pollutants and Green Synthesis of Biaryls. *Sep. Purif. Technol.* **2020,** *247*. DOI: 10.1016/j.seppur.2020.116987

Schrick, B.; Blough, J. L.; Jones, A. D.; Mallouk, T. E. Hydrodechlorination of Trichloroethylene to Hydrocarbons Using Bimetallic Nickel–Iron Nanoparticles. *Chem. Mater.* **2002,** *14* (12), 5140–5147.

Sen Gupta, S.; Chakraborty, I.; Maliyekkal, S. M.; Adit Mark, T.; Pandey, D. K.; Das, S. K. et al. Simultaneous Dehalogenation and Removal of Persistent Halocarbon Pesticides from Water Using Graphene Nanocomposites: A Case Study of Lindane. *ACS Sustain. Chem. Eng.* **2015,** *3* (6), 1155–1163.

Senthilnathan, J.; Philip, L. Photocatalytic Degradation of Lindane under UV and Visible Light Using N-doped TiO_2. *Chem.Eng. J.* 2010a, *161* (1–2), 83–92.

Senthilnathan, J.; Philip, L. Removal of Mixed Pesticides from Drinking Water System Using Surfactant-Assisted Nano-TiO_2. *Water Air Soil Poll.* 2010b, *210* (1), 143–154.

Shaoqing, Y.; Jun, H.; Jianlong, W. Radiation-Induced Catalytic Degradation of p-Nitrophenol (PNP) in the Presence of TiO_2 Nanoparticles. *Rad. Phys. Chem.* **2010,** *79* (10), 1039–1046.

Shen, J.; Zhou, Y.; Li, S.; Gu, P.; Xue, G. Hydrogel-Coated Fe_3O_4 Nanoparticles as an Efficient Heterogeneous Fenton Catalyst for Degradation of Phenol. *J. Mater. Sci.* **2019,** *54* (15), 10684–10694.

Shih, Y. H.; Chen, M. Y.; Su, Y. F. Pentachlorophenol Reduction by Pd/Fe Bimetallic Nanoparticles: Effects of Copper, Nickel, and Ferric Cations. *Appl. Catal. B: Environ.* **2011,** *105* (1–2), 24–29.

Shrivas, K.; Ghosale, A.; Nirmalkar, N.; Srivastava, A.; Singh, S. K.; Shinde, S. S. Removal of Endrin and Dieldrin Isomeric Pesticides through Stereoselective Adsorption Behavior on the Graphene Oxide-Magnetic Nanoparticles. *Environ. Sci. Poll. Res.* **2017,** *24* (32), 24980–24988.

Şimşek, B.; Sevgili, İ.; Ceran, Ö. B.; Korucu, H.; Şara, O. N. Nanomaterials Based Drinking Water Purification: Comparative Study with a Conventional Water Purification Process. *Period. Polytech. Chem. Eng.* **2019,** *63* (1), 96–112.

Sinha, A. K.; Pradhan, M.; Sarkar, S.; Pal, T. Large-Scale Solid-State Synthesis of Sn–SnO_2 Nanoparticles from Layered SnO by Sunlight: A Material for Dye Degradation in Water by Photocatalytic Reaction. *Environ. Sci. Technol.* **2013,** *47* (5), 2339–2345.

Soltani, T.; Entezari, M. H. Solar Photocatalytic Degradation of RB5 by Ferrite Bismuth Nanoparticles Synthesized via Ultrasound. *Ultrasonics Sonochem.* **2013,** *20* (5), 1245–1253.

Song, X.; Shi, X. Bioreductive Deposition of Highly Dispersed Ag Nanoparticles on Carbon Nanotubes with Enhanced Catalytic Degradation for 4-Nitrophenol Assisted by Shewanellaoneidensis MR-1. *Environ. Sci. Poll. Res.* **2017,** *24* (3), 3038–3044.

Su, J.; Lin, S.; Chen, Z.; Megharaj, M.; Naidu, R. Dechlorination of p-chlorophenol from Aqueous Solution Using Bentonite Supported Fe/Pd Nanoparticles: Synthesis, Characterization and Kinetics. *Desalination* **2011,** *280* (1–3), 167–173.

Tee, Y. H.; Grulke, E.; Bhattacharyya, D. Role of Ni/Fe Nanoparticle Composition on the Degradation of Trichloroethylene from Water. *Indus. Eng. Chem. Res.* **2005,** *44* (18), 7062–7070.

Tercan, M.; Demirci, S.; Dayan, O.; Sahiner, N. Simultaneous Degradation and Reduction of Multiple Organic Compounds by Poly (Vinyl Imidazole) Cryogel-Templated Co, Ni, and Cu Metal Nanoparticles. *New J. Chem.* **2020,** *44* (11), 4417–4425.

Tetteh, E. K.; Rathilal, S.; Naidoo, D. B. Photocatalytic Degradation of Oily Waste and Phenol from a Local South Africa Oil Refinery Wastewater Using Response Methodology. *Sci. Rep.* **2020,** *10* (1). DOI: 10.1038/s41598-020-65480-5.

Tian, H.; Chen, J.; He, J.; Liu, F. Pd-loaded Magnetic Mesoporous Nanocomposites: A Magnetically Recoverable Catalyst with Effective Enrichment and High Activity for DDT and DDE Removal under Mild Conditions. *J. Colloid Interf. Sci.* **2015,** *457*, 195–202.

Tian, H.; Li, J.; Shen, Q.; Wang, H.; Hao, Z.; Zou, L.; Hu, Q. Using Shell-Tunable Mesoporous Fe_3O_4@ HMS and Magnetic Separation to Remove DDT from Aqueous Media. *J. Hazard. Mater.* **2009,** *171* (1–3), 459–464.

Tripathi, A.; Narayanan, S. Impact of TiO_2 and TiO_2/g-C_3N_4 Nanocomposite to Treat Industrial Wastewater. *Environ. Nanotechnol. Monitor. Manage.* **2018,** *10*, 280–291.

Tso, C. P.; Kuo, D. T. F.; Shih, Y. H. Removal of Hexabromocyclododecane by Carboxymethyl Cellulose Stabilized Fe and Ni/Fe Bimetallic Nanoparticles: The Particle Stability and Reactivity in Water. *Chemosphere* **2020,** *250*. DOI: org/10.1016/j.chemosphere.2020.126155.

Ullah, R.; Dutta, J. Photocatalytic Degradation of Organic Dyes with Manganese-Doped ZnO Nanoparticles. *J. Hazard. Mater.* **2008,** *156* (1–3), 194–200.

Ulucan-Altuntas, K.; Debik, E. Dechlorination of Dichlorodiphenyltrichloroethane (DDT) by Fe/Pd Bimetallic Nanoparticles: Comparison with nZVI, Degradation Mechanism, and Pathways. *Front. Environ. Sci. Eng.* **2020,** *14* (1), 1–13.

Vinod, V. T. P.; Wacławek, S.; Senan, C.; Kupčík, J.; Pešková, K.; Černík, M.; Somashekarappa, H. M. Gum Karaya (Sterculia Urens) Stabilized Zero-Valent Iron Nanoparticles: Characterization and Applications for the Removal of Chromium and Volatile Organic Pollutants from Water. *RSC Adv.* **2017,** *7* (23), 13997–14009.

Vuong Hoan, N. T.; Thu, N. T. A.; Duc, H. V.; Cuong, N. D.; Khieu, D. Q.; Vo, V. 3Fe_3O_4/Reduced Graphene Oxide Nanocomposite: Synthesis and Its Application for Toxic Metal Ion Removal. *J. Chem.* **2016,** *2016*. DOI: 10.1155/2016/2418172.

Wan, C.; Li, J. Facile Synthesis of Well-Dispersed Superparamagnetic γ-Fe_2O_3 Nanoparticles Encapsulated in Three-Dimensional Architectures of Cellulose Aerogels and Their Applications for Cr(VI) Removal from Contaminated Water. *ACS Sustain. Chem. Eng.* **2015,** *3* (9), 2142–2152.

Wang, X.; Chen, C.; Chang, Y.; Liu, H. Dechlorination of Chlorinated Methanes by Pd/Fe Bimetallic Nanoparticles. *J. Hazard. Mater.* **2009,** *161* (2–3), 815–823.

Wang, X.; Ning, P.; Liu, H.; Ma, J. Dechlorination of Chloroacetic Acids by Pd/Fe Nanoparticles: Effect of Drying Method on Metallic Activity and the Parameter Optimization. *Appl. Catal. B: Environ.* **2010,** *94* (1–2), 55–63.

Wanjeri, V. W. O.; Sheppard, C. J.; Prinsloo, A. R. E.; Ngila, J. C.; Ndungu, P. G. Isotherm and Kinetic Investigations on the Adsorption of Organophosphorus Pesticides on Graphene Oxidebased Silica Coated Magnetic Nanoparticles Functionalized with 2-phenylethylamine. *J. Environ. Chem. Eng.* **2018,** *6* (1), 1333–1346.

Warang, T.; Patel, N.; Fernandes, R.; Bazzanella, N.; Miotello, A. Co_3O_4 Nanoparticles Assembled Coatings Synthesized by Different Techniques for Photo-Degradation of Methylene Blue Dye. *Appl. Catal. B: Environ.* **2013,** *132*, 204–211.

Wu, L.; Ritchie, S. M. Removal of Trichloroethylene from Water by Cellulose Acetate Supported Bimetallic Ni/Fe Nanoparticles. *Chemosphere*, **2008,** *63* (2), 285–292.

Wysocka, I.; Kowalska, E.; Trzciński, K.; Łapiński, M.; Nowaczyk, G.; Zielińska-Jurek, A. UV-Vis-induced degradation of phenol over magnetic photocatalysts modified with Pt, Pd, Cu and Au nanoparticles. *Nanomater.* **2018,** *8* (1). DOI: 10.3390/nano8010028

Xu, L.; Wang, J. Fenton-Like Degradation of 2, 4-dichlorophenol Using Fe_3O_4 Magnetic Nanoparticles. *Appl. Catal. B: Environ.* **2012,** *123*, 117–126.

Yang, B.; Zhang, Y.; Deng, S.; Yu, G.; Lu, Y.; Wu, J.; Shi, L. Reductive Degradation of Chlorinated Organic Pollutants-Contaminated Water by Bimetallic Pd/Al Nanoparticles: Effect of Acidic Condition and Surfactants. *Chem. Eng. J.* **2013,** *234*, 346–353.

Yin, J.; Zhan, F.; Jiao, T.; Deng, H.; Zou, G.; Bai, Z. et al. Highly Efficient Catalytic Performances of Nitro Compounds via Hierarchical PdNPs-Loaded MXene/Polymer Nanocomposites Synthesized through Electrospinning Strategy for Wastewater Treatment. *Chin. Chem. Lett.* **2020,** *31* (4), 992–995.

Yola, M. L.; Eren, T.; Atar, N. A Novel Efficient Photocatalyst Based on TiO_2 Nanoparticles Involved Boron Enrichment Waste for Photocatalytic Degradation of Atrazine. *Chem. Eng. J.* **2014,** *250*, 288–294.

Yu, B.; Zeng, J.-B.; GongL.; Zhang, M.; Zhang, L.; Chen, X. Investigation of the Photocatalytic Degradation of Organochlorine Pesticides on a Nano-TiO_2 Coated Film. *Talanta* **2007,** *72* (5), 1667–1674.

Zhang, S.; Zhao, X.; Niu, H.; Shi, Y.; Cai, Y.; Jiang, G. Superparamagnetic Fe_3O_4 Nanoparticles as Catalysts for the Catalytic Oxidation of Phenolic and Aniline Compounds. *J. Hazard. Mater.* **2009a,** *167* (1–3), 560–566.

Zhang, Z.; Cissoko, N.; Wo, J.; Xu, X. Factors Influencing the Dechlorination of 2, 4-dichlorophenol by Ni–Fe Nanoparticles in the Presence of Humic Acid. *J. Hazard. Mater.* **2009b,** *165* (1–3), 78–86.

Zhang, W.; Shi, X.; Zhang, Y.; Gu, W.; Li, B.; Xian, Y. Synthesis of Water-Soluble Magnetic Graphene Nanocomposites for Recyclable Removal of Heavy Metal Ions. *J. Mater. Chem. A* **2013,** *1* (5), 1745–1753.

Zhang, Z.; Yuan, Y.; Liang, L.; Cheng, Y.; Shi, G.; Jin, L. Preparation and Photoelectrocatalytic Activity of ZnO Nanorods Embedded in Highly Ordered TiO_2 Nanotube Arrays Electrode for Azo Dye Degradation. *J. Hazard. Mater.* **2008,** *158* (2–3), 517–522.

Zhu, J.; Wei, S.; Gu, H.; Rapole, S. B.; Wang, Q.; Luo, Z. et al. One-Pot Synthesis of Magnetic Graphene Nanocomposites Decorated with Core@ Double-Shell Nanoparticles for Fast Chromium Removal. *Environ. Sci. Technol.* **2012,** *46* (2), 977–985.

CHAPTER 9

Layer-by-Layer Assembly as a Powerful Nanofabrication Technique for the Design of Antimicrobial Surfaces in the Textile Industry

PEDRO J. RIVERO*

Materials Engineering Laboratory, Institute for Advanced Materials and Mathematics (INAMAT2), Department of Engineering, Public University of Navarre, Campus Arrosadía s/n, 31006, Pamplona, Spain

E-mail: pedrojose.rivero@unavarra.es

ABSTRACT

In this work, the fabrication of multilayer structures has been presented by using the layer-by-layer (LbL) nanoassembly technique with the aim to prevent the bacterial adhesion, pathogen colonization, or even contact-killing bacteria. The use of this nanofabrication technique is of great interest because the resultant thickness of the films can be perfectly controlled as a function of the electrostatic attraction between positive and negative charged compounds. In addition, this technique shows a great versatility because it can be implemented to a wide variety of substrates with any shape and composition (i.e., polymers, composites, medical devices, metallic substrates) without using any expensive equipment. In some cases, the polyelectrolytes can play a dual role for preparing multilayer antimicrobial surfaces. The first one is that most of them can be used for embedding active compounds, which can be released in a further step, while the second one, controlling the pH of the dipping cationic polyelectrolyte solutions can induce the presence of higher protonated charges in the outer surface, which

The Science of Nanomaterials: Basics and Applications. Suresh C. Ameta & Rakshit Ameta (Eds.)

can act as a better effective antimicrobial activity. Finally, a perspective in detail of these different approaches will be deeply analyzed against different bacterial strains.

9.1 INTRODUCTION

In recent years, the scientific community is continuously investigating in the design and development of functional coatings in the textile industry; thanks to the implementation of the nanotechnology (Rivero et al., 2015; Gottschalk and Nowack, 2011). The application of well-defined nanostructures or nanoparticles onto the yarns, or well directly to the final fabrics makes it possible to incorporate new functionalities to the textile (Yetisen et al., 2016). According to this, it is possible to engineer the properties of a fabric of different nature (natural or synthetic), and as a result, providing antibacterial activity (Vigneshwaran et al., 2007), flame retardant (Horrocks, 2011), UV-protection (Abidi et al., 2009), or even superhydrophobicity (Wang et al., 2011). Among all these additional functions, one of the most relevant is the antimicrobial or antibacterial activity because the textile fabrics are known to be very suitable environments for microbial growth under specific ambient conditions because humid conditions and warm temperatures are associated with establishing a desirable environment for the growth of bacteria, fungi, and molds. According to this, the ability for inhibiting or preventing the bacterial growth on the textiles is a critical factor to avoid fabric deterioration or discoloration (Mcmullan et al., 2001), the presence of unpleasant odors in a prolongated time (Callewaert et al., 2014), or the spread of infectious diseases through contaminated clothes (Sattar et al., 2001).

An important aspect to remark is that the applications of antimicrobial textiles can widely differ (e.g., infection control, wound healing, odor control), although in all the cases, the same active substances are used with a specific antibacterial activity. With this objective, the layer-by-layer (LbL) assembly appears to be one of the most promising techniques for the resultant functionalization of textiles because it is possible to fabricate well-ordered nanostructures with a great control over the chemical composition as well as the thickness coating. Finally, the wide range of disposable polyelectrolytes or substances chemically charged in the formation of a multilayer structure makes it possible that LbL can be considered as one of the most versatile nanofabrication techniques for the design of engineering surfaces, which can prevent bacterial adhesion and kill bacteria by direct contact-active or even by releasing specific antibacterial agents immobilized in the multilayer structure. Here, first the fundamentals of the LbL assembly, as well as their advantages in the field of nanotechnology, have been discussed and second, different strategies are presented for the design and development of highly effective antibacterial surfaces.

9.2 FUNDAMENTALS OF THE LAYER-BY-LAYER ASSEMBLY

The LbL nanoassembly technique, also known as electrostatic self-assembly multilayer, is a nanodeposition technique, which consists of the alternate immersion of a reference substrate into aqueous polyelectrolyte solutions with opposite electric charge, being the electrostatic attraction the main force to adsorb onto a surface (Decher, 1997; Choi and Rubner, 2005). Polyelectrolytes are polymers with ionizable functional groups that form charged polyions (with an overall positive or negative charge); thus, constituting either polycations or polyanions, respectively. According to this, a substance with positively charged functional groups (mostly quaternary ammonium or amino groups) is used as a polycation. The most significant examples are poly(allylamine hydrochloride) (PAH), polydiallyldimethylammonium chloride (PDDA), and poly(ethyleneimine) (PEI). The substance with negatively charged groups (mostly sulfonic acids or carboxylic acids) is the most common functional groups used in polyanions. The most common examples are poly(sodium 4-styrene sulfonate) (PSS) and poly(acrylic acid sodium salt) (PAA). In addition, another important aspect to remark related to these polyelectrolytes is that these charged polymers can be classified as strong or weak polyelectrolytes. According to this classification, polyelectrolytes that maintain a high and constant ionization ratio over a broad range of pH conditions are termed as "strong polyelectrolytes" (being the most representative examples PDDA or PSS), whereas polyelectrolytes that exhibit pH-dependent ionization degrees are called "weak polyelectrolytes" (being the most known examples PAH or PAA) (Rivero et al., 2019). However, the continuous advances in the LbL assembly technologies have explored different driving forces for the assembly process, when compared with the diffusion-driven kinetics of classical LbL assembly based on immersion steps in charged polyelectrolytes solutions (Richardson et al., 2015). According to this, this technology can be exploited into five broad categories for preparing functional materials denoted as immersive, spin, spray, electromagnetic, and fluidic assembly. A representative example for the preparation of functional coatings based on an immersive assembly based on the incorporation of silver nanoparticles (Richardson et al., 2015) which have an intrinsic antibacterial behavior is presented in Figure 9.1.

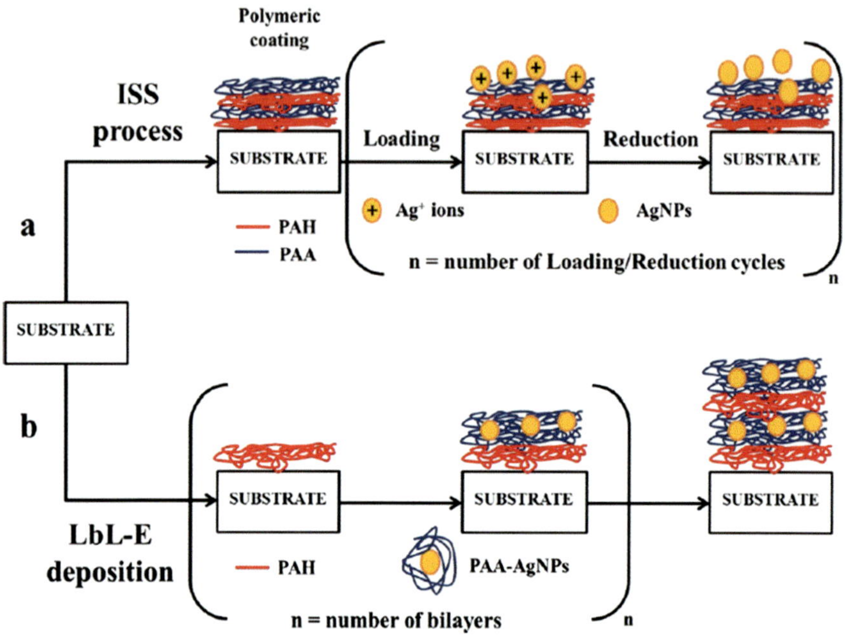

FIGURE 9.1 Schematic representation of the two alternative methods for the synthesis of AgNPs. (a) ISS process. (b) LbL-E deposition technique.

Source: Adapted from Rivero et al. (2014). With permission.

This technique offers a good number of advantages in comparison with other techniques. One of them is its simplicity because it does not require any special or expensive equipment and can be applied onto surfaces with any shape (cylindrical, conical, or plane). In addition, the thickness coating can be perfectly controlled with a high precision by just selecting the number of deposited layers. Finally, another great benefit of this technique is that it allows the deposition of different kinds of substances, such as polymers, fluorescent indicators, metallic or metal oxide nanoparticles, quantum dots, active compounds, or even colorimetric indicators.

9.3 ANTIMICROBIAL TEXTILES

As it was previously commented, the LbL assembly technique can be considered as a very effective nanofabrication technique for the design of antibacterial surfaces because it is possible to build highly controllable thin films with customizable compositions. In this section, different strategies will be presented for the design of effectively antibacterial surfaces based on contact-active antimicrobial coatings due to the presence of charged polyelectrolytes, or well by using the release-killing effect of specific antibacterial agents.

9.3.1 CONTACT-ACTIVE ANTIMICROBIAL COATINGS

One of the most known polyelectrolytes with an intrinsic antimicrobial behavior is chitosan, which is a natural linear polysaccharide comprised of N-acetyl-D-glucosamine and β-(1-4)-linked D-glucosamine with other interesting features, such as biodegradability, nonallergic, and nontoxicity. In addition, this polyelectrolyte in acidic conditions presents cationic groups, which exerts the corresponding antibacterial activity (Dash et al., 2011; Carmona-Ribeiro and Carrasco, 2013). The functionalization of linen fabric using LbL treatment with chitosan (CS) and green tea extract (GTE) has been presented Saini et al. (2020), showing multifunctional properties such as antibacterial activity in the range 75%–97% against *Staphylococcus aureus* bacteria, UV-protection, and antioxidant activity, respectively. An antibacterial finishing of cotton fabrics has been reported by Shirvan et al. (2014) with chitosan and pentasodium tripolyphosphate (TPP) self-assembled nanolayers, which can effectively increase the degree of inhibition on *Escherichia coli* and *S. aureus* bacteria. Another interesting approach is based on the design of bioinspired and green water repellent finishing of textiles using carnauba wax and LbL self-assembly approach (Bashari et al., 2020a). The use of fine Carnuba wax particles with a cubic shape makes possible a reduction of the surface energy of the fabrics, such as cotton, polyamide 6 (PA6), and cotton/PA6, and as a result, a nanosurface roughness with a water-repellent behavior was obtained. These resultant particles show an intrinsic negative charge (corroborated by zeta potential), and the chitosan is used as a positive charge component for the fabrication of the multilayer structure. The experimental results indicate that the treated fabrics have shown antibacterial effect against *E. coli* and *S. aureus* bacteria; thanks to the antibacterial behavior of the chitosan.

The same type of Carnauba wax nanoparticles (CWNs) has been also employed by Bashari et al. (2020b) for the design of eco-friendly dual-functional textiles with a water-repellent and antibacterial behavior. For this purpose, natural-based nanometer-sized particles were used to prepare the roughness with low-surface energy on fabric by using LbL assembly of CWNs and zinc oxide (ZnO) nanoparticles onto cotton, cotton/nylon6, and nylon6 fabrics. The antibacterial activities toward Gram-positive as well as Gram-negative bacterial strains due to the presence of ZnO on fabrics have been corroborated (Belay et al., 2020) ZnO NPs were synthesized at different temperatures and reaction media of water and 1,2-ethanediol by using precipitation and in situ methods on the surface of cotton fabrics (Figure 9.2). In addition, the antibacterial activities of ZnO NPs synthesized by these two methods possess very good bacteriostatic activity against *S. aureus* and *E. coli* bacteria demonstrated by the zone of inhibition (Figure 9.3).

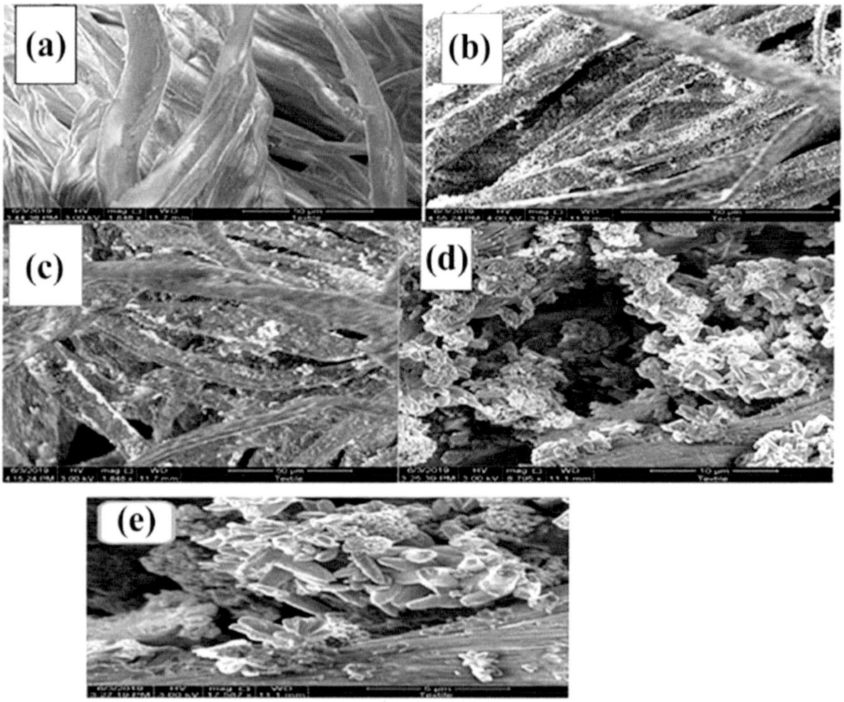

FIGURE 9.2 The SEM image of (a) uncoated cotton textiles, (b) cotton textiles coated with ZnO NPs synthesized using precipitation in a water medium, (c) cotton textiles coated with ZnO NPs synthesized using precipitation in 1,2-ethanediol medium, and (d, e) cotton textiles coated by ZnO NPs synthesized using an in situ method. *SEM*, scanning electron microscopy; *ZnO*, zinc oxide; *NP*, nanoparticle.

Source: Adapted from Belay et al. (2020). With permission. https://creativecommons.org/licenses/by/4.0/

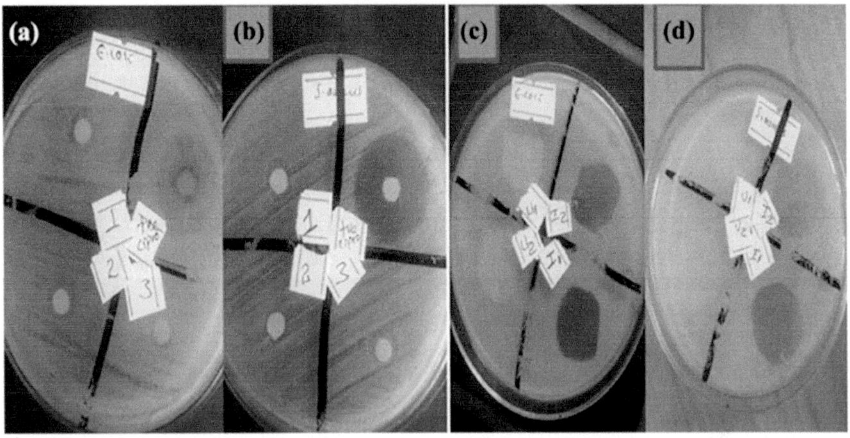

FIGURE 9.3 The antibacterial activities of ZnO NPs with different concentrations applied on *E. coli* and *S. aureus* synthesized using (a, b) precipitation method in water medium and (c, d) in situ method are shown. The numbers 1, 2, and 3 and U1, U2, and U3 show the concentrations of ZnO NPs and their corresponding zone of inhibition for ZnO NPs synthesized using precipitation method in water medium and in situ method, respectively. ZnO: zinc oxide; NP: nanoparticle.

Source: Adapted from Belay et al. (2020). With permission. https://creativecommons.org/licenses/by/4.0/

The use of other types of metal oxide nanoparticles such as titania (TiO_2) or silica (SiO_2) can be also implemented onto the textile fabrics. Photocatalytic multilayers composed of lignosulfonates (LS) and TiO_2 nanoparticles were deposited onto cellulose fiber surfaces (Li et al., 2013). The experimental results indicated that the degree of *E. coli* growth inhibition increased with the increasing bilayer number; this effect is more notorious on the modified cellulose fibers with TiO_2 NPs in the outmost layer than the sample modified with LS in the outmost layer. The sprayed LbL assembly of TiO_2 photocatalyst was used by Carré et al. (2015) to functionalize polyester fabrics and elaborate solar active antibacterial textiles. A synergetic effect resulting from both, biocidal photocatalytic activity of TiO_2 layers and biocidal activity of PEI layers within the multilayer structure, can explain the high self-decontaminating activity of the functionalized textiles. On this basis, the LbL spraying method was also employed, where multifunctional coatings on cotton fabric and leather surfaces are fabricated with the aim to endow them with fragrant sustained releasing behavior, antibacterial property as well as desired wear-resistance in a facile and benign way (Fan et al., 2018).

Here, the presence of chitosan in the top layer provides a more effective antibacterial action against *Aspergillus flavus*, while the use of SiO_2 particles embedded under the chitosan layer can produce an enhancement in the wear-resistance of the coating.

Other interesting work can be found by Safi et al. (2020), where cotton fabrics were modified by using LbL finishing with CS, sodium lignin sulphonate (SLS), and boric acid, showing multifunctional properties such as antioxidant, antibacterial, and UV-protective properties. More specifically, an antibacterial activity in the range of 70%–89% has been obtained. In addition, it is well-known that CS is a good antibacterial agent in textiles and the presence of boric acid and SLS can act in a synergetic way to provide a higher antibacterial activity against *S. aureus*. One sustainable biopolymer which has been emerged as an excellent candidate for antibacterial finishing of textiles is the alginate (Li et al., 2017). It is important to remark that alginate and its applications in medical biomaterials have been widely reported such as tissue engineering, drug delivery, or wound dressing (Augst et al., 2006; Lee and Mooney, 2012).

However, the number of research papers based on the design of antibacterial finishing of textiles has not been deeply studied. In this sense, a novel work can be found from Gomes et al. (2013), where multilayer structures composed of a positive polyelectrolyte (chitosan, CH) and negative polyelectrolyte (alginate, ALG) have been deposited onto cotton fabrics, and its corresponding antibacterial properties have been tested against *S. aureus* and *K. pneumonia*, respectively. The experimental results have demonstrated that the samples experimented with a bacteriostatic effect on the two bacteria tested, and the samples composed of five layers (CH/ALG/CH/ALG/CH) were found to be most effective in inhibition bacterial growth. In previous works, an increase of negative charges has been proposed to the cotton substrate by using sodium bromide (NaBr) and 2,2,6,6-tetramethylpiperidinyl-1-oxy free radical (TEMPO) to ensure the success of the LbL process with a better electrostatic interaction effect between the charged polyelectrolytes (Gomes et al., 2012).

Chen et al. (2017) demonstrated that it is possible to achieve antibacterial effects by the surface modification of cellulosic fibers via LbL technique by using different cationic polyelectrolytes, such as polyvinylamine (PVAm), poly(allylamine chloride) (PAH), and poly(diallyl dimethyl ammonium chloride (PDADMAC), whereas the anionic polyelectrolyte employed is the poly(acrylic acid) (PAA). The charge properties of the modified fibers have been determined by polyelectrolyte titration (PET) and zeta potential, indicating that there is a strong positive correlation between the charge density of the polyelectrolytes and the antibacterial activities concerning the bacterial removal and bacterial growth inhibition against *E. coli.* Cellulose pulp fibers have been modified by Chen and Ek (2018) via LbL deposition of polyvinylamine (PVAm) and wood-based cellulose nano-fibrils (CNF) showing a scheme of three-layer PVAm/CNF/PVAm. The CNF as the middle layer was adsorbed at varied pH and salt concentration, and the experimental results have demonstrated that the combination of a higher salt concentration with a lower assembly pH in the multilayer buildup resulted in a better bacterial reduction.

9.3.2 RELEASE-KILLING EFFECT OF ANTIBACTERIAL AGENTS

One of the most employed antibacterial agents used is silver because it presents a notorious broad-spectrum biocidal effect against bacteria, virus, and microorganisms (Lara et al., 2010). In addition, this excellent antibacterial activity can be implemented in different research fields, such as food preservation (Hoseinnejad et al., 2018), water treatment (Chou et al., 2005), wound dressing (Leaper, 2006), medical devices (Vasilev et al., 2009) or textile fabrics (Gao and Cranston, 2008). A representative example of this highly antibacterial activity related to the silver nanoparticles can be appreciated in Figure 9.4, where it has been compared for two different samples places onto agar slabs. The first one corresponds to a reference substrate (bare glass slide), where it is clearly seen that a high number of *Lactobacillus plantarum* colonies have grown up randomly throughout the whole agar slab. However, the second sample contains silver, where it can be clearly observed that there is no growth of colonies. In addition, another aspect to remark that the behavior is different in the uncoated area of the substrate and in the rest of the agar slab, where a high growth of colonies can be appreciated as it occurs in the reference substrate. These coatings with incorporated silver have reached more than 99.9% of killing efficiency (Rivero et al., 2011). Another interesting aspect is that there is a wide variety of chemical synthetic routes to obtain silver nanoparticles, being possible to synthesize AgNPs or silver clusters with a well-defined morphology (Liz-Marzán, 2004, 2006). A clear example is represented in Figure 9.5, where multicolored silver nanoparticles with different shapes (spherical, rod, hexagonal, triangular) and sizes (from nanometric up to micrometer) have been perfectly synthesized as a function of strict control of three operational parameters, such as an encapsulating agent (polyacrylic acid, PAA), a loading agent (silver nitrate, $AgNO_3$), and a reducing agent (dimethylamineborane, DMAB) (Rivero et al., 2011).

Another interesting aspect to remark is that the real mechanism of the antimicrobial action can be associated with two different types of hypotheses in the scientific community. The first one is associated with the antibacterial activity due to the release of silver ions (Ag^+) from silver nanoparticles (AgNPs), whereas the second one indicates that AgNPs are the responsible for cell death in the cell membrane. Although both cases (ionic or colloidal form) show low toxicity to human cells and a high degree of biocompatibility (Marambio-Jones and Hoek, 2010). Another aspect is that silver compounds can be easily implemented onto the fabrics by using different immersion deposition methods onto the fabrics surface (Rivero et al., 2015). The untreated and treated cotton fabrics against different strains are given in Figure 9.6, showing the samples doped with silver a strong inhibition effect on all the microorganisms, whereas the untreated samples have not shown any inhibition zone (Falletta et al., 2008).

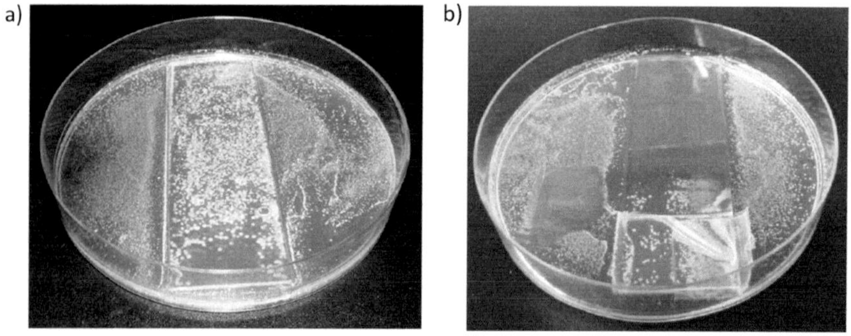

FIGURE 9.4 Bacteria growth on culture plates after 24 h in the (a) reference substrate and (b) coated substrate with silver where the coated area is clearly visible, as far as it inhibits completely the bacteria growth.

Source: Adapted from Rivero et al. (2011). With permission.

The LbL assembly can be used as an effective nanofabrication technique for the successive incorporation of the silver nanoparticles (LbL embedding) because one of the polyelectrolytes can be used as a polyelectrolyte-capped metal nanoparticles, making possible multilayer structures with a desired color as a function of the initial colloidal metal nanoparticles. However, the LbL can be also used as a host matrix for the

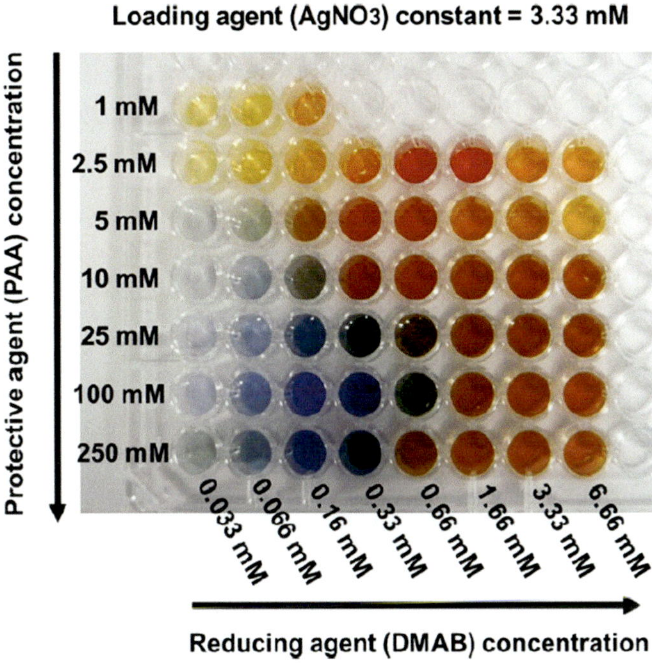

FIGURE 9.5 Photograph of multicolor silver map obtained as a function of variable protective (PAA) and reducing (DMAB) agents, maintaining a fixed molar concentration of the loading agent ($AgNO_3$).

Source: Adapted from Rivero et al. (2013). With permission.

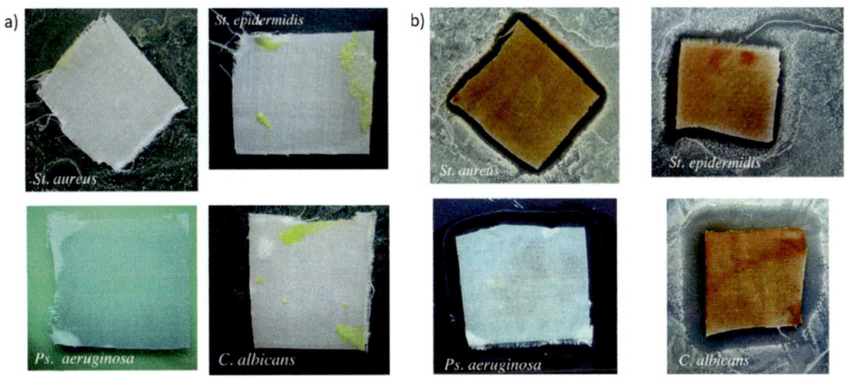

FIGURE 9.6 Effect of the (a) untreated and (b) Ag-treated cotton samples toward the growth of the investigated strains.

Source: Reprinted from Falletta et al. (2008). With permission. © 2008 American Chemical Society.

immobilization of silver ions, which can be reduced to AgNPs by using a chemical process denoted as in situ synthesis (ISS). A summary of both has been presented by Rivero et al. (2014), using poly(allylamine chloride) (PAH) as cationic polyelectrolyte and poly(acrylic acid) (PAA) as anionic polyelectrolyte. An example about the first approach can be found in Ahmed and Emam (2016) where it has been reported that the preparation of multifunctional cotton fabrics and silver nitrate reacts with carboxymethyl cellulose to prepare Ag NPs–carboxymethyl cellulose composite, which are applied onto the cotton fabrics. According to the second approach, a very interesting work has been reported by Meng et al. (2016), where a controllable ISS of silver nanoparticles on multilayered film-coated silk fibers is presented for antibacterial applications. The multilayer structure is built by poly(acrylic acid) (PAA)/poly(dimethyldiallylammonium chloride) (PDDA) which could act as both; 3D-matrix for the Ag NPs growth and stabilizing layer for Ag NPs protection and subsequently, in situ synthesizing AgNPs under UV irradiation. The as-prepared silk could effectively kill the existing bacteria (*E. coli*, *S. aureus*), inhibiting the bacterial growth and demonstrating the antimicrobial activity. Another very promising aspect is that the release of Ag^+ from the modified silk can last for 120 h, rendering the modified silk sustainable antimicrobial activity.

Another promising approach has been reported by Rubin et al. (2018), where surface anchored metal-organic framework (MOF)-cotton material was obtained for tunable antibacterial copper delivery. The LbL growth process has produced a highly uniform and evenly dispersed MOF around the entire cotton fibers.

The release killing effect of an active polyelectrolyte multilayer (PEM) structure of a polyethylene terephthalate (PET) textile fabrics support has been reported by Junthip et al. (2020), using both positive (cationic) and negative (anionic) cyclodextrines polyelectrolytes, denoted as PCD+ and PCD−, respectively. This PEM structure provides a reservoir property and prolonged release of the biocide compound of triclosan (TCS). It was revealed that the presence of quaternary ammonium groups carried by PCD provides intrinsic contact killing property, whereas it has been also appreciated an extrinsic release killing properties against *S. aureus* and *E. coli* once loaded with TCS up to 28 days. Very interesting work has been described by Aubert-Viard et al. (2019), where the surface modification of a polyester nonwoven textile using chitosan (CHT) crosslinked in mild conditions using genipin (Gpn) and the resultant cationic charge surface has been controlled by the reaction time and the concentration of both precursors. Then, a multilayer structure composed of both methyl-beta-cyclodextrin (PCD) (anionic) and CHT (cationic) was fabricated and after that, a specific thermal treatment has been performed in order to stabilize the thin film. Finally, the textiles are loaded with an antiseptic agent, known as chlorhexidine (CHX), and the antibacterial activity in vitro over several weeks has been demonstrated on two strains, such as *S. aureus* and *E. coli*.

Another novel approach was by Gomes et al. (2014) that because an antimicrobial agent such as L-cysteine (L-cys) is perfectly embedded between layers of chitosan (CH) and alginate (ALG), which are deposited onto cotton fabrics. The resultant functionalized CH/ALG cotton samples were investigated for their antibacterial properties toward two different strains such as *S. aureus* and *K. pneumonia*, showing an enhancement of the antibacterial, effect which is associated with the presence of L-cys in the cotton samples.

Finally, a summary about the different multilayer structures evaluated as well as the fabrics employed with their corresponding antibacterial tests against a specific strain is presented in Table 9.1.

9.4 CONCLUSION

The LbL assembly of nanofilms for preparing functional materials with antibacterial properties and their implementation in the textile industry has been studied as an effective easy and cost-effective method. The great versatility and simplicity of this technology combined with the possibility of incorporating a wide variety of charged polyelectrolytes or chemical substances of different nature (metal, semiconductors, and active compounds) makes it of great interest in the fabrication of nanomaterials with desired properties. On this basis, the design and development of novel antibacterial textiles can be performed by using the LbL assembly. Here, the use of multilayer structures has been evaluated, which can prevent the bacterial adhesion, pathogen colonization, or even contact-killing bacteria. The use of several strategies can be employed for this purpose, although the most common is the presence of charged polyelectrolytes in the outer surfaces or well by releasing antibacterial agents from a polymeric host matrix. Finally, a perspective in detail of these different approaches are analyzed against Gram-positive and Gram-negative bacteria.

TABLE 9.1 Summary of the Different LbL Coatings Implemented onto Different Types of Fabrics and the Corresponding Antibacterial Test (Type of Genus) Performed with the Resultant Bibliographic Reference.

Sensitive coating	Type of fabrics	Type of genus	References
CS/GTE	Linen	*Staphylococcus aureus*	Saini et al. (2020)
CS/TPP	Cotton	*Escherichia coli* *S. aureus*	Shirvan et al. (2014)
CS/Carnuba wax	Cotton, PA6 and cotton/PA6	*E. coli* *S. aureus*	Bashari et al. (2020a)
CWNs/ZnO	Cotton, cotton/nylon6 and nylon6	*E. coli* *S. aureus*	Bashari et al. (2020b)
TiO_2/LS	Cellulosic fibers	*E. coli*	Li et al. (2013)
PEI/TiO_2 PSS/TiO_2	Polyester	*E. coli* *Pseudomonas aeruginosa*	Carré et al. (2015)
PAA/Vanillin/SiO_2/CH	Cotton	*Aspergillus flavus*	Fan et al. (2018)
CS/SLS	Cotton	*S. aureus*	Li et al. (2017)
CH/ALG	Cotton	*S. aureus* *Klebsiella pneumonia*	Gomes et al. (2013)
PVAm/PAA PDADMAC/PAA PAH/PAA	Cellulosic fibers	*E. coli*	Chen et al. (2017)
PVAm/CNF/PVAm	Cellulosic fibers	*E. coli*	Chen and Ek (2018)
PDDA/PAA + Ag	Silk	*E. coli* *S. aureus*	Meng et al. (2016)
PCD+/PCD-+ TCS	Poly(ethylene terephthalate)	*E. coli* *S. aureus*	Junthip et al. (2020)
CHT/PCD + CHX	Poly(ethylene terephthalate)	*E. coli* *S. aureus*	Aubert-Viard et al. (2019)
CH/ALG + L-cys	Cotton	*S. aureus Klebsiella pneumonia*	Gomes et al. (2014)

KEYWORDS

- **layer-by-layer**
- **polyelectrolytes**
- **antibacterial agents**
- **antimicrobial surfaces**
- **textile industry**

REFERENCES

Abidi, N.; Cabrales, L.; Hequet, E. Functionalization of a Cotton Fabric Surface with Titania Nanosols: Applications for Self-Cleaning and UV-Protection Properties. *ACS Appl. Mater. Interf.* **2009,** *1* (10), 2141–2146.

Ahmed, H. B.; Emam, H. E. Layer by Layer Assembly of Nanosilver for High Performance Cotton Fabrics. *Fibers Polym.* **2016,** *17* (3), 418–426.

Aubert-Viard, F.; Mogrovejo-Valdivia, A.; Tabary, N.; Maton, M.; Chai, F.; Neut, C. Evaluation of Antibacterial Textile Covered by Layer-by-Layer Coating and Loaded with Chlorhexidine for Wound Dressing Application. *Mater. Sci. Eng. C*, **2019,** *100*, 554–563.

Augst, A. D.; Kong, H. J.; Mooney, D. J. Alginate Hydrogels as Biomaterials. *Macromol. Biosci.* **2006,** *6* (8), 623–633.

Bashari, A.; Salehi, K. A. H.; Salamatipour, N.; Bioinspired and Green Water Repellent Finishing of Textiles Using Carnauba Wax and Layer-by-Layer Technique. *J. Text. Inst.*; **2020a,** *111* (8), 1148–1158.

Bashari, A.; Koohestani, A. H. S.; Salamatipour, N. Eco-friendly Dual-Functional Textiles: Green Water-Repellent & Anti-Bacterial. *Fibers Polym.* **2020b,** *21* (2), 317–323.

Belay, A.; Mekuria, M.; Adam, G. Incorporation of Zinc Oxide Nanoparticles in Cotton Textiles for Ultraviolet Light Protection and Antibacterial Activities. *Nanomater. Nanotechnol.*; **2020,** *10*. DOI: 10.1177/1847980420970052.

Callewaert, C.; De Maeseneire, E.; Kerckhof, F.-M.; Verliefde, A.; Van de Wiele, T.; Boon, N. Microbial Odor Profile of Polyester and Cotton Clothes after a Fitness Session. *Appl. Environ. Microbiol.*; **2014,** *80* (21), 6611–6619.

Carmona-Ribeiro, A. M.; de Melo Carrasco, L. D. Cationic Antimicrobial Polymers and Their Assemblies. *Int. J. Mol. Sci.* **2013,** *14* (5), 9906–9946.

Carré, G.; Garnier, L.; Moeller-Siegert, J.; Gies, J. P.; Keller, V.; André, P. Antibacterial Textiles Functionalized by Layer-by-Layer Assembly of Polyelectrolytes and TiO_2 Photocatalyst. *RSC Adv.* **2015,** *5* (49), 38859–38867.

Chen, C.; Ek, M. Antibacterial Evaluation of CNF/PVAm Multilayer Modified Cellulose Fiber and Cellulose Model Surface. *Nord. Pulp Pap. Res. J.* **2018,** *33* (3), 385–396.

Chen, C.; Illergård, J.; Wågberg, L.; Ek, M. Effect of Cationic Polyelectrolytes in Contact-Active Antibacterial Layer-by-Layer Functionalization. *Holzforschung* **2017,** *71* (7–8), 649–658.

Choi, J.; Rubner, M. F. Influence of the Degree of Ionization on Weak Polyelectrolyte Multilayer Assembly. *Macromolecules* **2005,** *38* (1), 116–124.

Chou, W. L.; Yu, D.-G.; Yang, M. C. The Preparation and Characterization of Silver-Loading Cellulose Acetate Hollow Fiber Membrane for Water Treatment. *Polym. Adv. Technol.* **2005,** *16* (8), 600–607.

Dash, M.; Chiellini, F.; Ottenbrite, R. M.; Chiellini, E. Chitosan—A Versatile Semi-Synthetic Polymer in Biomedical Applications. *Prog. Polym. Sci.* **2011,** *36* (8), 981–1014.

Decher, G. Fuzzy Nanoassemblies: Toward Layered Polymeric Multicomposites. *Science* **1997,** *277* (5330), 1232–1237.

Falletta, E.; Bonini, M.; Fratini, E.; Lo Nostro, A.; Pesavento, G.; Becheri, A. Clusters of Poly(Acrylates) and Silver Nanoparticles: Structure and Applications for Antimicrobial Fabrics. *J. Phys. Chem. C* **2008,** *112* (31), 11758–11766.

Fan, Q.; Ma, J.; Xu, Q.; An, W.; Qiu, R. Multifunctional Coatings Crafted via Layer-by-Layer Spraying Method. *Prog. Org. Coat.* **2018,** *125*, 215–221.

Gao, Y.; Cranston, R. Recent Advaznces in Antimicrobial Treatments of Textiles. *Text. Res. J.* **2008,** *78* (1), 60–72, 2008.

Gomes, A. P.; Mano, J. F.; Queiroz, J. A.; Gouveia, I. C. Layer-by-Layer Deposition of Antimicrobial Polymers on Cellulosic Fibers: A New Strategy to Develop Bioactive Textiles. *Polym. Adv. Technol.* **2013,** *24* (11), 1005–1010.

Gomes, A. P.; Mano, J. F.; Queiroz, J. A.; Gouveia, I. C. Layer-by-Layer Deposition of Antibacterial Polyelectrolytes on Cotton Fibres. *J. Polym. Environ.* **2012,** *20* (4), 1084–1094.

Gomes, A. P.; Mano, J. F.; Queiroz, J. A.; Gouveia, I. C. New Biomaterial Based on Cotton with Incorporated Biomolecules. *J. Appl. Polym. Sci.* **2014,** *131* (15). DOI: 10.1002/a40519.

Gottschalk, F.; Nowack, B. The Release of Engineered Nanomaterials to the Environment. *J. Environ. Monit.* **2011,** *13* (5), 1145–1155.

Horrocks, A. R. Flame Retardant Challenges for Textiles and Fibres: New Chemistry Versus Innovatory Solutions. *Polym. Degrad. Stab.* **2011,** *96* (3), 377–392.

Hoseinnejad, M.; Jafari, S. M.; Katouzian, I. Inorganic and Metal Nanoparticles and Their Antimicrobial Activity in Food Packaging Applications. *Crit. Rev. Microbiol.* **2018,** *44* (2), 161–181.

Junthip, J.; Tabary, N.; Maton, M.; Ouerghemmi, S.; Staelens, J. N.; Cazaux, F. Release-Killing Properties of a Textile Modified by a Layer-by-Layer Coating Based on Two Oppositely Charged Cyclodextrin Polyelectrolytes. *Int. J. Pharm.* **2020,** *587*. DOI: 10.1016/j.ijpharm.2020.119730.

Lara, H. H.; Ayala-Núñez, N. V.; del Turrent, L. C. I.; Padilla, C. R. Bactericidal Effect of Silver Nanoparticles against Multidrug-Resistant Bacteria. *World J. Microbiol. Biotechnol.* **2010,** *26* (4), 615–621.

Leaper, D. J. Silver Dressings: Their Role in Wound Management. *Int. Wound J.* **2006,** *3* (4), 282–294.

Lee, K. Y.; Mooney, D. J. Alginate: Properties and Biomedical Applications. *Prog. Polym. Sci.* **2012,** *37* (1), 106–126.

Li, H.; Fu, S.; Peng, L. Surface Modification of Cellulose Fibers by Layer-by-Layer Self-Assembly of Lignosulfonates and TiO_2 Nanoparticles: Effect on Photocatalytic Abilities and Paper Properties. *Fibers Polym.* **2013,** *14* (11), 1794–1802.

Li, J.; He, J.; Huang, Y. Role of Alginate in Antibacterial Finishing of Textiles. *Int. J. Biol. Macromol.* **2017,** *94*, 466–473.

Liz-Marzán, L. M. Nanometals: Formation and Color. *Mater. Today* **2004,** *7* (2), 26–31.

Liz-Marzán, L. M. Tailoring Surface Plasmons through the Morphology and Assembly of Metal Nanoparticles. *Langmuir* **2006,** *22* (1), 32–41.

Marambio-Jones, C.; Hoek, E. M. V. A Review of the Antibacterial Effects of Silver Nanomaterials and Potential Implications for Human Health and the Environment. *J. Nanopart. Res.* **2010,** *12* (5), 1531–1551.

McMullan, G.; Meehan, C.; Conneely, A.; Kirby, N.; Robinson, T.; Nigam, P. Microbial Decolourisation and Degradation of Textile Dyes. *Appl. Microbiol. Biotechnol.* **2001,** *56* (1–2), 81–87.

Meng, M.; He, H.; Xiao, J.; Zhao, P.; Xie, J.; Lu, Z. Controllable in Situ Synthesis of Silver Nanoparticles on Multilayered Film-Coated Silk Fibers for Antibacterial Application. *J. Colloid Interf. Sci.* **2016,** *461*, 369–375.

Richardson, J. J.; Björnmalm, M.; Caruso, F. Technology-Driven Layer-by-Layer Assembly of Nanofilms. *Science* **2015,** *348* (6233). DOI: 10.1126/science.aaa2491.

Rivero, P. J.; Goicoechea, J.; Arregui, F. J. Layer-by-Layer Nano-Assembly: A Powerful Tool for Optical Fiber Sensing Applications. *Sensors* (Switzerland), **2019,** *19* (3). DOI: 10.3390/s19030683.

Rivero, P. J.; Goicoechea, J.; Matias, I. R.; Arregui, F. J. A Comparative Study of Two Different Approaches for the Incorporation of Silver Nanoparticles into Layer-by-Layer Films. *Nanoscale Res. Lett.* **2014,** *9* (1). DOI: 10.1186/1556-276X-9-301.

Rivero, P. J.; Goicoechea, J.; Urrutia, A.; Arregui, F. J. Effect of Both Protective and Reducing Agents in the Synthesis of Multicolor Silver Nanoparticles. *Nanoscale Res. Lett.* **2013,** *8* (1), 1–9.

Rivero, P. J.; Urrutia, A.; Goicoechea, J.; Arregui, F. J. Nanomaterials for Functional Textiles and Fibers. *Nanoscale Res. Lett.*; **2015,** *10* (1), 1–22.

Rivero, P. J.; Urrutia, A.; Goicoechea, J.; Zamarreño, C. R.; Arregui, F. J.; Matías, I. R. An Antibacterial Coating Based on a Polymer/Sol- Gel Hybrid Matrix Loaded with Silver Nanoparticles. *Nanoscale Res. Lett.*; **2011,** *6* (1). DOI: 10.1186/1556-276X-6-305.

Rubin, H. N.; Neufeld, B. H.; Reynolds, M. M. Surface-Anchored Metal-Organic Framework-Cotton Material for Tunable Antibacterial Copper Delivery. *ACS Appl. Mater. Interf.* **2018,** *10* (17), 15189–15199.

Safi, K.; Kant, K.; Bramhecha, I.; Mathur, P.; Sheikh, J. Multifunctional Modification of Cotton Using Layer-by-Layer Finishing with Chitosan, Sodium Lignin Sulphonate and Boric Acid. *Int. J. Biol. Macromol.* **2020,** *158*, 903–910.

Saini, S.; Gupta, A.; Singh, N.; Sheikh, J. Functionalization of Linen Fabric Using Layer by Layer Treatment with Chitosan and Green Tea Extract. *J. Ind. Eng. Chem.* **2020,** *82*, 138–143.

Sattar, S. A. Springthorpe, S.; Mani, S. Transfer of Bacteria from Fabrics to Hands and Other Fabrics: Development and Application of a Quantitative Method Using *Staphylococcus aureus* as a Model. *J. Appl. Microbiol.* **2001,** *90* (6), 962–970.

Shirvan, A. R.; Nejad, N. H.; Bashari, A. Antibacterial Finishing of Cotton Fabric via the Chitosan/TPP Self-Assembled Nano Layers. *Fibers Polym.* **2014,** *15* (9), 1908–1914.

Vasilev, K.; Cook, J.; Griesser, H. J. Antibacterial Surfaces for Biomedical Devices. *Expert Rev. Med. Devices* **2009,** *6* (5), 553–567.

Vigneshwaran, N.; Kathe, A. A.; Varadarajan, P. V.; Nachane, R. P.; Balasubramanya, R. H. Functional Finishing of Cotton Fabrics Using Silver Nanoparticles. *J. Nanosci. Nanotechnol.* **2007,** *7* (6), 1893–1897.

Wang, L.; Zhang, X.; Li, B.; Sun, P.; Yang, J.; Xu, H. Superhydrophobic and Ultraviolet-Blocking Cotton Textiles. *ACS Appl. Mater. Interf.* **2011,** *3* (4), 1277–1281.

Yetisen, A. K.; Qu, H.; Manbachi, A.; Butt, H.; Dokmeci, M. R.; Hinestroza, J. P. Nanotechnology in Textiles. *ACS Nano*, **2016,** *10* (3), 3042–3068.

CHAPTER 10

Prospect of Utilizing Novel Nanomaterials for Effective Food Security, Safety, Packaging, and Preservation

C. O. ADETUNJI[1*], O. A. ANANI[2], O. T. OLANIYAN[3], A. INOBEME[4], R. E. BODUNRINDE[5], J. T. MATHEW[6], J. B. ADETUNJI[7], and D. I. HEFFT[8]

[1]Applied Microbiology, Biotechnology and Nanotechnology Laboratory, Department of Microbiology, Edo University Iyamho, PMB 04, Auchi, Edo State, Nigeria

[2]Laboratory of Ecotoxicology and Forensic Biology, Department of Biological Science, Faculty of Science, Edo University, Iyamho, Edo State, Nigeria

[3]Laboratory for Reproductive Biology and Developmental Programming, Department of Physiology, Edo University Iyamho, Nigeria

[4]Department of Chemistry, Edo University Iyamho, Nigeria

[5]Department of Microbiology, Federal University of Technology Akure, Nigeria

[6]Department of Chemistry, Ibrahim Badamasi University Lapai, Niger State, Nigeria.

[7]Nutritional and toxicological Research Laboratory, Department of Biochemistry Sciences, Osun State University, Osogbo, Nigeria

[8]Edgbaston Campus, University of Birmingham, School of Chemical Engineering, Birmingham B15 2TT, UK

[*]Corresponding author. E-mail: adetunji.charles@edouniversity.edu.ng

---

The Science of Nanomaterials: Basics and Applications. Suresh C. Ameta & Rakshit Ameta (Eds.)

ABSTRACT

Food commodities are normally wrapped and hygienically transferred after processing to safeguard and preserve them from any unpalatable changes, especially in the level of quantity before it is delivered to the consumer. Moreover, food packaging has been identified as one of the current innovations in material science and technology, which also constitutes a greater factor that could influence the choice of consumer's demand. It has been observed that the consumers of nowadays normally request for food with high level of food product having the following essential features, such as extended shelf life, natural quality, minimal processing, assured safety, and so forth. The application of innovative packaging system also constitutes effective distribution, preservation of food during transportation, and improvement in adequate communication with the consumer level. Therefore, there is a need to search for an innovative and sustainable technology that could lead to tremendous advancement in the area of food packaging especially through science of nanomaterials and nanotechnology. The advances in nanotechnology have drastically led to changes in numerous areas of food science, such as storage, processing, functionality, packaging, and food safety. Several nanomaterials, which include nanoorganic material incorporated into bioactive substances, metal oxides, and inorganic metal as well as their nanocomposites have been utilized in food industry. Moreover, despite all the benefits of nanotechnology, the high level of buildup of these nanomaterials in the environment and human health has constituted several health hazards. Hence, there is a need to put necessary policies regarding safety and health-related issues as one of the greatest priorities that need to be considered during processing, ingestion of nanomanufactured food products, production stage, intelligently and actively packaging, and processing. This chapter intends to provide a detailed information on recent advances in the application of nanotechnolgy in processing and food packaging industries. Moreover, detailed information on the application of sensors and biosensor in food packaging is also highlighted.

10.1 INTRODUCTION

Nanotechnology is a field of science which integrates several aspects of research in chemistry, physics, engineering, and biotechnology. It entails the utilization of nanomaterials within the range of 1–100 nm to serve various purposes. When these materials are modified to these sizes, they evolve into unique characteristics, which normally are not found in its original form. When the principles of these unique properties are known, such knowledge can be adopted in the improvement of novel and better products. This has been the goal of nanoscientist in recent times around the world.

Nanocomposite of unique characteristics can be formed by the combination of diverse nanostructure materials with other polymers. Nanostructure materials can have several dimensions for their crystallites, structural elements, molecules, clusters, and dimension, such as its one, two, and three dimensions within the range of 1–100 nm. Nanoclusters, quantum dots, nanorods, nanotubes, nanothin films, and nanomaterials are typical examples of 0, 1, 2, 3 dimensions. Nanomaterials with relatively high-value surface ratio showed promising physicochemical properties. Some of such properties include strength, optics, magnetism, solubility, thermodynamics, and color.

Nanomaterials are gaining increased applications in medicine, agriculture, industrial and domestic goods, and food. This is added to their unique properties of improved functional and physiochemical properties in production processes (Pathkoti et al., 2017; Fu, 2014; Hu et al., 2017). They also find usefulness against pathogenic microbes and thereby they are used in preservation and food safety (Fu, 2014; Baranwal et al., 2018). Other benefits are also known in food application. Nanomaterials are applied as nanosensors, encapsulation agent and used in new packaging materials. In a broad spectrum, this chapter focuses on the use of novel nanomaterials for use in preservation and packaging in the food sector.

The role of nanotechnology in food-related application has been reported previously. In a study by Sandoval (2009), certain food properties were developed when a specific nanomaterial type was integrated into the food product of a specific brand. Likewise, nanomaterials have been reported to improve food safety through the improvement of shelf life, packaging materials, and nutritional value without affecting the characteristics of the food product, a very important characteristics of fortifying agent (Das et al., 2017; He and Hwang, 2016).

In spite of their potential, the economic cost of deploying nanotechnology is still a challenge to its successful use in the food industry. Due to safety concerns, there are ongoing research in developing nanomaterials that are suitable for food grade ingredients using strategies that are cost-effective as well as simple, environment-friendly, and layer-by-layer technique (Adabi et al., 2017).

In light of the advantages as well as the limitation of nanomaterials on the food industry, the discussion in this chapter has been geared toward addressing the current status as regard the use of nanomaterials in food processing, safety, and preservation. The recent use of nanomaterials in the food packaging industry has also been discussed.

10.2 PRESERVATION OF FRUITS AND VEGETABLES

Singh et al. (2015) stated that the utilization of nanoparticles and nanomaterials in agriculture increased drastically over the years due to accelerated increase in research output concerning nanobiotechnology's safety mechanisms, beneficial role in plant growth, development, and storage for sustainable development. It was revealed that nanoparticles are transporters or carriers of agrochemicals that can be utilized for drug delivery, effective enhancement of storage, and preservation to increase availability. Recently, the utilization of nanodevices and nanobiosensors in crop preservation, protection, genetic manipulation, and storage has grown tremendously. Studies have shown that the physiological effectiveness of the nanoparticles are more determined by their size, chemical configuration, concentration, and reactivity.

Gutiérrez et al. (2011) reported that the treatment of various plant with nanoparticles increased the productivity, nutrient availability and enhances dehydrogenase activity. Scrinis and Lyons (2007) also demonstrated that nanoparticles can be utilized for the discovery and analysis of plant/crop pathogens through dissimilar nanobiosensor, immune sensor and nanochips so as to monitor crop quality and yield, minimize the utilization of antibiotics or pesticides and preserve agricultural products. The concept of nanobiosensor is to rapidly detect biological agents, such as nucleic acids, antibiotics, heavy metal ions, preservatives, microbial load, toxins, antibodies, metabolites, and pathogens.

The working principles of these nanobiosensors are based on binding of bioanalytes onto bioreceptors for modulation of the physicochemical signal linked with the binding. The surface transducer traps and converts the physicochemical events into an electrical signal, which can be further monitored by the signal variations, such as current, electric potential, conductance, intensity/phase of electromagnetic radiation, impedance, mass, viscosity, and temperature. Analysis of the signal variations in these parameters represents the absence or presence of bioagents. It was also revealed that nanobiosensors can be used in the monitoring of environmental conditions, such as temperature, humidity, moisture best fit for crop perseveration, to prevent spoilage, and increase the shelf life so as to support sustainable agriculture through improved crop productivity and preservation. They showed that nanobiosensors, immune sensors, and nano chips, which are the new class of nanotechnology biosensor comprises a recognized biomolecules immobilized on the surface as surface transducer for reaction between the analyte and the biorecognition molecule at ultra-low concentration for greater sensitivity, response time, effective data generation, processing, real-time decision-making, and rapid manipulation for change.

Fernando et al. (2018) revealed that nanoparticles are beginning to gain considerable attention among the scientists as antimicrobial agents due to physicochemical properties and safety for human health. Recently, due to the advances in the development of nanotechnology, the confluence of physiology and nanoparticles has brought to fore metallic nanoparticles with vast biological roles, such as antimicrobial, anti-inflammatory, and antioxidant agents. Nanoparticles have unique physicochemical features, which can be manipulated to suit biomedical applications (Wang et al., 2008).

Fernando et al. (2018) also discovered that many of the nanomaterials are beneficial as an alternative to many multidrug resistance bacteria, such as *Pseudomonas aeruginosa, Staphylococcus aureus, Escherichia coli, Mycobacterium tuberculosis, Streptomyces fradia,* and *Acinetobacter baumanii.*

Tarafdar et al. (2013) demonstrated that the health attributes of fruit and vegetable consumption is the high content of antioxidants, such as anthocyanins, flavonoids, polyphenols, resveratrol, curcumin, vitamins, lycopene, catechins, and carotenoids with health-promoting benefits. It was revealed that many of these bioactive molecules or phytochemicals in fruits and vegetables could be enhanced through the application of nano-biotechnology. This is generally achieved by increasing the bioavailability of these biomolecules in the fruit and vegetable matrix trapped in lipid-based nanocarriers or facilitators, such as nanoliposomes, nanoemulsions, micelles, and nanocochleates. They noted that the bioactive molecules can be enhanced through increases in biochemical stability particularly during storage or preservation stage—even with harsh weather conditions, such as high temperature and UV light exposure.

Studies have demonstrated that nanoparticles can be utilized as alternative in facilitating food flavorings in bulk amounts and also to increase the shelf life of tomatoes and other food products (Nile et al., 2020). These nanotechnology-linked shelf life extension strategies have received massive acceptance from many stakeholders through the addition of bio-nanoencapsulated quercetin, and now this approach has been extended to other vegetables and fruits so as to increase the shelf life (Liu et al., 2019). Nanoemulsions are also commonly utilized in fruits and vegetables packaging and sanitization of packaging equipment. Again, glycerin added with nanomicelle-based products removes pesticide residues in vegetables and fruits, oil/dirt from cutlery and equipment possibly by acting as surface decontamination during food processing and packing or by reduction of surface contamination (Sekhon, 2010). Gerardino et al. (2005) revealed that glycerol may alter the surfactants micellization through an indirect mechanism like solvent-mediated mechanism or salting-out effect, thus, lowering the medium dielectric constant, and subsequently enhancing the electrostatic interactions and glycerol competing for water molecules, causing a dehydration effect on the surfactant ethoxylic head-group.

10.3 PRESERVATION OF DAIRY FOODS

The major reason for many research activities as alternative strategies on improving meat, fish, and dairy packaging system is to eliminate spoilage, prevent contamination, improve tenderness through enzymatic activity, reduce weight loss, and retaining the color, and texture. Radha et al. (2014) have revealed that nanotechnology in dairy food production and preservation has received massive attention in developed countries, particularly the US. The livestock industry, which represents a sustainable and self-sufficient manufacturing of diverse range of products in food industry has been revealed to face a lot of challenges in productivity and quality, such as nutritional deficiencies and pathogenic attack. Nano/microencapsulation as nanofeeds have been shown to respond to these factors, which enables active biomolecules or potential antioxidant supplements to be delivered to the small intestine to treat nutritional deficiency and improve the quality of production and yield (Trinchese et al., 2015). Nanofeeds and nanominerals are now being adopted as alternatives to chemical antibiotics for bacteria treatment in dairy foods production and packaging. Also, various nutritional supplements needed for improving the quality and yield of dairy products are being deployed through the utilization of nanotechnology in many parts of the developed countries (Chen and Yada, 2011).

Curing or salting is one of the oldest technologies utilized in meat processing in many parts of world to enhance flavor, texture, taste, and prevent microorganisms or microbial load. In the developed countries, different methods are combined for dairy food preservation such as modified atmosphere packaging, high pressure or curing, which may alter the nature and quality of food (Mastromatteo et al. 2010). Different studies have shown the beneficial role of salting in food processing, such as protein solubilization, hydration, protein extraction activation, water-holding-capacity enhancement, increase in food product cooking yield, increase in viscosity of meat, enhancing the formation of heat-stable emulsions, and decrease fluid loss (William et al., 2011). Increased concerns about the health implications, such as carcinogens, high blood pressure, and other cardiovascular risks due to high dietary salt intake due to curing have been witnessed in the last few decades (Petit et al., 2019). Hence, other advanced methods like nanotechnology are now attracting more attention as an alternative in dairy food products preservation.

Tarafdar et al. (2013) and Singh (2016) revealed that during postharvest food processing, nanotechnology has been utilized for security, packaging, and production of dairy food products with immense potential to enhance food bioavailability, texture, taste, consistency, or conceals unpleasant odor or taste with edible nanocoatings for meat products.

Again, nanofilters and edible nanocoatings have been reported to be widely utilized to eliminate browning compounds from different dairy products while retaining the flavor content or to remove lactose content from milk or other dairy products to be resubstituted with other brands of sugars so as to make the milk appropriate for patients with lactose intolerance (Nile et al., 2020). Another advantage of the nanofilter and nanosieves is to help in producing milk free of bacteria pathogens and prevent boiling that may denature the dairy product. Fortified nanotechnology-based products with dietary supplements, active oil, vitamins (A, C, D, E, and K), nutraceuticals and drugs are becoming commercially available in international market across the globe (Nile et al., 2020).

The development of many devices and products in nanotechnology has massively assisted in the food industry so as to meet up with the global food demand and satisfying consumers by preventing microbial contamination, food spoilage, detecting pesticides and toxins. Addition of nanoemulsified bioactive compounds to food products will not change the product appearance but different food pests and pathogens, such as Gram-negative bacteria can be controlled massively by nanoemulsions as an active and intelligent packaging system (Liu et al., 2019).

10.4 PRESERVATION OF BAKERIES AND CONFECTIONARIES

Mustafa and Andreescu (2020) documented the advancement in nanotechnology has made remarkable provision for new methods for sensing and packaging of various kinds of foods, which thereby aid in addressing one of the major problems in the food sector. Restuccia et al. (2010) revealed that in the preservation of bakeries and confectionaries, the application of nanotechnology for processing, production, packaging and safety is now an emerging technology. A nanocomposite coating system for placing antimicrobial agent to prevent infection and enhance storage has also attracted serious attention in bakeries and confectionaries. It was also discovered that in bakeries and confectionaries, nanotechnology in the production of functional foods through removal, addition, concentration, improvement or alteration of the bioavailability of a particular active molecule in the food can be achieved through novel food processing techniques with long shelf life (Nile et al., 2020).

Studies have also revealed that edible nanocoatings can be utilized in cheese, fast food bakery, and many confectionery products as gas or moisture barriers. Antibacterial nanocoatings in bakery food products can be utilized to prevent pathogens, boost the antioxidant content, flavor, taste, and prolonged shelf life (Nile et al., 2020). Shivraj et al. (2020) revealed that packaging industry constitutes a major global economy yielding billions of dollars for the packaging of food and beverages to prevent contamination, to increase foods shelf life, enhance food safety, reduce spoilage, repair tears in packaging, improve the general health of individual, and reduce the problem of food shortage. Food packages inserted with nanoparticles can alert the consumer about the safety and quality of product. These purposes are achieved by the adoption of nanoparticles or nanocomposites as part of the processes involved in the production, packaging of food and the restriction of store-product pathogens or pests (Pal, 2017).

10.5 SPECIFIC APPLICATIONS OF NANOTECHNOLOGY IN PACKAGING

There are different ways of packaging foodstuffs, such as bags, cans, boxes, and cartoons to ensure preservation and food safety. However, many of these methods might have potential health impact, thus, granting an unsafe medium for the preservation of food and reduce shelf life. Mandal et al. (2009) reviewed the utilization of nanotechnology as a better way of food packaging. It was pointed out that nanotechnology as molecular-scale functional engineering has the prospective to alteration of the world's system of packaging food. Currently, nanotechnology can be employed by engineers to change the structure of material packaging on a structural scale to offer the materials a perfect asset example is the nanoencapsulation of nutraceuticals. Besides, enhancing food safety and quality, an intelligent way of food packaging using various nanostructures has been shown to improve the permeability of plastics carrying water/gas, thus, utile for the safeguarding of beverages, vegetables, fruits, and other perishable foods.

This asset of nanostructure used in packaging has been tested to be efficient and has a longer shelf life in the preservation of the natural contents, color, and flavor of the stored food for easy transportation. Mandal et al. (2009) opined that the nano-enhanced plastics have higher thermal and gas absorption assets, stronger resistance, and mechanical properties. Besides, these properties, nanomaterials can also efficiently prevent the incursion of microorganisms in the package, thus, ensuring food quality and safety. More so, with nanosensors embedded in the package, the consumers can read the items of ingredients of different food listed. The nanosensors can give an alarm on food spoilage or provide information on the nutritional contents of the food. They stated that the utilization of this technology has replaced the traditional way of packaging for a smart packaging system. This will provide a more customized and fresher outlook of products.

Sujithra and Manikkandan (2019) also reviewed the utilization of nano-technology in food packaging. Nanostructures used in packaging foods have gained considerable ground over several decades because of the potentials to preserve, protect, and improve the shelf life of packaged materials. The role of coatings, comestible films, and barriers can viably decrease the difficulty, thus, enhancing the recycling of materials used in food packages, when related to the traditional method of packaging that is not green. This novel form of the package can gradually substitute the synthetic types of polymers that have possible lesser protection, preservation, and shelf life for packing materials. The nanomaterials have remarkable antimicrobial repellant and coating also that ensure food security and safety of packed consumables. In conclusion, it was opined that nanotechnology could permit a more sustain-able and efficient food manufacturing by the way of safeguarding food materials and preserving their natural quality.

Alfadul and Elneshwy (2010) gaged in a review of the utilization of nanotechnology in the safety, packaging, and processing of food materials with special effects on the regulations. They opined that nanotechnology can be applied in all sectors of food technology, such as food color modifica-tions, production, monitoring, packaging, and processing of nutraceutical foods as well as in the sustenance of nutritional needs. Some other potentials in employing nanotechnology in food preservation are the ability to prolong the shelf life of the packaged product and the reduction of the economic cost for purchasing ingredients and other food flavors. Special effects of the utilization of nanofoil in food packaging are the ability to signal biochemical and microbiological changes, the creation of active sites for antifungal and antimicrobial interactions, enhancing the heat and mechanical resistance chattels, and barrier chattels. More so, it aids in the detection of a minute quantity of chemical pollutants as well as bacteria and viruses in the food system. So, this technology has what it takes to transform the food industries into a life-sustaining structure most especially with the fast-growing human population and need for effective food storage. Based on all these properties of nanofood packaging, there is a clarion need to ascertain the potential risks linked to their utilization in the food sector.

Agriopoulou (2016) evaluated the role of nanotechnology in the packaging of food materials. They stated that nanopackaging is a radical technology, which has come to substitute the traditional method of packaging food because of the smart system of operation. The chemical and physical properties of nanomaterials make a lot of commercial and production industries invest in different food processing and analytical packaging techniques. Nanoparticles also have high antimicrobial and gas barrier properties. Besides, nanopackaging also offers different latent benefits, which can be recommended for further investigations. However, there are growing worries about consumers' acceptance because of the possibility of toxicity. Be as that may, nanopackaging offers an assortment of stirring benefits such as an increase in the shelf life of the food material, enhancement of the hygiene, quality of the food, and many other innovative chattels. There are some main gaps in the terms of knowledge on the evaluation of the influence, behaviors, and properties of nanoparticles. Extant issues on the hazards resulting from various areas, such as their bioaccumulation, biodistribution, bioavailability, behavior, fate of the chemicals, noxiousness, and so on should be ascertained as a recommendation for subsequent studies.

The use of nanotechnology as a sensor, antimicrobial, and barrier substance in food safety and packaging was also reviewed by Duncan (2011). It was stated that Ag-nanoparticles have the potency as nanosensors and antimicrobial messengers, for the discovery of food analytes, such as pathogenic microbes, minutes organic substances, and gases. He reported that nanomaterials provide some stirring benefits, such as antimicrobial chattels, safer and better ways of the packaging of food materials, and possible reduction in the risk of adulterant chemicals. However, nanopackaging is still at the nascent stage, and that is the main reason that consumers have not yet fully accepted its utilization for the preservation and safety of food because of perceived noxiousness. Nonetheless, the current application of this technology will see the light at the tunnel because of the exciting benefits it portends. In conclusion, it was recommended that continuing investigation of the probable dangers that loom in the utilization of this technology and address stunning questions tactfully and scientifically regarding the health and safety that pervade the persistence use of this novel nanofood packaging techniques. This will negate the already perceived attitudes of many intent consumers and promotes their appeal for use.

Ravichandran (2010) examined the uncertainties, opportunities, and ground-breaking green methods in the utilization of nanotechnology in packaging and food processing. It was recounted that nanotechnology and science are at the forefront of modern food technology. Their use in several industrial sectors specifically in the pharmaceuticals and in biomedicine for drug delivery has yielded astonishing breakthroughs. Of recent, many engineers and scientists have documented the possible usage of nanotechnology in the food engineering sector. In this sector, nanotechnology can be used as a sensor and for packaging food materials. In the latter, food safety and quality are guaranteed by employing this technology. Meanwhile, in the former, microfluidics, microelectromechanical, and microarrays of DNA technological systems are used as smart nanosensors in the delivery of colors, chemical, biological, and food nutrients, the bioseparation of proteins, solubilization, nutraceuticals, and nanoencapsulation in food systems. It was suggested that the adverse and positive influences in the utilization of nanopackaging should be reviewed continuously to forestall their safety and quality use.

Yu et al. (2018) reviewed the safety, practical utilization, and preparative techniques of nanopackaging of food materials. Recently, the advancement of nanotechnology in different industrial sectors of food technology and science has been on increase. This technology can be used for the quality management, control, storage, and processing of food. The technology has been utilized to design nanosensors for the discovery of toxic compound in food as well as a smart system of packaging that enables the sensitive and rapid recognition of food adulteration. It was recounted that nanoencapsulation plays a pivotal role in the design and target transfer of bioactive substances and flavor. More so, it controls the discharge of flavor at the point of storage and processing. In conclusion, it was suggested that the current information on food packaging should be made public to sensitize the public and end-users of the potentials and safety in the utilization of nanomaterials.

Hellarama et al. (2014) evaluated the importance of nanotechnology in the food sector. They described nanotechnology as a powerful tool that can be used to revolutionize different industrial sectors of the world. The application of nanoencapsulation can be used to transfer substances to target spots, improving the potency of antibacterial influence, food flavor, sensing adulteration, improving the shelf life and storage of the food, tracing, tracking, and product protection. Nanofood products and processing can alter the sensory properties, flavor, and color of package materials, thus, improving food quality, safety, and repairs. Nanobarcodes, another advanced type of nanopackaging, can also be used to label, monitor, distribute food products, and also to ensure the safety of food. More so, nanosupplements, a technique of nanoencapsulation can be used as an effective drug transfer system and nutritional entity.

For the last two decades, nanotechnology has been re-engineered to better the society of humans. In the food industries, several issues permeate the safety, processing, and package of food materials, which have caused excruciating wanton destruction of valuable and edible consumables consequents by bacteria, fungi invasion, and reduce shelf life. These destructive episodes are a result of the inefficiency of the traditional methods of packaging. Goyal and Goyal (2012) reviewed and evaluated special potential effects in the use of nanotechnology in the preservation of food materials as well as their application in effective food packaging. It was stated that the use of nanopackaging can make the food materials cheaper and more efficient in the reduction of energy and wastes generation. However, many novel technologies carry different ethical responsibilities for their utilization such as general acceptance and appeal to the consumers choice. So, their recognition in the market system and use by the consumers might not pose low patronage as well as any unforeseen hazard. The ideal of nanomaterials signifies a motivating pathway for creating novel and creative materials, more so in the natural polymers sector. The use of renoprotective coatings for the packaging food materials has become an interesting area of stimulating interest and because of the chattels such prolongation of the shelf life of the food products is possible.

Goyal and Goyal (2012) recounted that the R and D (Research and Development) of nanocomposite substances used for packaging and food contact exteriors is predicted to improve in the next couple of years with the coming of novel polymeric composites and materials with mineral nanoparticles. Nanomaterials sourced by combining crystalline clay (LDHs) and natural novel polymers offer a countless variety of great asset profiles. Despite the potentials of extant packaging using biobased nanoparticles, it is still vague to forecast future situations of what environmental risk they portend? It is expected that in the nearest decades, the native form of packaging will gradually be substituted with a multifunctional next-generation smart-packaging substance that will able to accommodate the preservation of food materials, dairy, wine, beverages, vegetables, and fruit products. The introduction of nanomaterials into these products will possibly improve the thermal barrier, and mechanical performance of the products. More so, it will act as antimicrobial security, forestall the safety and quality of the packaged food. Nanosensors fixed in the packaging materials may also sensitize the consumers if the materials in the package are deteriorating.

Food technology using nanomaterials has become one of the emerging sectors in the food industries. Sekhon (2010) in a review, looked at the use of novel nanotechnology techniques in the packaging and preservation of food materials. The utilization of nanoscience and technology has improved the functionalities of food packaging and preservation by the introduction of active substances that can transport useful attributes yonder those of traditional dynamic method of packaging, as well as signaling and sensing of related information that will enhance the food quality and preserve their safety. Nanofood packaging can be used to mend tears and wears in materials packaged, extend the food life by the discharge of preservatives, signal the consumers if there is food spoilage, and ensure food safety. The use of nanotechnology in the food industries can also aid in the detection of microbes in the package, produce and maintain the natural colors and flavors, and improve the barriers of the properties therein. It was reported that this technology holds immense potentials that can benefit the health of the products and the consumers at large. The technology has also created novel chances in the food sector at a great speed. But health and uncertainty concerns remain the issues to be tackled. Global regulation by legislation will help to manage any unforeseen future episodes.

Fabulous benefits have been discovered by the use of nanomaterials in the quality management, safety, and packaging of food materials, which are reducing wastes and increasing the shelf life. Berekaa (2015) evaluated the use of nanomaterials in food safety, processing, and packaging of food. They pinpointed that nanotechnology can be used to modify and create a better way of packaging materials against microbes via the presence of bioactive substances, enhancement of the heat and mechanical resistance, and the improvement of the barrier chattels. Besides, this technology can also be used in the detection of chemical pollutants and pathogens via a nanosensor method. However, there is a need to evaluate the possible toxicological concerns that may permeate from the utilization of this technology. Frameworks that are capable of regulating the risks like dust and fire burst, ingestion and skin exposures linked with the use of this technology should be set into urgent deployment when the need arises.

10.6 BIOSENSORS IN FOOD PACKAGING

The loss of food as a result of crop contagions from pathogens, which include viruses, fungi, and bacteria is a tenacious issue in food production for years throughout the globe. Foodborne diseases affected through bacteriological pathogens are dangers to human well-being. There are conservative principles, which established microbial techniques for investigating microbial food infection and these are time-consuming, laborious, in addition to entail precise services. Ruyuan et al. (2020) used several outstanding nanomaterials for the production of new biosensors for foodborne pathogen detection. It was as a result of their findings, they were able to establish that the novel biosensors applied in the study consist of numerous advantages, such as specificity, rapidity, accuracy, sensitivity as well as simplicity using nanomaterials properties. They also accounted for an innovative biosensor which includes colorimetric, electrochemical, along with outward improved Raman scattering as well as fluorescent biosensors for detecting key foodborne pathogens. However, they opined that the traditional methods for foodborne pathogen detection have several limitations, such as low sensitivity, low selectivity, and long detection time.

Kumar et al. (2020) reported that nanomaterials were applied for innovative skills in biosensors, which was relocation of signals to improve their sensitivity along with efficiency. Although microbial polluted food poses a lot of health concerns as a result of epidemics of severe foodborne illnesses, and for that reason, there is need for specific identification and revealing of pathogenic toxins in addition to microbes in food to avoid some of these concerns. Consequently, they were able to establish the concept of biosensing, which allowed them to form nanobiosensors by means of diverse composites along with nanomaterials to increase the specificity and the sensitivity of pathogen detection. Nanomaterials used for developing biosensors are gold and magnetic, carbon nanotubes, graphene nanomaterials, quantum dots as well as dendrimers, which predominantly improved sensitivity and specificity for detection as a result of their limited magnetic, chemical, mechanical, physical, and optical properties. Also, they opined that new composites and nanoparticles utilized in biosensors required to be categorized and classified for the quick detection, improved productivity, effective as well as unobtrusive application in foodborne investigation.

Zhaolui et al. (2017) used a modest, fast, as well as sensitive detection of *E. coli,* which was established based on urease-catalyzed signal amplification and antimicrobial peptide functionalized magnetic nanoparticles. It was revealed that antimicrobial peptide functionalized magnetic nanoparticles were found to capture *E. coli* by means of a high effectiveness through hydrophobic as well as electrostatic interfaces. So, as the microbes are netted by means of antimicrobial peptides AMP, the binding sites for urease on the exterior of antimicrobial peptide functionalized magnetic nanoparticles are blocked resulting in a huge quantity of urease in the supernatant after attractive magnetic separation. In addition, this novel podium was effectively useful to detect *E. coli* in ground beef and pointed apple nectar samples with a limit of detection ranging between 84 and 233 CFU/mL, respectively.

Neha et al. (2017) provided a general idea of the application of diverse nanomaterials using fluorescent in the improvement of nanobiosensors through distinct highlighting on fundamental sensitivity, revealing scheme, capability of multiplexed analysis, and specificity. Polymerase chain reaction-based molecular and immunologic techniques lack specificity, they are costly and might produce false results. They also have established fluorescent nanobiosensors which demonstrate that they have become alternative tools for routine and rapid detection of foodborne microorganisms.

Fang and Ramasamy (2015) observed that in order to reduce the infection-induced damage to yields all through growth, postharvest and harvest processing, and to safeguard agricultural sustainability to maximize efficiency, advanced disease detection and prevention in crops are imperative. It was revealed by research laboratory-based methods. Immunofluorescence, enzyme-linked immunosorbent assay, polymerase chain reaction, gas chromatography-mass spectrometry, fluorescence *in situ* hybridization, as well as flow cytometry are some of the direct detection methods, while incidental techniques include fluorescence imaging, hyperspectral techniques as well as thermography. Similarly, they make available a complete view of biosensors established on exceedingly discriminating biorecognition elements, such as antibody, enzyme, bacteriophage, and RNA/DNA as a novel device for the primary documentation of farm product infections.

Jia et al. (2019) while studying immunofiltration band technique had centered on the photothermal influence of gold nanoparticles and established a rapid, simple low-cost, sensitive as well as handheld technique for the detection of *E. coli*. Nevertheless, the band analysis is usually less sensitive and small compared with other immunological methods. They used photothermal effect of gold nanoparticles to obtain the intensification of bands to advance the sensitivity of the strip technique. They noted that thermal differences are caused by the photothermal effect, which were related to the microbes' yield as well as concentrations in virtually logarithmic relation. When subjected to optimum conditions, this technique has the detection strength up to about 1.95×10^4 CFU/mL, where the sensitivity was enhanced approximately 10 times when compared with that of the conventional optical strip technique. They opined that the study showed that photothermal-based immunofiltration band might be utilized for the sensitive and rapid detection of additional pathogens as a pronounced prospective for food quality device practice.

Umesha and Manukumar (2015) expressed that some foodborne pathogens, such as *Campylobacter, Aspergillus* spp., *Penicillium* spp., *Fusarium* spp., *Salmonella* spp. and some of the other pathogens are detected in adulterated food substances that could lead to continuously sicknesses in mankind in one way or the other. In their own case, they used short period of time to identify foodborne pathogens in food technology. Some level of contamination of food through main pathogens involves specific sensitive detection methods, where the present research is making way forward to novel nanomolecular systems for nanomaterials, making them appropriate for the improvement of analyses with portability, detection of pathogen in short period of time as well as high sensitivity.

Cho and Ku (2017) explored present analytical skills, which includes magnetic and metallic nanomaterials, electrochemistry, spectroscopy as well as optics. They based their findings on the existing techniques on the early pathogens detection by means of improved analytical specificity and sensitivity. However, in order to acquaint with the prospective combination and comparative investigation of numerous advanced techniques, they also referred a new specimen preparation procedure that uses recovery technologies and microbial concentration. This technology has the prospective to advance the pre-enhancement stage that paved the way for detection process. The improvement of high-tech along with novel solutions for accurate, nonlaborious, and rapid bacterial detection techniques is imperative to increase the supplying of food globally.

Moreover, Yang and Bashir (2008) intensified on the application of impedance as a novel technique for the enumeration of foodborne pathogenic microorganisms. In the recent years, the incorporation of impedance by means of biological recognition skill for bacterial detection has steered to the development of impedance biosensors that have found wide-spread use. They were trying to work on the applications and progress of impedance microbiology for foodborne pathogenic detection of bacteria which entails the utilization of interdigitated microelectrodes, the use of corresponding circuits for the investigation of the impedance systems, and the improvement of chip-based impedance microbiology. Also, the study opined the importance of improvements of impedance biosensors for the detection of bacteria in the past 5 years, focusing on microfluidic-based Faradaic electrochemical impedance biosensors and microfabricated microelectrodes-based, non-Faradaic impedance biosensors, and the incorporation of impedance biosensors through other methods, such as electropermeabilization and dielectrophoresis.

Recently, Luo et al. (2020) made use of lateral flow strip assays and have gained an extensive attention as a substitute technique for rapid diagnosis of foodborne pathogens to the unadventurous culture-based analyses, which includes polymerase chain reaction (PCR) and plate counting. They noted some main drawback of the present lateral flow strip assays for the pathogenic recognition of bacteria is its limitation in its applied uses in point of care testing (POCT), which is the result of low sensitivity. However, the study showed existing advances in the lateral flow strip assays for the pathogenic detection of bacteria in terms of analyte types and chromatic agents. In addition, current methods for signal modifications and enhancement of the lateral movement band analysis architecture for complex detection of pathogenic bacteria were also a part of the study.

Base on the potential dangers related through the globalization of the food production, attention has been shifted to investigation on the improvement of novel sensing methods to make available the means of contamination detection at various steps in the supply of food chain. This study revealed the modern developments in platforms that have the highest prospective for low-cost identification and real-time detection of foodborne pathogens. Particular emphasis has been made on the development methods/techniques that use nano- and microscale technology (Carlson et al., 2018).

Rubab et al. (2018) emphasized on the detection of foodborne pathogens most especially from *Staphylococcus aureus,* which has been identified as a significant pathogen responsible for the cause of foodborne diseases globally. Nevertheless, several of these techniques are laborious and are not appropriate for on-site uses. It is extremely significant to advance fast and more approachable detection methods. This study stressed on the novelty entails in the application of biosensor techniques toward the quantification and detection of *S. aureus*.

Bonah et al. (2019) performed a study to evaluate some essential information about hyperspectral imaging, including data acquisition, instrumentation, data analysis and image processing.The existing application of hyperspectral imaging serves as a potential tool for the classification, discrimination, and detection of numerous foodborne pathogens.

Sharma et al. (2015) played an indispensable role for effective monitoring, packaging, and assessing the contaminants in food. Nanomaterials-based aptasensors hold pronounced prospective to overwhelm the shortcomings of conservative analytical procedures. Aptamers consist of a new class of extremely precise biorecognition elements that are shaped through efficient evolution of ligands through exponential enhancement process. Several nanomaterials including, metal nanoparticles, carbon and metal quantum dots, metal nanoclusters, metal oxide nanoparticles, carbon nanotubes, nanocomposites, and graphene permit higher sensitivity through signal intensification and present numerous new transduction schemes, such as fluorescence, improved chemiluminescence, Raman signals, electrochemical signals, super-paramagnetic properties as well as enhanced catalytic activity to the biosensor.

Sangeeta et al. (2019) opined that there are shortcomings associated with the conventional procedures, such as requirement of costly sophisticated devices and multistep sample processing, which limit their applications at research laboratories. To address the restrictions associated with the traditional approaches, novel techniques have been established by several study groups. Recent improvements in the nanomaterial-based detection of histamine in diverse food products have exposed important measurement as a result of high sensitivity, specificity, ease of operation, field deployability as well as cost. Therefore, the improvement of nanomaterials-based histamine detecting strategies/assays, where the detection is based on optical and electrochemical results.

10.7 MAINTAINING FOOD SAFETY

The utilization of nanomaterials for food packaging could help to reduce toxicity and effective circulation (Slootweg et al., 2002). Through the use of NPs, there is possible reason for the development of mechanical properties that resist heat in food packaging, thereby extending the shelf life of such food. Polymers are naturally occurring complexes consisting of large molecules that are not naturally penetrable to gas or water vapor but they do have the potential of developing barriers for gas, resistance to heat, and strength for food packaging attributes (Cho et al., 2008; Morais et al., 2014). Nanolaminates are utilized as thin layer of covering materials on food surface due to their increased relatively small nature, which makes the materials so fragile (Pawar, 2016).

When silver is deposited on titanium dioxide nanoparticles, it leads to increase in bactericidal efficiency against *E. coli* (Kim et al., 2006). Nanoparticles are easily fastened for food packing due to their increased superficial area. Many food producing industries now utilize nanomaterials in preserving food products from microbial spoilage. Materials that are mostly used are nanotubes, as there are great analytical importance for the separation of complexes, such as proteins, vitamins, minerals, peptides, and in the detection of enzymes, antibodies and DNA (Lee and Martin, 2002). Zinc oxide nanoparticles were prepared by Tayel et al. (2011) and were used against foodborne bacteria, such as *Salmonella typhimurium* and *S. aureus*. They discovered that the higher is the zinc oxide NPs, the higher will be the likelihood of the NPs to attach to the microbial cell, thereby, resulting in cell death.

The innovative nanoparticles could be useful in food package (Zhao et al., 2008). This material can be applied as a food packaging material via the application of penetrable amount of silver nanoparticle fraction of sodium benzoate, which is eco-friendly and cost-efficient (Munish et al., 2015). Nanoparticles synthesized from nature are good form of food preservative. The ions inherent in Ag NPs possess the ability to hinder the biological process in bacterial cell (De Azeredo, 2009).

Silver nanoparticles are mostly utilized as antibacterial and antifungal agents in packing food, containers for storing, chopping board, and as supplements to improve human health (Wang et al., 2008; Chen and Mikecz, 2005). Zinc oxide is used as antimicrobial agent, as food product container, and also as food preservative (Brunner et al., 2006). Silicon dioxide NPs are also used in preserving food. Nanocombinations are globally utilized in food wrapping because they are economically friendly and easily degradable (Nam et al., 2003).

10.8 DETECTION OF FOOD SPOILAGE MICROORGANISMS BY NANOPACKAGING MATERIALS

Enzymes impregnated with Au NPs are mainly used in determining the presence of microorganisms to ascertain the state of the food substance. Nanofibril shares a crucial role by showing the extent of meat deterioration by indicating the presence of amine gas. Barcodes from nanoparticles are utilized for the purpose of labeling and security (Avella et al., 2005). Piezolectric biosensor was synthesized from the electric affinity of Au NPs, which are applied during the detection of pathogenic microorganism that could cause food spoilage, such as *E. coli* O157:H7 (Gordon et al., 2011). Quantum dots were utilized as fluorescent tags in several assays in detecting the presence of foodborne pathogenic microbes, such as *C. jeuni, E. coli* O157:H7, *S. aureus, S. typhimurium* and to what extent, they could possibly cause harm (Xie et al., 2011; Chaudhry et al., 2008)?

Nanocombination is utilized in the recognition of gas produced by microorganism that causes food deterioration. These packages were found to detect the rate of spoilage exhibited by the change in color. The application of these nanomaterials as food wrappers is economical in the sense that it helps to monitor early-stage spoilage in food and also in reducing the stress of sending food products to the laboratory for test (Tang and Sauceda, 2009; Meetoo, 2011; Zhao et al., 2008). Electric tongues are a type of sensors that was designed to spot early spoilage of food. They reveal the unfit nature of a food product upon contact with the food (Qasim et al., 2008). Normally, food spoilage is caused by the influence of environmental and anthropogenic activities. The advent of nanomaterials and nanocombinations has therefore been developed into food wrapper that can help to preserve food and also to detect the presence of microbial infiltration. These materials can also be called nanobiosensors due to their ability to detect explicitly chemical complexes, hydrogen sulfide, pathogenic microbes, ammonia, and the manner of occurrence in the food product (Mylvaganam and Rathnayake, 2020; Qureshi and Karthikeyan, 2012). These nanosensors are highly sensitive and selective, making them much more effective than the synthetic sensors. The gas metals involve metals, such as gold and palladium (Mannino and Scampicchio, 2007). Nanobiosensors are nanofood packaging materials that are utilized for the affirmation of the presence or absence of viruses and bacteria in food substances, they are also used in detecting the presence of mycotoxins and other toxic complexes (Mason et al., 2006; Coles and Frewer, 2013). Cerium oxide immunosensors and chitosan-based nanocombination are used in food packaging for detecting Ochratoxin A and carbon nanotubes in detecting the presence of *staphylococcal enterotoxin B* (Mousavi and Rezaei, 2011).

10.9 PROSPECT OF UTILIZING NOVEL NANOMATERIAL FOR EFFECTIVE FOOD SECURITY, SAFETY, PACKAGING, AND PRESERVATION

Foltynowicz et al. (2017) fabricated iron nanoparticles of zero-valency to enhance the packaging of food. The nanoparticles act as a scavenger for the prevention of oxidation in food. Cindi et al. (2015) reported a reduction in the cases and extent of rot in peach fruit caused by *Monilinia laxa*. This was observed by the use of chitosan nanocomposite films containing thyme oil and punnets of polyethylene terephthalate. They used a copolymer of biodegradable chitosan–lactide for the transport of pyrachlostrobin, a foliar fungicide that is broad spectrum in its action.

Swaroop and Shukla (2018) combined polylactic acid biopolymer and nanostructure of magnesium oxide in the fabrication of a material for the packaging of food and observed that the nanomaterial produced was highly efficient in the protection of the food against the biofilms of bacteria.

Nishiyama (2009) produced silver nanoparticles functionalized with sodium benzoate using silver nitrate. A significant activity was observed against various foodborne pathogenic bacteria such as *E. coli, S. typhimurium*, and *S. aureus*. This was relatively high when compared with pure sodium benzoate. It was concluded that this is useful for packaging of food through the application permutable content of sodium benzoate–silver nanoparticle composite.

Bajpai et al. (2018) opined that the advancement in the field of nanotechnology has positively affected different areas of food science, most especially in the aspect of packaging, storage, processing, and overall food safety. Various nanomaterials from metallic oxides, inorganics, as well as nanocomposites have been employed in the area of food technology.

Mihindukulasuriy and Lim (2014), explained that nanotechnology has proved to be highly efficient in the area of food packaging and security with a remarkable target on indicators for targeting freshness, oxygen scavenging, and humidity as well as nanocoatings, which increase the features of nanofilms used for packaging.

Yu et al. (2017) produced a chitosan and polyvinyl alcohol-based polymer film that is economical and environment-friendly using a silica nanocomposite. This was revealed to show a remarkable decrease in moisture and oxygen permeability and showed that the film increased the life span of the cherries when compared with the usual packaging material.

Arfat et al. (2017) fabricated a nanocomposite film utilizing silver-copper bimetallic nanoparticles and gelatin from fish skin. A remarkable antibacterial potency was observed against *Salmonella enteric and Listeria* spp. They also reported the use of various nanoparticles based on silver for the retardation of *E. coli* and *Campylobacter* spp.

Hosseini et al. (2015) successfully entrapped nutraceuticals with low solubility in aqueous media, such as curcumin, folic acid, ergocalciferol, and carotene in electrostatically stabilized nanocomplexes from sodium alginate combination resulting in a nano level green delivery system in liquid food.

Lingamdinne et al. (2017) reported nanoparticles of iron oxide in the reduction of heavy metals present in food without a loss of the stability of the food. The iron oxides were shown to be recoverable and reusable.

Yang et al. (2010) worked on the enhancement of physiological and physicochemical properties and qualities of straw berry fruits using nanotechnology. They fabricated a new material for nanopackaging, which had high strength, lower humidity, and oxygen transmission speed. The nanoparticles were also capable of enhancing the absorption and decomposition of ethylene, thereby bringing about an extension of the shelf life of the fruits.

Sarwar et al. (2018) formulated a nanocomposite film of silver and polyvinyl alcohol for the main purpose of applying them as food packaging materials and prevention of microbial attack. The films were observed to show significant potency against *S. aures* and *E. coli*.

Saleem et al. (2017) developed nickel oxide nanoparticles and then investigated anti-biofilm potential. These nanoparticles were proved to have antitumor and antibacterial potential. The approach for the synthesis of the nanoparticle was a green approach in which *Eucalyptus globules* was used.

Zhao et al. (2008) emphasized on the utilization of nanoclay in the area of bottling formulation which aid in stiffening the material used in packaging and makes the beer relatively fresher, reducing the escape of carbon dioxide from them, increasing the shelf life of the products, and bringing about a reduction in the permeability of gas.

Verma et al. (2012) opined that nanotechnology would play an indispensable role even for vegetarians who have preference to consume a high-protein food with the animal not killed, in forms such as meat grown in the laboratory, cultured meat, and in vitro meat. Hence, it would be a reliable weapon for managing the fast-growing global population in the nearest future. The use of nanoparticles would further enhance the easy transport of products from livestock while ensuring their freshness throughout the process.

Nam et al. (2003) emphasized that nano particles such as zinc oxide and titanium oxide that are metallic in nature are capable of removing pathogenic organisms and toxic chemicals from food through the reduction of transport of oxygen within the packaging materials, moisture leakages, thereby increasing the freshness and shelf life of the food. Table 10.1 illustrates the various types of nanomaterials used in food packaging.

TABLE 10.1 Different Types of Nanomaterials Used in Food Packaging.

Nanomaterial and systems	Food material	Nature of packaging/role	References
Bio-nanoencapsulated quercetin	Fruits and vegetables	Increase shelf life	Tarafdar et al. (2013)
Nanofilters	Juice and milk	Elimination of colors in juice and elimination of microbes	Fermando et al. (2018)
Silica nanocomposite	Cherries	Reduction of oxygen and moisture permeability	Yu et al. (2017)
Magnesium oxide NPs and biopolymer of polylactic acid		Potency against bacterial biofilms	Swaroop and Shukla (2018)
ZnO and iron oxide NPs	Fuji apples	Acting as oxygen scavenger	Foltynowicz et al. (2017)
Au NPs	Milk	Detection of aflatoxin B1	Mao et al. (2006)
Nanobiosensor with polyaniline	Milk	Detection of foodborne pathogen	Sozer and Kokini (2009)
Carbon nanotubes and zinc oxide NPs		Significant enhancement of coating	Mylvaganam and Rathnayake (2020)
Zinc oxide NPs	Food materials	Potency against *Salmonella typhimurium*	Nishima et al. (2011)

10.10 CONCLUSION AND FUTURE RECOMMENDATION

A detailed information on several nanomaterials has been provided, which includes information on nanoorganic material incorporated into bioactive substances, metal oxides, and inorganic metal, as well as their nanocomposite that have been utilized as packaging material in food industry. Moreover, despite all the benefits of nanotechnology, the high level of buildup of these nanomaterials in the environment and human health has constituted several health hazards. There is a need to provide detailed information on the advantages of nanotechnology on human health when compared with the conventional techniques. Moreover, there is a need to provide more facts on the relevance of function foods and neutraceuticals for effective release of active constituents from packaging materials, bioavailability, and stability that could lead to improvement of human health.

KEYWORDS

- **nanotechnology**
- **nanomaterials**
- **processing**
- **storage**
- **functionality**
- **packaging**
- **food safety**

REFERENCES

Abbas, K. A.; Saleh, A. M.; Mohamed, A.; Azhan, N. M. The Recent Advances in the Nanotechnology and Its Applications in Food Processing: A Review. *J. Food Agric. Environ.* **2009,** *7* (3 & 4), 14–17.

Adabi, M.; Naghibzadeh, M.; Adabi, M.; Zarrinfard, M. A.; Esnaashari, S. S.; Seifalian, A. M. Biocompatibility and Nanostructured Materials: Applications in Nanomedicine. *Artif. Cells Nanomed. Biotechnol.* **2017,** *45*, 833e42.

Agriopoulou, S. Nanotechnology in Food Packaging. *EC Nutrition* **2016,** *5*, 1137–1141.

Alfadul, S. M.; Elneshwy, A. A. Use of Nanotechnology in Food Processing, Packaging and Safety—Review. *Afr. J. Food Agric. Nutr. Dev.* **2010,** *10*, 2719–2739.

Arfat, Y. A; Ahmed, J.; Hiremath, N.; Auras, R.; Joseph, A. Thermo-Mechanical, Rheological, Structural and Antimicrobial Properties of Bionanocomposite Films Based on Fish Skin Gelatin and Silver-Copper Nanoparticles. *Food Hydrocoll.* **2017,** *62*, 191–202.

Avella, M.; De Vlieger, J. J.; Errico, M. E.; Fischer, S.; Vacca, P.; Volpe, M. G. Biodegradable Starch/Clay Nanocomposite Films for Food Packaging Applications. *Food Chem.* **2005,** *93*, 467–474.

Bajpai, V.; Kamle, M.; Shukla, S.; Han, Y. Prospects of Using Nanotechnology for Food Preservation, Safety, and Security. *J. Food Drug Anal.* **2018,** *26* (4). DOI: 10.1016/j.jfda.2018.06.011.

Baranwal, A.; Srivastava, A.; Kumar, P.; Bajpai, V. K.; Maurya, P. K.; Chandra, P. Prospects of Nanostructure Materials and Their Composites as Antimicrobial Agents. *Front. Microbiol.* **2018,** *9*. DOI: 10.3389/fmicb.2018.00422.

Berekaa, M. M. Nanotechnology in Food Industry; Advances in Food Processing, Packaging and Food Safety. *Int. J. Curr. Microbiol. App. Sci*, **2015,** *4* (5), 345–357.

Bezbaruah, A. N.; Kalita, H.; Almeelbi, T. Ca–Alginate-Entrapped Nanoscale Iron: Arsenic Treatability and Mechanism Studies. *J. Nanopart. Res.* **2014,** *14*. DOI: 10.1007/s11051-013-2175-3.

Bonah, E.; Huang, X.; Aheto, J. H.; Osae, R. Application of Hyperspectral Imaging as a Nondestructive Technique for Foodborne Pathogen Detection and Characterization. *Foodborne Pathog Dis.* **2019,** *16* (10), 712–722.

Brunner, T.; Piusmanser, P.; Spohn, P.; Grass, R.; Limbach, L.; Ruinink, A. B.; Stark, W. In Vitro Cytotoxicity of Oxide Nanoparticles: Comparison to Asbestos, Silica, and the Effect of Particle Solubility. *Environ. Sci. Technol.* **2006,** *40*, 4374–4381.

Carlson, K.; Misra, M.; Mohanty, S. Developments in Micro- and Nanotechnology for Foodborne Pathogen Detection. *Foodborne Pathog. Dis.* **2018,** *15* (1), 16–25.

Chaudhry, Q.; Scotter, M.; Blackburn, J.; Ross, B.; Boxall, A.; Castle, L. et al. Applications and Implications of Nanotechnologies for the Food Sector. *Food Add. Contam.* **2008,** *25*, 241–258.

Chen, M.; Mikecz, A. V.; Formation of Nucleoplasmic Protein Aggregates Impairs Nuclear Function in Response to SiO^2 Nanoparticles Experiment. *Cell Res.* **2005,** *305*, 51–62.

Chen, H.; Yada, P. Nanotechnologies in Agriculture: New Tools for Sustainable Development. *Trends. Food Sci. Technol.* **2011,** *22* (11), 585–594.

Cho, I. H.; Ku, S. Current Technical Approaches for the Early Detection of Foodborne Pathogens: Challenges and Opportunities. *Int. J. Mol. Sci.* **2017,** *18* (10). DOI: 10.3390/ijms18102078.

Cho, K.; Wang, X. U.; Nie, S.; Shin, D. M. Therapeutic Nanoparticles for Drug Delivery in Cancer. *Clin. Cancer Res.* **2008,** *14* (5), 1310–1316.

Cindi, M. D.; Shittu, T.; Sivakumar, D.; Bautista-Banos. Chitosan Boehmite-Alumina Nanocomposite Films and Thyme Oil Vapour Control Brown Rot in Peaches (Prunus persica L.) during Postharvest Storage. S. *Crop Protect.* **2015,** *72*, 127–131.

Coles, D.; Frewer, L. J. Nanotechnology Applied to European Food Production—A Review of Ethical and Regulatory Issues. *Trends Food Sci. Technol.* **2013,** *34*, 32–43.

Coma, V. Bioactive Packaging Technologies for Extended Shelf Life of Meat-Based Products. *Meat Sci.* **2008,** *78*, 90–103.

Das, M. P.; Livingstone, J. R.; Veluswamy, P.; Das, J. Exploration of *Wedelia chinensis* Leaf-Assisted Silver Nanoparticles for Antioxidant, Antibacterial and in Vitro Cytotoxic Applications. *J. Food Drug Anal.* **2017,** *26*, 917–925.

De Azeredo, H. M. Nanocomposites for Food Packaging Applications. *Food Res. Int.* **2009,** *42*, 1240–1253.

Dingman, J. Nanotechnology-Its Impact on Food Safety. *J. Environ. Health*, **2008,** *70* (6), 47–50.

Duncan, T. V. Applications of Nanotechnology in Food Packaging and Food Safety: Barrier Materials, Antimicrobials and Sensors. *J. Colloid Interf. Sci.* **2011,** *363*. DOI: 10.1016/j.jcis.2011.07.017.

Elgadir, M. A. Application of Nanotechnology in Meat Packaging, Review. *Int. J. Innov. Sci. Eng. Technol.* **2017,** *4* (4), 267–275.

Errico, D.; Ciccarelli, G. D.; Ortona, O. Effect of Glycerol on Micelle Formation by Ionic and Nonionic Surfactants at 25°C. *J. Colloid Interf. Sci.* **2005,** *286*, 747–754.

Fang, Y.; Ramasamy, R. P. Current and Prospective Methods for Plant Disease Detection. *Biosensors*, **2015,** *5* (3), 537–561.

Fernando, S. S.; Gunasekara, T.; Holton, J. Antimicrobial Nanoparticles: Applications and Mechanisms of Action. *Sri Lankan J. Infectious Diseases*. **2018.** DOI: 10.4038/sljid.v8i1.8167.

Foltynowicz, Z.; Bardenshtein, A.; S€angerlaub, S.; Antvorskov, H.; Kozak, W. Nanoscale, Zero Valent Iron Particles for Application as Oxygen Scavenger in Food Packaging. *Food Packag. Shelf Life* **2017,** *11*, 74–83.

Fu, P. P. Introduction to the Special Issue: Nanomaterials Toxicology and Medical Applications. *J. Food Drug Anal.* **2014,** *22*. DOI: 10.1016/j.jfda.2014.01.013.

Gordon, T.; Perlstein, B.; Houbara, O.; Felner, I.; Banin, E.; Margel, S. Synthesis and Characterization of Zinc/Iron Oxide Composite Nanoparticles and Their Antibacterial Properties. *Colloids Surf. A. Physicochem. Eng. Aspects.* **2011,** *374*. DOI: 10.1016/j.colsurfa.2010.10.015.

Goyal, S.; Goyal, G. K.; Nanotechnology in Food Packaging—A Critical Review. *Russ. J. Agric. Socio-Econ. Sci.* **2012,** *10* (10), 14–20.

Gutiérrez, F. J.; Mussons, M. L.; Gatón, P.; Rojo, R. *Nanotechnology and Food Industry. Scientific, Health and Social Aspects of the Food Industry*; In Tech: Croatia, 2011; pp 95–128.

He, X.; Hwang, H. M. Nanotechnology in Food Science: Functionality, Applicability, and Safety Assessment. *J. Food Drug Anal.* **2016,** *24*, 671–681.

Hellarama, C. C.; Murugaboopathib, G.; Johna, A. A.; Sivakumarc, R.; Ganesand, S.; Krithikae, S. Significance of Nanotechnology in Food Industry. *APCBEE Procedia* **2014,** *8*, 109–113.

Hosseini, S. M. H.; Emam-Djomeh, Z.; Sabatino, P.; Van der Meeren, P. *Food Hydrocoll.* **2015,** *50*, 16–26.

Hu, B.; Liu, X.; Zhang, C.; Zeng, X. Food Macromolecule Based Nanodelivery Systems for Enhancing the Bioavailability of Polyphenols. *J. Food Drug Anal.* **2017,** *25*, 3–15.

Jia, M.; Liu, J.; Zhang, J.; Zhang, H. An Immunofiltration Strip Method Based on the Photothermal Effect of Gold Nanoparticles for the Detection of *Escherichia coli* O157:H7. *Analyst*. **2019,** *144*, 573–578.

Kim, W.; Zide, J.; Gossard, A.; Klenov, D.; Stemmer, S.; Shakouri, A. et al. Thermal Conductivity Reduction and Thermoelectric Figure of Merit Increase by Embedding Nanoparticles in Crystalline Semiconductors. *Phys. Rev. Lett.* **2006,** *96* (4). DOI: 10.1103/PhysRevLett.96.045901.

Kumar, H.; Kuca, K.; Bhatia, S. K.; Saini, K.; Kaushal, A.; Verma, R. et al. Applications of Nanotechnology in Sensor-Based Detection of Foodborne Pathogens. *Sensors* **2020,** *20* (7). DOI: 10.3390/s20071966.

Lee, S. B.; Martin, C. R. Electromodulated Molecular Transport in Gold Nanotube Membranes. *J. Am. Chem. Soc.* **2002,** *124*, 11850–11851.

Lingamdinne, L.P.; Chang, Y.Y.; Yang, J.K.; Singh, J.; Choi, E.H.; Shiratani, M.; Koduru, J.R.; Attri, P. Biogenic Reductive Preparation of Magnetic Inverse Spinel Iron Oxide Nanoparticles for the Adsorption Removal of Heavy Metals. *Chem. Eng. J.* **2017,** *307*, 74–84.

Liu, W.; Zhang, M.; Bhandari, B. Nanotechnology—A Shelf Life Extension Strategy for Fruits and Vegetables. *Crit. Rev. Food Sci. Nutr.* **2019,** *60*, 1706–1721.

Luo, K.; Kim, H. Y.; Oh, M. H.; Kim, Y. R. Paper-Based Lateral Flow Strip Assay for the Detection of Foodborne Pathogens: Principles, Applications, Technological Challenges and Opportunities. *Crit. Rev. Food Sci. Nutr*. **2020,** *60* (1), 157–170.

Mandal, P. K.; Choil K.; Min, S. G.; Lee, C. H. Application of Nanotechnology in Food Packaging: An Overview Orean. *J. Food Sci. Ani. Resour.* **2009,** *29*, 403–408.

Mannino, S.; Scampicchio, M. Nanotechnology and Food Quality Control. *Vet. Res. Commun.* **2007,** *31*, 149–151.

Mao, X.; Huang, J.; Leung, M. F. Novel Core-Shell Nanoparticles and Their Application in High-Capacity Immobilization of Enzymes. *Appl. Biochem. Biotechnol.* **2006,** *135*, 229–246.

Mason, T. G.; Wilking, J. N.; Meleson, K.; Chang, C. B.; Graves, S. M. Nanoemulsions: Formation, Structure, and Physical Properties. *J. Phys. Condensed Matter.* **2006,** *18*, R635–R666.

Mastromatteo, M.; Conte, A.; Nobile, M. A. Combined Use of Modified Atmosphere Packaging and Natural Compounds for Food Preservation. *Food Eng. Rev*. **2010,** *2* (1), 28–38.

Meetoo, D. D. Nanotechnology and the Food Sector: From the Farm to the Table. *Emir. J. Food Agric.* **2011,** *23*, 387–407.

Mihindukulasuriya, S. D. F.; Lim, L. T. Nanotechnology Development in Food Packaging: A Review. *Trends Food Sci. Technol*. **2014,** *40*, 149–167.

Mills, A. Oxygen Indicators and Intelligent Inks for Packaging Food. *Chem. Soc. Rev.* **2005,** *34*, 1003–1011.

Morais, M. G. D.; Martins, V. G.; Steffens, D.; Pranke, P.; Costal, J. A. V. D. Biological Applications of Nanobiotechnology. *J. Nanosci. Nanotechnol.* **2014,** *14*, 1007–1017.

Mousavi, S. R.; Rezaei, M. Nanotechnology in Agriculture and Food Production. *J. Appl. Env. Biol. Sci.* **2011,** *1*, 414–419.

Munish, G.; Paramjeet, S.; Shveta, R. Optimizing Physical Layer Energy Consumption for Reliable Communication in Multi-Hop Wireless Sensor Networks. *Ind. J. Sci. Technol.* **2015,** *8*, 1–7.

Mustafa, F.; Andreescu, S. Nanotechnology-Based Approaches for Food Sensing and Packaging Applications. *RSC Adv.* **2020,** *10*, 19309–19336.

Mylvaganam, H.; Rathnayake, P. Application of Nanotechnology in Food Packaging: A Review. *J. Food Technol. Pres.* **2020,** *4* (4), 1–4.

Nam, J. M.; Thaxton, C. S.; Mirkin, C. A. Nanoparticle-Based Bio-Bar Codes for the Ultrasensitive Detection of Proteins. *Science* **2003,** *301*, 1884–1886.

Neha, B.; Sanjeev, B.; Manoj, K. N.; Akash, D. Fluorescent Nanobiosensors for the Targeted Detection of Foodborne Bacteria. *Trends Anal. Chem.* **2017,** *97*. DOI: 10.1016/j.trac.2017.09.010.

Nile, S. H.; Baskar, V.; Selvaraj, D.; Nile, A.; Xiao, J.; Kai, G. Nanotechnologies in Food Science: Applications, Recent Trends, and Future Perspectives. *Nano-Micro Lett*. **2020,** *12*. DOI: 10.1007/s40820-020-0383-9.

Nishiyama, Y. Structure and Properties of the Cellulose Microfibril. *J. Wood Sci*. **2009,** *55*, 241–249.

Pal, M. Nanotechnology: A New Approach in Food Packaging. *J. Food Microbiol. Saf. Hyg.* **2017,** *02* (02). DOI: 10.4172/2476-2059.1000121.

Pathkoti, K.; Manubolu, M.; Hwang, H. M. Nanostructures: Current Uses and Future Applications in Food Science. *J. Food Drug Anal.* **2017,** *25* (2), 245–253.

Pawar, A. Nanotechnology: The Multidisciplinary Technology—A Review. *J. Pure Appl. Sci. Technol.* **2016,** *6*, 11–16.

Prombutara, P.; Kulwatthanasal, Y.; Supaka, N.; Sramala, I.; Chareonpornwattana, S. Production of Nisin-Loaded Solid Lipid Nanoparticles for Sustained Antimicrobial Activity. *Food Control.* **2012,** *24*, 184–190.

Petit, G.; Jury, V.; Lamballerie, M. D.; Duranton, F.; Pottier, L.; Martin, J. L. Salt Intake from Processed Meat Products: Benefits, Risks and Evolving Practices. *Compr. Rev. Food Sci. Food Saf.* **2019,** *18* (5), 1453–1473.

Qasim, C.; Scotter, M.; Blackburn, B.; Ross, B.; Boxall, A.; Castle, L. et al. Applications and Implications of Nanotechnologies for the Food Sector. *Food Addit. Contam.* **2008,** *25*, 241–258.

Qureshi, M. A.; Karthikeyan, S. Application of Nanotechnology in Food and Dairy Processing: An Overview. *J. Food Sci. Technol.* **2012,** *22*, 23–31.

Radha, K.; Thomas, A.; Sathian, C. T. Application of Nano Technology in Dairy Industry: Prospects and Challenges—A Review Received. *J. Dairy Sci.* **2014,** *67* (5). DOI: 10.5146/IJDS.V67I5.45669.G19818.

Ravichandran, R. Nanotechnology Applications in Food and Food Processing: Innovative Green Approaches, Opportunities and Uncertainties for Global Market. *Int. J. Green Nanotechnol. Phys. Chem.* **2010,** *1*, 72–96.

Restuccia, D.; Spizzirri, U. G.; Parisi, O. I.; Cirillo, G.; Curcio, M.; Iemma, F. New EU Regulation Aspects and Global Market of Active and Intelligent Packaging for Food Industry Applications. *Food Control.* **2010,** *21* (11), 1425–1435.

Rubab, M.; Shahbaz, H. M.; Olaimat, A. N.; Oh, D. H. Biosensors for Rapid and Sensitive Detection of *Staphylococcus aureus* in Food. *Biosens. Bioelectron.* 2018, *105*, 49–57.

Ruyuan, Z.; Tarun, B.; Li, L.; Xingyu, L.; Yangun, L.; Zisheng, L. Nanomaterial-Based Biosensors for Sensing Key Foodborne Pathogens: Advances from Recent Decades. *Compr. Rev. Food Sci. Food Saf.* 2020, *19*, 1465–1487.

Saleem, S.; Ahmed, B.; Khan, M. S.; Al-Shaeri, S.; Musarrat, J. Inhibition of Growth and Biofilm Formation of Clinical Bacterial Isolates by NiO Nanoparticles Synthesized from Eucalyptus Globulus Plants. *Microb. Pathog.* **2017,** *111*, 375–387.

Sandoval, B. Perspectives on FDA's Regulation of Nanotechnology: Emerging Challenges and Potential Solutions. *Compr. Rev. Food Sci. Food Saf.* **2009,** *8*, 375–393.

Sangeeta, Y.; Sheethal, S. N.; Sai, V. V. R.; Jitendra, S. Nanomaterials Based Optical and Electrochemical Sensing of Histamine: Progress and Perspectives. *Food Res. Int.* **2019,** *119*, 99–109.

Sarwar, M. S.; Niazi, M. B. K.; Jahan, Z. T. Ahmad, A.; Hussain, A. Preparation and Characterization of PVA/Nanocellulose/Ag Nanocomposite Films for Antimicrobial Food Packaging. *Carbohydr. Polym.* **2018,** *184*, 453–464.

Scrinis, G.; Lyons, K. The Emerging Nano-Corporate Paradigm: Nanotechnology and the Transformation of Nature, Food and Agri-Food Systems. *Int. J. Sociol. Food Agric.* **2007,** *15* (2), 22–44.

Sekhon, B. S. Food Nanotechnology—An Overview. *Nanotechnol. Sci. Appl.* **2010,** *3*, 1–15.

Sharma, R. Ragavan, K. V.; Thakur, M. S.; Raghavarao, K. S. Recent Advances in Nanoparticle Based Aptasensors for Food Contaminants. *Biosens. Bioelectron.* **2015,** *15*, 612–627.

Shivraj, N. H.; Baskar, V.; Selvaraj, D. et al. Nanotechnologies in Food Science: Applications, Recent Trends, and Future Perspectives. *Nano-Micro Lett.* **2020,** *12*, DOI: 10.1007/s40820-020-0383-9.

Singh, A.; Singh, N. B.; Hussain, I.; Singh, H.; Singh, S. C. Plant-Nanoparticle Interaction: An Approach to Improve Agricultural Practices and Plant Productivity. *Int. J. Pharm. Sci. Invent.* **2015,** *4*, 25–40.

Singh, H. Nanotechnology Applications in Functional Foods: Opportunities and Challenges. *Prev. Nutr. Food Sci.* **2016,** *21*. DOI: 10.3746/pnf.2016.21.1.1

Slootweg, P. J.; Hordijk, G. J.; Schade, Y.; van Es, R. J.; Koole, R. Treatment Failure and Margin Status in Head and Neck Cancer. A Critical View on the Potential Value of Molecular Pathology. *Oral Oncol.* **2002,** *38*, 500–503.

Sozer, N.; Kokini, J. Nanotechnology and Its Applications in the Food Sector. *Trends Biotechnol.* **2009,** *27*, 82–89.

Sujithra, S.; Manikkandan, T. R. Application of Nanotechnology in Packaging of Foods-A Review. *Int. J. Chem. Tech. Res.* **2019,** *12*, 7–14.

Swaroop, C.; Shukla, M. Nano-Magnesium Oxide Reinforced Polylactic Acid Biofilms for food Packaging Applications. *Int. J. Biol. Macromol*, **2018,** *113*, 729–736.

Tang, D.; Sauceda, Z. Lin Magnetic Nano Gold Microspheres-Based Lateral-Flow Immunodipstick for Rapid Detection of Afflation B2 in Food. *Biosens. Bioelectron.* **2009,** *25*, 514–518.

Tarafdar, J. C.; Sharma, S.; Raliya, R. Nanotechnology: Interdisciplinary Science of Applications. *Afr. J. Biotechnol.* **2013,** *12* (3), 219–226,

Tayel, A. A.; El-Tras, W. F.; Moussa, S.; El-Baz, A.F.; Hoda, M.; Salem, M.F. et al. Antibacterial Action of Zinc Oxide Nanoparticles against Foodborne Pathogens. *J. Food Safety*. **2011,** *31*, 211–218.

Trinchese, G.; Cavaliere, G.; Berni, R.; Matamoros, S.; Bergamo, P.; De Filippo, C. et al. Human, Donkey and Cow Milk Differently Affects Energy Efficiency and Inflammatory State by Modulating Mitochondrial Function and Gut Microbiota. *J. Nutr. Biochem*. **2015,** *26*. DOI: 10.1016/j.jnutbio.2015.05.003.

Umesha, S.; Manukumar, H. M. Advanced Molecular Diagnostic Techniques for Detection of Food-Borne Pathogens: Current Applications and Future Challenges. *Crit. Rev. Food Sci. Nutr.* **2015,** *2*, 58, 84–104.

Verma, A. K.; Singh, V. P.; Vikas, P. Application of Nanotechnology as a Tool in Animal Products Processing and Marketing: An Overview. *Am. J. Food Technol.* **2012,** *7* (8), 445–451.

Wang, B.; Feng, W.; Wang, M.; Wang, T.; Gu, Y.; Zhu, M. et al. Acute Toxicological Impact of Nano- and Submicro- Scaled Zinc Oxide Powder on Healthy Adult Mice. *J. Nanopart. Res.* **2008,** *10*, 263–276.

William, A.; Sanchez, I.; Grau, R.; Barat, J. Salt in Food Processing; Usage and Reduction: A Review. *Int. J. Food Technol.* **2011,** *46* (7), 1329–1336.

Xie, Y.; He, Y.; Irwin, P. L.; Jin, T.; Shi, X. Antibacterial Activity and Mechanism of Action of Zinc Oxide Nanoparticle against Campylobacter *jejuni. Appl. Environ. Microbiol.* **2011,** *77*, 2325–2331.

Yang, F. M.; Li, H. M.; Li, F.; Xin, Z. H.; Zhao, L. Y.; Zheng, Y. H. J. Effect of Nano-Packing on Preservation Quality of Fresh Strawberry (Fragaria ananassa Duch. cv Fengxiang) during Storage at 4°C. *Food Sci.* **2010,** *75*. DOI: 10.1111/j.1750-3841.2010.01520.x.

Yang, L.; Bashir, R. Electrical/Electrochemical Impedance for Rapid Detection of Foodborne Pathogenic Bacteria. *Biotechnol. Adv*. **2008,** *26* (2), 135–150.

Yu, H.; Park, J. Kwon, C. W.; Hong, S.; Park, K. M.; Pahn-Shick, C. An Overview of Nanotechnology in Food Science: Preparative Methods, Practical Applications, and Safety. *J. Chem.* **2018,** 2018. DOI: 10.1155/2018/5427978.

Yu, Z. B.; Lib, J.; Chud, P.; Zhang Silica in Situ Enhanced PVA/Chitosan Biodegradable Films for Food Packages. *Carbohydr. Polym.* **2017,** *184*, 214–220.

Zhao, R.; Torley, P.; Halley, P. J. Emerging Biodegradable Materials: Starch- and Proteinbased Bio-Composites. *J. Mater. Sci.* **2008,** *43*, 3058–3071.

Zhaolui, Q.; Chunyang, L.; Yingchun, F.; Yanbin, L. Rapid and Sensitive Detection of *E. coli* O157:H7 Based on Antimicrobial Peptide Functionalized Magnetic Nanoparticles and Urease-Catalyzed Signal Amplification. *Anal. Methods*, **2017,** *9*, 5204–5210.

CHAPTER 11

Bio- and Synthetic Nanocomposites for Food Packaging

AMRA BRATOVCIC*

Department of Physical Chemistry and Electrochemistry, University of Tuzla, Faculty of Technology, Urfeta Vejzagica 8, 75000 Tuzla, Bosnia and Herzegovina

**E-mail: amra.bratovcic@untz.ba*

ABSTRACT

Today, the packaging industry is growing and expanding very fast. Approximately, 40% of the materials used for packaging is plastics. However, in the last decade, we have witnessed the presence of a very huge amount of nondegradable synthetic plastic materials dumped in oceans and rivers and they are mainly coming after removing food packaging materials, plastic bags, and foils. This chapter aims to point out the recent achievements of nanotechnology in terms of food packaging with focus on bio- and synthetic nanocomposites materials that enhance the antimicrobial, mechanical, thermal, as well as barrier properties against the migrating element in the food packaging system. Biodegradable biopolymers are often synthesized from agricultural resources, such as polysaccharides and proteins, from biotechnology, such as polylactic acid (PLA) and polyhydroxyalkanoate (PHA), or biopolymers from petrochemical sources like polycaprolactone (PCL).

Nanoadditives in food packaging materials are nanoparticles, nanotubes, nanoclays and nanofibers. Among them, nanoclays and silicate are the most studied nanoadditives/fillers due to their layered structure, high availability, low cost, and good performance. Therefore, nanotechnology can be used to

The Science of Nanomaterials: Basics and Applications. Suresh C. Ameta & Rakshit Ameta (Eds.)

address the main problems of conventional packaging. The effort of scientist in the nanotechnology is focused on the production of edible and biodegradable films, which preserve the freshness of the food and are extending its shelf life. Finally, these green, safe, recyclable, and eco-friendly materials are attractive products that may replace conventional materials in food packaging and consequently, reducing the contamination of the environment and improved human health.

11.1 INTRODUCTION

In recent years, numerous studies have shown that the presence of micro and nanoplastics in large quantities in the river, lakes, seas, and oceans is ubiquitous in the environment. (Bratovcic, 2019). According to Plastics Europe, 2019 (https://www.plasticseurope.org/application/files/9715/7129/9584/FINAL_web_version_Plastics_the_facts2019_14102019.pdf), packaging (39.9%) and building and construction (19.8%) by far represent the largest end-use markets. The third biggest end-use market is the automotive industry (9.9%). Several thermoplastics used in packaging, such as polyethyelene (PE), polypropylene (PP) (i.e., packaging hinged caps), polystyrene (PS) (i.e., yogurt pots), poly(ethylene terephthalate) (PET) (water bottles), and poly(vinyl chloride) (PVC) are very often found in rivers worldwide (Xiong et al., 2018; Scherer et al., 2020) and seas and oceans as well as in sediment.

The presence of such small dimensions of plastics in the environment may be dangerous since there are not enough studies available on understanding of its interaction with living organisms. It may enter into the food chain. In the last decades, use of biopolymer materials as renewable and biodegradable polymeric materials has appeared worldwide with an intention to replace conventional plastics. In the domain of food packaging, the technological advances have been achieved by the introduction of nanotechnology or by the creation of nanomaterials that has significantly contributed to overcoming problems related to safety and stability as well as food quality. Besides, nanotechnology-based packaging intimates consumers about the real-time quality of food product (Sharma et al., 2017).

In the last several years, one of the fastest growing materials sectors has been the production of bio-based and biodegradable polymers from renewable resources. Their development has attracted attention because they may replace fossil fuel-based polymers. They are based on plants, produced by biological pathways, and generally they are biodegradable. However, not all bio-based polymers are biodegradable. In this context, both bio-based polymers, not biodegradable and biodegradable ones, contribute to the protection of the environment. Due to the increased demand for bioplastics from the packaging industry, the market for bioplastics and biopolymers has experienced a major boom. Bottles, film, clamshell cartons, waste collection bags, carrier bags, and food serviceware are mainly produced from bioplastics.

11.1.1 PACKAGING

Packaging is one of the fastest growing industries globally and its task is to protect goods during distribution, transportation, storage, and use. Among numerous materials (glass, metals, wood, and polymers), about 40% of the polymer is used for packaging of different types of products to provide protection and containment. Plastic bags are typically composed of polyethylene and these are distributed in a very huge amount, more than 500 billion plastic bags, out of which only less than 3% of bags are recycled. The main problem is that their degradation process can take up to thousand years in landfills that emit toxic greenhouse gases (Heap, 2009). Increased use of plastics is due to its specific properties, such as high durability, plasticity, and water resistance. However, the use of conventional polymers for the production of carry bags and packages has been restrained by increasing environmental regulations which aroused great interest in the development of new bioplastics and biopolymers. Therefore, the need for sustainable solutions has encompassed food and beverage industry, by enhancing industrial and manufacturing technologies enabling them to incorporate bioplastics in packaging. Bioplastics and biopolymers have found different applications in food, pharmaceutical, cosmetics packaging, and shopping bags. Bioplastics are often blended with synthetic polymers to reduce the cost of the material.

The evident increased use of PLA, PHA, and starch- and chitosan-based polymers has arisen due to the growing demand for eco-friendly packaging. Europe is one of the major producers of plastics and the largest bioplastics and biopolymers market. Bioplastics have become identified as one of the most important products at the markets according to the EU Commission (https://www.marketsandmarkets.com/Market-Reports/biopolymers-bioplastics-market-88795240.html?gclid=Cj0KCQjw9ZzzBRCKARIsANwXaeKABH7G4izM9F115y4hEsbVH92cJbd67YrCXkuflWndvTIvrO66oVwaApXvEALw_wcB).

Currently, 3.59 million tons of bioplastics are produced annually and its market continuously grows and diversify. According to the latest research institute nova-Institute and market data compiled by European Bioplastics, global bioplastics production capacity should reach 2.43 million tons in 2024 and it would represent an increase of 0.32 million tons from 2019 to 2024 (https://www.european-bioplastics.org/market/). In economic terms, that would mean the growth from USD 10.5 billion in 2020 and USD 27.9 billion by 2025 (https://www.marketsandmarkets.com/Market-Reports/biopolymers-bioplasticsmarket88795240.html?gclid=Cj0KCQjw9ZzzBRCKARIsANw-XaeKABH7G4izM9F115y4hEsbVH92cJbd67YrCXkuflWndvTIvrO66oVwaApXvEALw_wcB).

11.2 BIOPOLYMERS

Biopolymers are monomer molecules interconnected by covalent bonds and have become very attractive due to their properties, such as renewability and biodegradability and represent an excellent substitute for conventional petroleum-derived plastic materials (Muratore et al., 2005). Their final degradation products are carbon dioxide (CO_2), water (H_2O), and organics (Rhim et al., 2013; Tang et al., 2012).

Biodegradable and nonbiodegradable are the main class of biopolymers and alternatively biopolymers can be classified into bio-based and nonbio-based. Generally, bio-based polymers are gained from biological sources. More specifically, they can be produced within the producing organisms (microorganisms, algae, or plants), and in an external environment from bio-based monomers.

The simplest bio-based polymers are cellulose and starch, while novel bio-based polymers are PLA and PHA. In this context, the bioplastics and biopolymers market has been divided into biodegradable and nonbiodegradable bioplastics. It is predicted that Bio-PET as nonbiodegradable bioplastics will be the largest and the fastest growing type of bioplastics and biopolymers due to low cost and easy production process. Biodegradability together with sustainability is the main property of biopolymers which differs from fossil-fuel-derived polymers.

11.2.1 BIODEGRADABLE POLYMERS

Biodegradable products are mostly generated from biopolymers. Starch, sugar, natural fibers, or other organic biodegradable components in different compositions are frequently used as starting materials for the synthesis of biodegradable biopolymers. These systems include biopolymers from agricultural resources, such as polysaccharides and proteins, from biotechnology (PLA, PHA), or biopolymers from petrochemical sources, such as PCL. Recently, all of them have been extensively reviewed (Bari et al., 2016), making reference to fundamental industrial sectors, such as food packaging (Rouf and Kokini, 2018) or the biomedical field (Tripathy, 2017).

According to Gabor and Tita (2012), biodegradable polymers used in food packaging can be classified based on their composition (Figure 11.1).

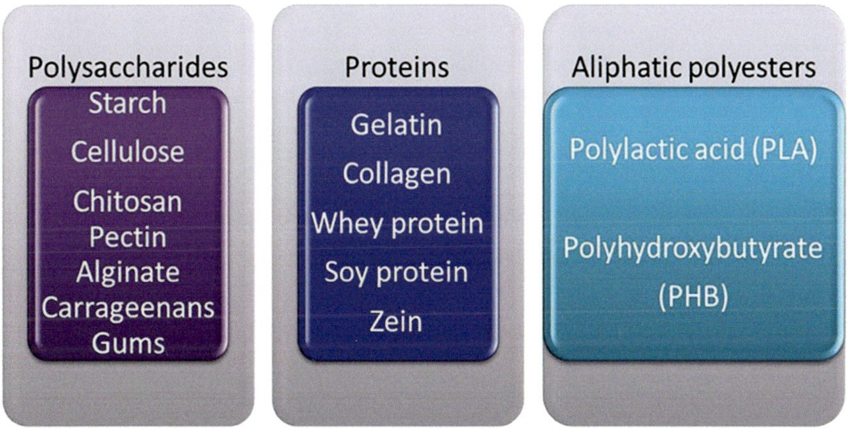

FIGURE 11.1 Classification of biodegradable polymers used in food packaging.

PLA is becoming increasingly popular owing to its high mechanical strength and high recyclability. However, Auras et al. (2004) reported that PLA is sensitive to thermal degradation, which is regarded as one of the main drawbacks for PLA in industrial processing. Static water contact angle of nearly 80° makes PLA relatively hydrophobic material, (Rasal et al., 2010), PLA shows higher wetting properties compared with low density polyethylene (LDPE) and PP which exhibitthe static water contact angle around 100° (Fombuena et al., 2013; Balart et al., 2016).

Starch-based material has attracted much attention in the food packaging sector owing to its low cost and biodegradability. However, they possess some shortcomings like strong hydrophilic behavior (poor moisture barrier) and also weaker mechanical properties than the conventional nonbiodegradable plastic film used in the food packaging industry (Avella et al., 2005).

A growing trend of the number of published papers on ScienceDirect by searching the keywords "Biopolymers" on (07.09.2020) is presented in Figure 11.2.

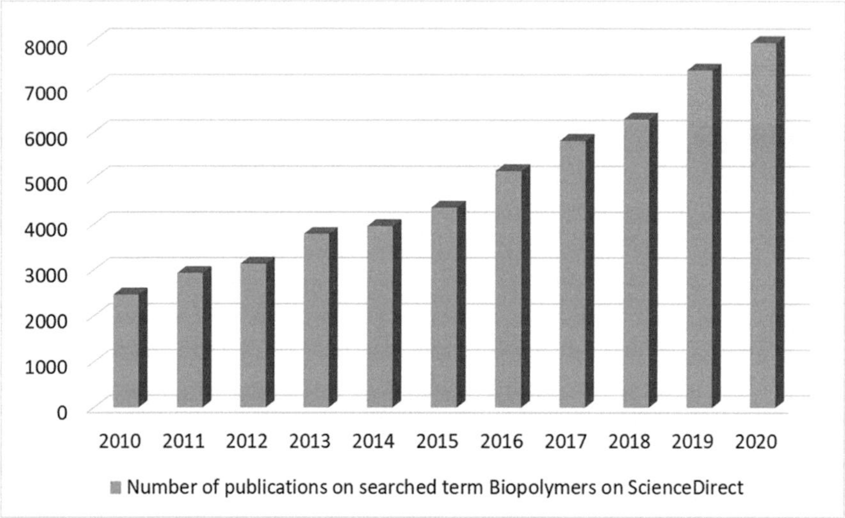

FIGURE 11.2 The number of research articles in ScienceDirect for the keywords "Biopolymers."

The rate of biodegradation of biopolymers depends on several things, such as the type and ratios of the components as well as the environment of disposal. Biodegradability property of biodegradable biopolymer product is decreasing on increasing the thickness.

Blending biopolymers with other polymers or additives are often used to improve inferior properties of many biopolymers. However, the drawback of many biopolymer blends or composites is that they do not have uniform mechanical properties. For example, the aliphatic polyesters, most known biodegradable polymers have low softening temperatures (T_m), which prevents their use in various areas. Cost reduction of biopolymers production, wider availability, improvement of their barrier and thermomechanical properties, speed of biodegradability, and availability are still some challenges for researchers. It is quite reasonable to expect that with the increased demand for biopolymers, their production capacity will also increase, then prices will decrease, as well as a better and more widespread infrastructure of industrial composting plants. However, the question arises as to whether the performance properties and processability of biopolymers will ever be comparable and compete with nonrenewable polymers (Niaounakis, 2013).

11.2.2 EDIBLES COATINGS

Edible films and coatings as their name suggests are made of edible components and they belong to the group of biodegradable packaging. The packaging consists of a thin layer of edible material by which a food product can be coated without affecting the packed ingredients. Thin coatings, about five nanometers thick, are usually made of edible components and used as moisture and gas barriers for food packaging (vegetables, fruits, meat) and food products, such as pastries and confectionery (Bratovcic, 2020a).

Simple cellulose derivatives are most often used for the production of edible coatings, and cellulose is one of the most widespread and naturally available polymers in abundant quantities in plants. Flexibility and toughness are the most common properties that characterize polymer edible coatings.

11.2.2.1 SOURCE OF EDIBLE POLYMERS

Polysaccharides, proteins, and lipids are the most common constituents of edible polymers. Among all the coatings, polysaccharide coatings should have barrier properties for gases and moisture. However, these coatings have weaker barrier properties for moisture. It may be due to their hydrophilicity. Yadav (2017) has shown in his research that it is possible to increase the shelf life of tomato using bionanoencapsulated quercetin (biodegradable poly-D, L-lactide). However, similar studies in different types of fruits and vegetables should be investigated to determine whether the bionanoencapsulated quercetin used has the same or similar effect.

Starch-based coatings containing potassium sorbate were applied on the surface of fresh strawberries just to reduce growth of microorganism and also to extend their storage life. Antifungal activity against the growth of *Cladosporium* and *Rhizopus* on fresh strawberries may be increased by chitosan coatings containing potassium sorbate (Park et al., 2005).

Proteins with similar properties are also hydrophilic and these include cornzein, wheat gluten, peanut, soy, collagen, gelatin, egg, whey, and casein.

Lipids and waxes tend to be more permeable to gases but they present a better barrier to water vapors. These include petroleum-based waxes, beeswax, vegetable oils, and so forth.

Encapsulated fats and oils, aroma compounds, oleoresins, colorants, vitamins, minerals, and enzymes are embedded in food products. Films of coating contains encapsulated oils, aroma compounds, antimicrobials, and enzymes. Moreover, the microencapsulation of nutrients like vitamins and minerals within the film matrix can augment the nutritional content of the foodstuff (Pascall and Lin, 2013; Bratovcic and Suljagic, 2019a). The main drawback of incorporated nutraceuticals in food is their stability, because they can lose their activity due to oxidation or other chemical reactions. The major advantage of edible films and encapsulation is the preservation of bioactive effect of active biomolecules present in edible coatings or encapsulated and protection from contact with foods. Milk proteins, maltodextrines, and other modified starches are largely used as encapsulation matrices, and as a result, it will increase the bioavailability of molecule.

11.2.2.2 PROPERTIES OF EDIBLE COATINGS

The low water vapor permeability should have films or coatings for fresh vegetables and fruits in order to decrease desiccation rates, while oxygen permeability should be low enough to retard respiration, but it should be not too low to create anaerobic conditions favorable to ethanol production and off-flavor formation (Lin and Zhao, 2007). Films or coatings for nuts should have low oxygen permeability in order to reduce lipid oxidation rates and low water vapor to reduce water absorption and loss of crunchy texture.

Usually, transparency and flavorless, not interfering with food sensory properties are the main characteristics of edible films and coatings. McHugh et al. (1996) produced first edible films based on fruit purees. Fruit purees such as apple, apricot, pear, peach (McHugh et al., 1996), mango (Sothornvit and Rodsamran, 2008; Azeredo, 2009), and banana were used for film production (Martelli et al., 2015). So far, the most used has been pectin-binding agent in edible films based on fruits and vegetables (Otoni et al., 2017).

Some milk proteins, such as whey protein and caseinates, have also been extensively used because of their high-quality nutritional value and their multiple functional properties which are significant for the creation of edible films (MCHugh and Krochta, 1994). Chen (1995) reported that edible films based on whey protein were flavorless, tasteless, flexible and have desirable film-forming and barrier properties, which can be well compared with petrol-based products. However, they have the low tensile strength and high-water vapor permeability (Zhao et al., 2008; Malathi et al., 2019).

A growing trend of published papers in the last 10 years on ScienceDirect by searching the keywords "Edible polymers" (07.09.2020) is presented in Figure11.3.

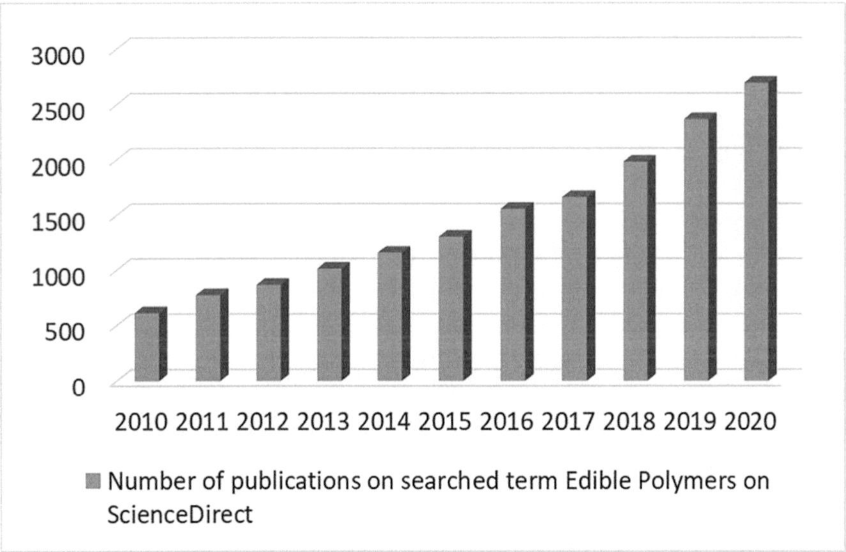

FIGURE 11.3 The results of published papers on ScienceDirect by searching the keywords "edible polymers"

Apart from these polymers, edible-polymer-based hydrogel films are used for food packaging, they include some plasticizer and additives also. Plasticizers are based on glycerol, sorbitol, sodium benzoate, glutaraldehyde, gelatin, polyethylene glycol (PEG), glycerine, agar, and olive oil.

Various parameters, such as the type and structure of the monomer, the functionality of the monomer, the concentration of the monomer, and the degree of cross-linking can affect the structure of the hydrogel. The films of dried hydrogels that are used for food, pharmaceuticals products, and cosmetics are known as biodegradable packaging materials (Langmaier et al., 2008). It is worth noting that environmental contamination may be reduced by increased use of edible polymers.

11.2.2.3 COMMERCIALLY AVAILABLE NANOPROCESSED FOOD

Nanogreen tea, Neosino capsules (dietary supplements), Canola active oil, Aquanova (micelle to intensify the solubility of vitamins E, K, D, C, and A), β-carotene and omega fatty acids and Nutralease (strengthening nanocarriers to carry nutraceuticals and drugs) are common commercial nanotechnology-based products within the market. In addition, oat nutritional drinks, fortified fruit juices, nanoteas and nanocapsules which contain tuna fish oil in breads are few commercially offered nanoprocessed foods within the market present in Australia, Japan, China, and the US (Nile et al., 2020; Poças et al., 2008).

11.2.3 BIOPOLYMER-NANOCOMPOSITES

Nanocomposites have attracted plenty of attention in academic research and the industry because of their distinctive properties as compared with standard composites, even at low filler content. A good interaction between the polymer matrix (continuous phase) and the nanofiller (discontinuous phase) is desired. The nanofillers have sturdy reinforcing effects and it has been additionally shown that they have to attain these modifications with positive impact on barrier properties. Development of nanocomposites (inserting nanomaterials in polymers) may be a new strategy to enhance the physical properties of polymers, together with mechanical strength, thermal stability, and barrier properties to gases and biodegradation (Bratovcic et al., 2015). Thus, bio-based nanocomposites and biodegradable nanocomposites become a subject of interest in recent years and several other appropriate bio-based and biodegradable polymer matrices are developed and their properties enhanced after the incorporation of various nanoscale materials.

The impact of the nanoscale in materials and systems is usually associated with their large surface area-to-volume ratio magnitude relation, resulting in the enhanced surface area, distinct optical behavior, chemical and kinetic stability, and low density versus high mechanical properties.

Recent studies involving the use of nanotechnology in food packaging showed an excellent potential to enhance packaging materials' main properties in addition as adding new functionalities like active (antimicrobial) and intelligent (controlled release) features. Therefore, nanotechnology can be used to overcome most of the issues of standard packaging (low gas barrier and controlled delivery of bioactive compounds). Polymer blending and compositing are well-used techniques whenever modification of properties is required and as a result, it uses conventional technology at low cost.

11.2.3.1 NANOFILLERS IN FOOD PACKAGING

Nanofillers applied in food packaging can improve the thermal, barrier, and mechanical properties and moreover they exhibit different desired functions, such as antimicrobial agents, biosensors, and oxygen scavengers. A promising alternative to conventional plastics, particularly those made up of fossil fuel used for food packaging are bionanocompsite materials.

Nanofillers, which are used in food packaging materials are composed of nanoparticles (Cheng et al., 2006), nanotubes (Kang et al., 2007), nanoclays (Sothornvit et al., 2009) and nanofibers. Among them, nanoclay and silicate are the most studied nanofillers/additives, because of their layered structure, high availability, low cost, and good performance.

There are reports of different particle fillers being employed, together with silver nanoparticles (Moura et al., 2012), copper, copper oxide, zinc oxide (Duncan, 2011), titanium dioxide (Kadam et al., 2013), carbon nanotubes, graphene nanoplates, carvacrol/thymol/linalool (Kuorwel et al., 2013), lignin (Bhat et al., 2013), potassium sorbate (Sayanjali et al., 2011), clove oil (Muppalla et al., 2014), graphene (Demitri et al., 2016), cinnamaldehyde, propolis (Siripatrawan and Vitchayakitti, 2016), tea extract (Siripatrawan and Harte, 2010), enzyme (lysozyme) (Bayarri et al., 2014), silica (Farris et al., 2012), montmorillonite (Weiss et al., 2006; Kumar et al., 2010; Introzzi et al., 2012), garlic oil (Pranoto et al., 2005), oregano oil (Wu et al., 2014), cellulose fibers (Jensen et al., 2015), flaxseed oil (Hopkins et al., 2015), and so forth. Incorporation of those nanofillers in polymer or biopolymer matrices has an inhibiting effect on the growth of microorganisms, and thus contribute to the reduction of food spoilage (Honarvar et al., 2016).

11.2.3.1.1 Nanoparticles as Additives

To date, several nanoparticles are known as fillers for creating polymer nanocomposites to reinforce their packaging performances. The addition of nanoparticles to develop nanocomposites has been established as a good strategy to enhance the barrier properties of polymer-based packaging materials (Sanchez-Garacia and Lagaron, 2010).

Among all metal oxide nanoparticles silver oxide (Ag_2O), titanium dioxide (TiO_2), magnesium oxide (MgO), copper and copper oxide (CuO), zinc oxide (ZnO), lead oxide (PbO), cadmium selenide/telluride, chitosan, carbon nanotubes (CNTs), and silver nanoparticles (AgNPs) are the most powerful wide spectrum antimicrobials (Emamifar et al., 2011; Bratovcic, 2020d).

Recently, Bratovcic (2020b) has synthetized green silver nanoparticles from orange peel extracts with possible antimicrobial effects. Common antibiotics eradicate only five to six disease-causing microbial pathogens, while silver molecules can kill more than 650 pathogens in 6 min of contact (Han and Li, 2008).

11.2.3.1.2 Carbon Nanotubes as Additives

CNTs are another promising nanomaterials in packaging applications.

In order to enhance the inherent properties of polymeric hydrogels, CNTs were embedded in the polymer matrix as nanofillers obtaining nanocomposite hydrogels (Bratovcic, 2020c). The obtained nanocomposite improves polymer properties in multiple aspects, such as tensile strength, stiffness, toughness, thermal stability, electrical conductivity, and chemical resistance (Shahbazi et al., 2017). However, due to high cost and the difficulty in processing of polymer nanocomposites containing CNTs, their use has been difficult (Arora and Padua, 2010). Furthermore, large surface area, chemical inertness, large number of antibacterial and antifungal properties may be used as protein carriers, because these contain exposed functional groups, which make them enormously attractive in numerous biosensor applications (Baughman et al., 2002).

11.2.3.1.3 Nanoclay as Additives

Nanoclay is typically used nanomaterial in food packaging to boost the physical, thermal and mechanical properties, therefore the moisture and gas barrier properties of plastic packaging.

Nanocomposite obtained by mixing bioplastics and nanoclay particles exhibit improved barrier properties in comparison to the pure bioplastics. After their useful life, these can be composted and returned back to the soil.

Academic and industrial researchers showed a great interest in montmorillonite (MMT, $MMT\text{-}Na^+$) and organic-modified MMT application in the packaging area due to high surface area with a quite large aspect ratio (50–1000) and fine compatibility with many organic thermoplastics (Golebiewski et al., 2008; Farhoodi, 2016). Many studies have shown that application of montmorillonite clay as an additive offers resistance to gases and water vapor permeability (Lotti et al., 2008; Koh et al., 2008; Adame and Beall, 2009). Furthermore, the mechanical strength of biopolymers

(Ghanbarzadeh et al., 2015; Gutierrez et al., 2017), the glass transition temperature (Petersson and Oksman et al., 2006), and thermal degradation temperatures (Cyras et al., 2008) may be improved by the addition of clays.

11.2.3.1.4 Fibers as Additives

Fiber reinforcement has been widely utilized in developing polymeric composites thanks to their specific properties, such as high strength, stiffness, and low weight. Typical example of natural polymer composites is starch reinforced by cellulose fiber. The reinforced composites may be described as being renewable, biodegradable, abundantly available, and inexpensive (Jiang et al., 2020). Natural fibers in the composition of composites in relation to inorganic additives offer a number of advantages, such as a large variety of fillers available worldwide, they are cheap, do not consume much energy, have low density, high sound attenuation, high specific strength and modulus, and they are easy to process and have a fairly reactive surface (Dufresne et al., 2000). The starch films reinforced by 7% cellulose nanocrystals could achieve the strength and stiffness quite close to that of polyolefins (Panaitescu et al., 2015).

11.2.3.1.5 Graphene as Additives

For the development of sensors for intelligent packaging, graphene has found its application as a detector of food spoilage (e.g., detecting the presence of a gas), thus informing the consumer about food quality. It is reported that nanocomposites made from graphene nanoplates (GNPs) can be an excellent option for food-packaging applications by ameliorating heat resistance and barrier properties (Ramanathan et al., 2008).

11.2.3.1.6 Reactive Nanofillers for Biosensors

Biopolymer containing reactive nanofillers can act as a sensor. Biosensors or bionanocomposites are characterized by the ability to respond to changes in the environment, such as oxygen concentration, humidity, temperature, as well as the decomposition and contamination of packaged food and food products (Bouwmeester et al., 2009). All these parameters are especially important to monitor during their storage to ensure fresh and healthy food (Othman, 2014).

Nanofillers with oxygen scavenger properties: The oxygen scavenger properties are exhibited by some of the nanofillers. Monitoring of the oxygen concentration in packaged foods is of particular importance because its high concentrations can lead to oxidation and rapid spoilage of food. For this reason, there is a need to incorporate oxygen scavenger nanofillers into food packaging in order to monitor the oxygen concentration, maintain it at the permitted concentration and thereby to contribute to easier monitoring of product quality (Azeredo et al., 2009). For example, direct oxidation of fruits and vegetables can result in browning while oxidation of fats results in rancid odors, flavors, and stale of food that contains the fat. The change in color and appearance of the food product makes the product less attractive, but at the same time, it may indicate that oxidation of pigments in the food has occurred. Packaging materials with oxygen scavenger properties can be used to prevent oxidation and keep food fresh. One such material has been developed by Xiao-e et al. (2004). They used TiO_2 as nanofiller, which exhibits oxygen scavenger properties on irradiation with ultraviolet (UV) light. They observed that nanocrystalline TiO_2 as a photocatalyst when deposited on glass or acetate film has the ability to create a deoxygenated closed environment and this nanocomposite is effective in removing ethylene vapor and thus delaying fruit ripening.

Another TiO_2-based material, more precisely TiO_2-coated polypropylene films were developed by Maneerat and Hayata (2006) and were used to remove ethylene vapor from packaged horticultural products.

Nanofillers with antimicrobial properties:Tee et al. (2017) have studied physical, chemical, and barrier properties of flax seed mucilage-based films for its potential for coating or packaging of bioproducts. They prepared the film samples through casting by the addition of glycerol up to 5 wt.%. They discovered possible interactions between the mucilage and alcohol by Fourier transform infrared (FTIR) spectra because of the appearance of the - OH stretching vibration at 2883.1 cm^{-1}. With increasing alcohol concentration, the water activity decreased, but the moisture content and water solubility increased. Their analysis additionally reveals that regarding barrier properties, the water vapor permeability (WVP) and oxygen permeability (OP) notably increased with increasing alcohol content. The obtained results of high WVP and OP values suggests that prepared films can be potentially extended for coating or packaging fresh products.

Toxicity of nanoparticles in packaging:The small particle size of the nanomaterials is actually a major fact that raises concerns about their application in food packaging. Thus, the point is that nanoparticles can be potentially dangerous to human health because they can have different physical and chemical properties compared with their chemical counterparts from the macroscale (Huang et al., 2015), giving them the potential for causing health problems. In addition, their transformation and bioavailability to the environment can also vary considerably because of their physicochemical properties as well as their large surface area (Bratovcic, 2019b). As a result, nano-based packaging materials are not evaluated as new chemicals, but their permitted concentrations that may come into contact with food through migration have not been defined either (Busolo or Lagaron, 2012). Thus, at this time, there are no international legal regulations or standards that address the above-mentioned issues related to the application of nanomaterials.

11.2.4 POLYSACCHARIDE-BASED COMPOSITES

11.2.4.1 STARCH

Among natural polymers, native starch is taken under consideration as one among the foremost promising materials owing to its attractive combination of availability, price, performance, good film-forming ability, and a good barrier to lipids and oxygen. However, it has poor water resistance. Starch-based edible films may be used to package candy and bakery products (Bravin et al., 2006). Applications of starch-based materials in food packaging include films for food wrapping and containers, such as bowls, plates, cups, and egg trays. As a packaging material, however, starch-based systems are very brittle, and therefore to enhance their mechanical responses, they are typically plasticized, blended with other materials, chemically modified (Adeyeye et al., 2019).

The research carried out by Kristo and Biliaderis (2007) and Arora and Padua (2010) has shown that starch crystalline particles with thickness of 6–8 nm improve the tensile strength and modulus of pullulan films, with decreased stretching property. Arora et al. (2016) have observed increased antimicrobial activity of metals adsorbed or incorporated on polysaccharides due to enhanced surface area.

11.2.4.2 CELLULOSE

Cellulose is the most abundant natural polymer on earth consisting of β-1,4-linked glucose monomer units. The glucose units in cellulose are joined together via β-1,4 glycosidic linkages, the glucose units are joined by strong hydrogen bonds. Cellulose is a very cheap natural material due to its wide presence in all plants. However, it is not easy to use cellulose in packaging due to its hydrophilicity, poor solubility, and highly crystalline structure. Cellulose-based polymers are used in the packaging of bread, fruits, meat, dried products, and so forth.

11.2.4.2.1 Nanocelluloses

Nanocelluloses belong to renewable biomaterials, which contain lignocellulose fibers that have characteristic strengthening properties and therefore they are used as reinforcement phases for the production of nanocomposites (Azeredo et al., 2018). Pure nanocellulose crystalline structure are usually 3–10 nanometers wide, and aspect ratio higher than 5 and frequently lower than 50% cellulose nanocrystals (CNCs), while cellulose nanofibrils (CNFs) contain both crystalline and amorphous regions with 5–30 nm in width and aspect ratio is usually higher than 50. In aqueous suspension, CNCs has the following properties, such as high tensile strength (approximately 7.5 GPa), high axial stiffness (approximately 100 and 50 GPa), thermal stability up to 300°C, low thermal expansion coefficient (approximately 1 ppm/K), low density (approximately 1.6 g/cm^3), shear-thinning rheology, and lyotropic liquid crystalline behavior. The tendency for gel formation at low consistency, high strength and stiffness, good elastic properties, and surface area between 100 and 200 g/m^2 are properties of CNFs (Moon et al., 2013).

Khan et al. (2012) observed that chitosan films containing 1 wt.% CNC increased tensile modulus (43%), and it was ascribed to the formation of a percolated network of nanocellulose. Azeredo et al. (2010) have found that for the preparation of chitosan edible films, the optimal concentration of glycerol and nanocellulose was 18 wt.% and 15 wt.%, respectively for mechanical, thermal, and barrier properties.

11.2.4.2.2 Hemicelluloses

Hemicelluloses are also polysaccharides available in plant cell walls. However, hemicellulose contains diverse sugars, while cellulose contains only glucose. In hemicelluloses, in addition to glucose, other sugars, such as xylose and arabinose, mannose and galactose, and rhamnose can also are found. The hemicellulose contains mostly D-pentose sugars with small amounts of L-sugars (Ebringerová et al., 2005). Shorter chains of 500–3000 sugar units are exhibited by hemicelluloses, whereas 7–15,000 glucose molecules comprise each polymer of cellulose (Gibson, 2013). The most common polysaccharides in the plant kingdom are xylans.They are characterized by resistance to degradation in the stomach and small intestine as well as being degraded in the colon by anaerobic microflora, and these properties make them suitable for the development of hydrogels (Oliveira et al., 2010; Kumar and Negi, 2012). Xylans are the second most abundant polysaccharides in the plant kingdom and the most common hemicelluloses. Xylans are resistant to degradation at the stomach and small intestine, being degraded in colon by anaerobic microflora (Oliveira et al., 2010; Kumar and Negi, 2012). This property has motivated studies on hydrogels (Sun et al., 2013) and xylan nanoparticles for controlled release of drugs to the colon.

11.2.4.2.3 Lignin

After cellulose, lignin is the second most available biopolymer (Duval and Lawoko, 2014). It is an irregularly highly branched polyphenolic polyether, where the main monomers (monolignols) being 4-hydroxyphenyl, guaiacyl, and syringyl derivatives are connected by aromatic and aliphatic ether bonds (Calvo- Calvo-Flores and Dobado, 2010). The cell wall is impermeable to water due to its hydrophobic nature. Depending on the plant source, its structure is considerably variable, because of different proportions of the monolignols, which determine the type of interunit bonds, the degree of branching, and thus the reactivity of the lignin.

11.2.4.2.4 Chitosan

Chitosan is linear polysaccharide composed of β-1,4-linked glucosamine (2-amino-2-deoxy-β-D-glucose) units. Chitosan biopolymer has been used most commonly in food industries due to its exceptional properties, such as effective antimicrobial activity, high biodegradability, and low toxicity. Lately, new chitosan-based materials were produced by introducing nanoparticles, organic substances, such as aldehyde (cinnamaldehyde), acids (acetic, lauric and propionic acids), nanoparticles, and metal oxides into chitosan matrix with extra ordinary applications in food packaging industries (Hooda et al., 2018). Antimicrobial properties of chitosan are mainly due to its polycationic nature (Arora et al., 2016).

Chitosan-based antimicrobial film developed by Tripathi et al. (2009) [112] has been efficiently used against *E. coli*, *S. aureus*, and *B. subtilis* and exhibits the extended shelf life of tomato. Chitosan-based edible films and coatings were prepared and applied to fresh-cut vegetables and fruits (Ce et al., 2012; Ramos et al., 2012). Furthermore, chitosan coatings may improve both the gloss and oxygen barrier properties of paper. By adding chitosan, the gloss value was increased. Nanoclays incorporated into chitosan may also improve the barrier properties. Barrier properties against oxygen, water vapor, grease, and UV-light transmission have been enhanced by biohybrid nanocomposite films and multilayer coatings.

11.2.5 PROTEIN-BASED COMPOSITES

In recent years, biodegradable polymers and bio-based nanocomposite formulations have received great attention for its production and utilization for food packaging. The source of proteins used were taken from different plants, such as corn-zein, sunflower, wheat gluten, and soy protein. Especially, small portions of beans, nuts, and cashew nuts were packed by using protein-based edible films (Wittaya, 2012).

Natural protein-based polymers have shown some drawbacks, such as poor barrier properties (high sensitivity to humidity) and weak mechanical properties, which limit their final practical application. In order to eliminate the disadvantages of the properties of natural protein-based polymers and the possibility of expanding their application in more complex conditions, they are blanded with other synthetic polymers, but they are rarely chemically modified. Greater use of edible and biodegradable films based on nanocomposites could certainly lead to a reduction in packaging waste.

Safety control and migration possibilities of the nanopackaged foods are also crucial. Today, unfortunately, the current legal acts do not distinguish materials produced using nanotechnologies from those produced by conventional production processes.

11.2.5.1 GELATIN

Among hydrocolloids, gelatin is unique in the creation of a thermoreversible substance with a melting point close to body temperature, which is distinctly important in pharmaceutical and edible applications. A unique sequence of amino acids with the high content of glycine, proline, and hydroxyproline makes up its content.

11.2.5.2 COLLAGEN

Collagens contain a significant amount of triple helical structure. Collagen is used for the production of the most commercially successful edible protein films. They are biocompatible and nontoxic to most tissues.

11.2.6 ALIPHATIC POLYESTERS

11.2.6.1 POLYLACTIC ACID

Due to excellent mechanical properties, transparency, and commercial availability, PLA has become a very promising material (Jonoobi et al., 2010; Jamshidian et al., 2012). However, its poor gas barrier properties, poor thermal stability, and high cost need to be solved in order to achieve its fabrication for wider industrial applications.

11.2.6.2 POLYHYDROXYBUTYRATE

Polyhydroxy butyrate (PHB) is a PHA, an ester that can be found in nature (Erich et al., 2018). Its properties are similar to the properties of PP. In fact, because of its particularly similar properties, PHB can be used as a substitute for PP, which is otherwise a polymer of fossil origin. Meat salad can be packed by using PHB films (Levkane et al., 2008), as well as fatty products (mayonnaise, cream cheese, and margarine) (Bucci et al., 2007). Moreover, there is no difference in stability when the orange juice was packed in PHB and in HDPE indicating that the material is convenient for liquid acidic and fatty foodstuff (Haugaard et al., 2003).

11.3 SYNTHETIC BIODEGRADABLE POLYMER-BASED NANOCOMPOSITES

Adding MMT, silver, and ZnO nanofillers to the biopolymers, such as starch and PLA, opens up potential to create a new and innovative bionanocomposite materials with improved properties and performance.

11.3.1 PLA COMPOSITES

Added nanofillers may contribute to enhance the barrier to gases due to the synergistic tortuosity, crystal nucleation, and chain immobilization effects (Fortunati et al., 2014b). Recently, Arrieta et al. (2017) have summarized the advantages of reinforcing PLA matrix with CNCs for the development of bionanocomposites for biodegradable food packaging applications. The addition of low amounts of CNC, from 1 to 3 wt.%, does not affect the wettability properties of PLA (Fortunati et al., 2014a). However, the addition of 5 wt.% of CNC causes an increase in the PLA surface wettability due to the high amount of hydroxyl groups of CNC (Arrieta et al., 2014). Sanchez-Garcia et al. (2008) reported that incorporated alpha purified cellulose microfibers may improve barrier properties of PLA biocomposites. Sanchez-Garcia and Lagaron (2010) found that the the water permeability may be reduced by up to 82% and the oxygen permeability by up to 90% with only 3 wt.% of nanocellulose content.

11.3.2 PLA/LIGNOCELLULOSIC NANOFILLERS

Nanocellulose and nanolignin are the most abundant biodegradable natural bio-based nano reinforcements, which can be easily extracted from various biomasses.

Water vapor and oxygen barrier effects as well as antimicrobial properties were improved by PLA-nanocellulose composites in combination with silver nanoparticles (Fortunati et al., 2012a) or essential oils (Salmieri et al., 2014). PLA/cellulose films act as a promising bioactive packaging to preserve fresh food products against foodborne pathogens. PLA–lignin renewable and biodegradable composites have also been prepared by blending lignin-g-rubber-g-poly(D-lactide) copolymer particles and commercial poly(L-lactide) (PLLA) in chloroform (Sun et al., 2015). The rubbery layer made of poly(ε-caprolactone-co-lactide) (PCLLA) was formed via lignin-initiated ring opening copolymerization of ε-caprolactone/L-lactide mixture, followed by the formation of poly(D-lactide) (PDLA) outer segments via polymerization of D-lactide. The prepared composites exhibited sixfold increase of elongation at break and a simultaneous improvement in their tensile strength and Young's modulus, though to a lesser extent.

11.3.3 PLA/NANOCLAY

The compatibility of clay nanoparticles (layered double hydroxides and layered silicates) and its homogeneous distribution into polymer matrix are two critical points, which should be satisfied for the successful formation of its nanocomposites. However generally, natural layered silicates are hydrophilic due to maintaining Na^+ or K^+ on their surfaces, and as a result, they are not much miscible with hydrophobic polymers.. In order to achieve better miscibility of silicates with hydrophobic polymers, the surface of the silicates can be modified with various organic cations such as PLA. In this way, the interaction between the functional groups derived from the organic cation on the modified silicate and the polymer matrix is improved. The surface energy of the silicate surface is reduced in the presence of organic cations, which results in a larger interlayer spacing, enabling better-intercalated nanocomposite formation.

11.3.4 PLA/METAL OR METAL OXIDE NANOPARTICLES

Among all metals, such as Ag, Cu, Zn, Ti, or its metal oxide nanoparticles, silver nanoparticles have been one of the most commonly used for the development of innovative packaging materials. The unique electrical, optical, catalytic, thermal stability, and particularly antimicrobial activity against a broad spectrum of bacteria, viruses, and fungi have been demonstrated by silver nanoparticles (Russell and Hugo, 1994). The antimicrobial mechanism of silver nanoparticles works on the principle that positively charged silver ions are released from the surface of the particles and react with negatively charged biomacromolecular components, which leads to membrane damage. Silver nanoparticles may aggregate in the bacterial cytoplasmic membranes, leading to membrane disruption and a major increase of permeability, resulting in cell death (Feng et al., 2000; Sondi and Salopek-Sondi, 2004; Kim et al., 2007).

Nanocomposites prepared from PLA and silver nanoparticles have shown a strong antibacterial activity against both Gram-negative and Gram-positive bacteria (Shameli et al., 2010). High potential application in PLA-based films for food packaging exhibit even more complexed nanocomposites prepared from silver nanoparticles in PLA matrix mixed with cellulose (Fortunati et al., 2012b), chitosan (Yu et al., 2007), nanoclay (Busolo and Lagaron, 2013), and zeolite (Fernández et al., 2010) showing excellent antimicrobial activity.

Nanocomposites made from PLA-modified TiO_2 have shown an excellent mechanical and barrier properties and improved anti-UV performance (Zhu et al., 2011; Man et al., 2012). TiO_2 used as a packaging material can protect the food against against radiation of relatively low wavelength near the UV region, the foodborne microorganism, odor, staining, and allergens.

11.3.5 BARRIER PROPERTIES OF SYNTHETIC BIOPOLYMERS

Packaging must satisfy low permeability to gases, such as oxygen and water vapor in order to increase the shelf life of food. However, bio-based food packaging materials due to their hydrophilicity do not have good barrier properties for water vapor and therefore special attention should be paid, when using them for packaging (Vartiainen et al., 2014). Moisture resistance with hydrophilic polymers is inadequate in many cases; thus, excessive water vapor transmission through packaging dilutes the quality of foods resulting in shorter shelf-lives, increased costs and eventually more waste. One of the most promising new commercially available bioplastic is polyglycolic acid (PGA) with excellent barrier properties. Härkki et al. (2011) mentioned that bioplastics include PLA, polyhydroxyalkanoates, polyhydroxybutyrates, and cellophane which may enhance barrier properties by adding fillers such as nanoclays.

Pectin is a structural heteropolysaccharide found in sugar beet. Since pectin is an anionic polyelectrolyte, which means that ionic interactions will be of great importance when creating a film and behavior in dispersions with nanoclays. The biohybrid nanocomposite films were grease proof and also showed improved barrier properties against oxygen and water vapor.

Glycerol, sorbitol, or xylitol are known as plasticizers and these are typically used for the reduction of brittleness. It has been shown that films plasticized with xylitol–sorbitol mixture have lower water sorption and water vapor permeability compared with the films plasticized with glycerol–xylitol and glycerol–sorbitol mixtures at constant plasticizer contents (Talja et al., 2007). The water content in the films significantly affects the oxygen permeability. In this context, at low humidity conditions, the films are excellent barriers to oxygen transfer, however, if the water content is higher than 20%, the barrier is mostly lost. Increased polymer chain mobility can increase permeability as a result of an easier transport process (Forssell et al., 2002). Using sorbitol below 21% can reduce the permeability of starch films to oxygen.

11.4 CONCLUSION

Recently, bio-based and biodegradable polymers, which may replace fossil-fuel-based polymers have become an important class of materials for food-packaging applications. Since they are natural polymers, there is no concern about the materials directly getting in contact with food because they do not pose any threat to health. The increasing demand for eco-friendly packaging made from bio-based polymers is driving the demand for products, such as PLA, PHA, and starch- and chitosan-based polymers in different applications. Biopolymers have been considered as an alternate for plastic material prepared from petroleum due to their advance properties like renewability and biodegradability. PLA is a hydrophobic material and it is popular owing to its high mechanical strength and high recyclability. Starch- and chitosan-based materials are characterized with low cost and biodegradability, but they posses strong hydrophilic behavior leading to poor moisture barrier and poor mechanical properties. Chitosan-based materials have shown good antibacterial properties. The insertion of nanomaterials into polymer matrix may be used as a new strategy to enhance the physical properties of polymers, together with mechanical strength, thermal stability, barrier properties to gases and biodegradation. CNTs as nanofillers may improve tensile strength, stiffness, toughness, thermal stability, electrical conductivity, while graphene may detect food spoilage and consequently be used for intelligent packaging. TiO_2 as nanofiller may be used as a packaging material against foodborne microorganism odor, staining, as well as an oxygen scavenger. In terms of the safety of using nanofillers to prepare nanocomposites for food packaging, despite the many benefits of bionanocomposite materials, there are still not enough studies on the toxicological effects and migration of nanofillers into food from the packaging films. To conclude, packaging must satisfy low permeability to gases, such as oxygen and water vapor in order to increase the shelf life of food.

KEYWORDS

- **biopolymers**
- **edible polymers**
- **synthetic bio-based polymers nanocomposites**
- **food packaging**
- **biodegradable**

REFERENCES

Adame, D.; Beall, G. W. Direct measurement of the Constrained Polymer Region in Polyamide/Clay Nanocomposites and the Implications for Gas Diffusion. *Appl. Clay Sci.* **2009,** *42*, 545–552.

Adeyeye, O. A.; Sadiku, E. R.; Reddy, A. B.; Ndamse, A. S.; Makgatho, G.; Sellamuthu, P. S. The Use of Biopolymers in Food Packaging. In *Green Biopolymers and their Nanocomposites. Materials Horizons: From Nature to Nanomaterials*; Gnanasekaran, D., Ed.; Springer: Singapore, 2019.

Arora, A.; Padua, G. W. Review: Nanocomposites in Food Packaging. *J. Food Sci.* **2010,** *75*, *1*, R43–R49.

Arora, D.; Sharma, N.; Sharma, V.; Abrol, V.; Shankar, R.; Jaglan, S. An Update on Polysaccharide-Based Nanomaterials for Antimicrobial Applications. *Appl. Microbiol. Biotechnol.* **2016,** *100*, 2603–2615.

Arrieta, M. P.; Fortunati, E.; Dominici, F.; Rayón, E.; López, J.; Kenny, J. M. PLA-PHB/Cellulose Based Films: Mechanical, Barrier and Disintegration Properties. *Polym. Degrad. Stab.* **2014,** *107*, 139–149.

Arrieta, M. P.; Peltze,r M. A.; López, J.; Peponi, L. PLA-Based Nanocomposites Reinforced with CNC for Food Packaging Applications: From Synthesis to Biodegradation. In *Industrial Applications of Renewable Biomass Products*; Goyanes, S., D'Accorso, N., Eds.; Springer: Berlin, 2017.

Auras, R.; Harte, B.; Selke, S. An Overview of Polylactides as Packaging Materials. *Macromol. Biosci.* **2004,** *4* (9), 835–864.

Avella, M.; Vlieger, J.; Errico, E. M.; Fischer, S.; Vacca, P.; Volpe, G. M. Biodegradable Starch/Clay Nanocomposite Films for Food Packaging Applications. *Food Chem.* **2005,** *93* (3), 467–474.

Azeredo, H. M. C.; De Fritas Rosa, M.; de Figueirêdo, M. C. B. Lignocellulosic-Based Nanostructures and Their Use in Food Packaging. In *Nanomaterials for Food Packaging Materials***.** Elsevier: Amsterdam, 2018; pp 47–69.

Azeredo, H. M. C. D. Nanocomposites for Food Packaging Applications. *Food Res. Int.* **2009,** *42*, 1240–1253.

Azeredo, H. M. C.; Mattoso, L. H. C.; Wood, D.; Williams, T. G.; Avena-Bustillos, R. J.; McHugh, T. H. Nanocomposite Edible Films from Mango Puree Reinforced with Cellulose Nanofibers. *J. Food Sci.* 2009, *74* (5), N31–N35.

Azeredo, H.; Mattoso, L. H. C.; Avena-Bustillos, R. J.; Munford, M. L.; Wood, D.; Mchugh, T. H. Nanocellulose Reinforced Chitosan Composite Films as Affected by Nanofiller Loading and Plasticizer Content. *J. Food Sci.* **2010,** *75*, N1–N7.

Balart, J. F.; Fombuena, V.; Fenollar, O.; Boronat, T.; Sánchez-Nacher, L. Processing and Characterization of High Environmental Efficiency Composites Based on PLA and Hazelnut Shell Flour (HSF) with Biobased Plasticizers Derived from Epoxidized Linseed Oil (ELO). *Compos. Part B* **2016,** *86*, 168–177.

Bari, S. S.; Chatterjee, A.; Mishra, S. Biodegradable Polymer Nanocomposites: An Overview. *Polym. Rev.* **2016,** *56*, 287–328.

Baughman, R. H.; Zakhidov, A. A.; de Heer, W. A. Carbon Nanotubes–the Route toward Applications. *Science* **2002,** *297* (5582), 787–792.

Bayarri, M.; Oulahal, N.; Degraeve, P.; Gharsallaoui, A. Properties of Lysozyme/Low Methoxyl (LM) Pectin Complexes for Antimicrobial Edible Food Packaging. *J. Food Eng.* **2014,** *131*, 18–25.

Bhat, R.; Abdullah, N.; Din, R. H.; Tay, G-S. Producing Novel Sago Starch Based Food Packaging Films by Incorporating Lignin Isolated from Oil Palm Black Liquor Waste. *J. Food Eng.* **2013,** *119*, 707–713.

Bioplastics & Biopolymers Market, Bioplastics & Biopolymers Market by Type (Non-biodegradable, Biodegradable), End-use Industry (Packaging, Consumer Goods, Automotive, Textiles, Agriculture) and Region (APAC, North America, Europe, and RoW)—Trends & Forecast to 2023. https://www.marketsandmarkets.com/Market-Reports/biopolymers-bioplastics-market 88795240.html?gclid=Cj0KCQjw9ZzzBRCKARIsANwXaeKABH7G4izM9F115y4hEsbVH92cJbd67YrCXkuflWndvTIvrO66oVwaApXvEALw_wcB (accessed Mar 10, 2020).

Bioplastics & Biopolymers Market, Bioplastics & Biopolymers Market by Type (Non-Biodegradable/Bio-Based, Biodegradable), End-Use Industry (Packaging, Consumer Goods, Automotive & Transportation, Textiles, Agriculture & Horticulture), Region - Global Forecast to 2025. https://www.marketsandmarkets.com/Market-Reports/biopolymers-bioplastics-market-88795240.html?gclid=Cj0KCQjw9ZzzBRCKARIsANwXaeKABH7G4izM9F115y4hEsbVH92cJbd67YrCXkuflWndvTIvrO66oVwaApXvEALw_wcB (accessed Sep 8, 2020)

Bouwmeester, H.; Dekkers, S.; Noordam, M. Y.; Hagens, W. I.; Bulder, A. S.; De Heer, C.; Review of Health Safety Aspects of Nanotechnologies in Food Production. *Regul. Toxicol. Pharmacol.* **2009,** *53*, 52–62.

Bratovcic, A.; Suljagic, J. Micro- and Nano-Encapsulation in Food Industry. *Croat. J. Food Sci. Technol.* **2019a,** *11*, 113–121.

Bratovcic, A. Different Applications of Nanomaterials and Their Impact on the Environment. *Int. J. Mater. Sci. Eng. (IJMSE)*, **2019b,** 5, 1, 1–7.

Bratovcic, A. Degradation of Micro- and Nano-Plastics by Photocatalytic Methods. *J. Nanosci. Nanotechnol. Appl.* **2019c,** *3*. DOI: 10.18875/2577-7920.3.304.

Bratovcic, A. Nanomaterials in Food Processing and Packaging, Its Toxicity and Food Labeling. *Acta Sci. Nutr. Health.* **2020a,** *4* (9), 07–13.

Bratovcic, A. Biosynthesis of Green Silver Nanoparticles and Its UV-Vis Characterization. *IJISET—Int. J. Innov. Sci. Eng. Technol.* **2020b,** *7* (7), 170–176.

Bratovcic, A. Nanocomposite Hydrogels Reinforced by Carbon Nanotubes. *IJERA—Int. J. Eng. Res. Appl.* **2020c,** *10* (5), (Series-IV), 30–41.

Bratovcic, A. Synthesis, *Characterization, Applications, and Toxicity of Lead Oxide Nanoparticles*; IntechOpen: London, 2020d.

Bratovcic, A.; Odobasic, A.; Catic, S.; Sestan, I. Application of Polymer Nanocomposite Materials in Food Packaging. *Croat. J. Food Sci. Technol.* **2015,** *7*, 86–94.

Bravin, B.; Peressini, D.; Sensidoni, A. Development and Application of Polysaccharide—Lipid Edible Coating to Extend Shelf-Life of Dry Bakery Products. *J. Food Eng.* **2006,** *76*, 280–290.

Bucci, D. Z., Tavares, L. B. B.; Sell, I. Biodegradation and Physical Evaluation of PHB Packaging. *Polym. Test.* **2007,** *26* (7), 908–915.

Busolo, M. A.; Lagaron, J. M. Oxygen Scavenging Polyolefin Nanocomposite Films Containing an Iron Modified Kaolinite of Interest in Active Food Packaging Applications. *Innov. Food Sci. Emerg. Technol.* **2012,** *16*, 211–217.

Busolo, M. A.; Lagaron, J. M. Antimicrobial Biocomposites of Melt-Compounded Polylactide Films Containing Silver-Based Engineered Clays. *J. Plast. Film Sheet.* **2013,** *29*, 290–305.
Calvo-Flores, F. G.; Dobado, J. A. Lignin as Renewable Raw Material. *Chem. Sus. Chem.* **2010,** *3* (11), 1227–1235.
Ce, N.; Norena, C. P. Z.; Brandelli, A. Antimicrobial Activity of Chitosan Films Containing Nisin, Peptide P34, and Natamycin. *CyTA J. Food* **2012,** *10*, 21–26.
Chen, H. Functional Properties and Application of Edible Films Made of Milk Protein. *J. Dairy Sci.* **1995,** *78*, 2563–2583.
Cheng, Q.; Li, C.; Pavlinek, V.; Saha, P.; Wang, H. Surface-Modified Antibacterial TiO_2/Ag^+ Nanoparticles: Preparation and Properties. *Appl. Surf. Sci.* **2006,** *252*, 4154–4160.
Cyras, V. P.; Manfredi, L. B.; Ton-That, M. T.; Vazquez, A. Physical and Mechanical Properties of Thermoplastic Starch/Montmorillonite Nanocomposite Films. *Carbohydr. Polym.* **2008,** *73*, 55–63.
Moura, M. R. D.; Mattoso, L. H. C.; Zucolotto, V. Development of Cellulose-Based Bactericidal Nanocomposites Containing Silver Nanoparticles and Their Use as Active Food Packaging. *J. Food Eng.* **2012,** *109*, 520–524.
Demitri, C.; De Benedictis, V. M.; Madaghiele, M.; Esposito Corcione, C.; Maffezzoli, A. Nanostructured Active Chitosan-Based Films for Food Packaging Applications: Effect of Graphene Stacks on Mechanical Properties. *Measurement* **2016,** *90*, 418–423.
Dufresne, A.; Dupeyre, D.; Vignon, M. R. Cellulose Microfibrils from Potato Tuber Cells: Processing and Characterization of Starch—Cellulose Microfibril Composites. *J. Appl. Polym. Sci.* **2000,** *76* (14), 2080–2092.
Duncan, TV. Applications of Nanotechnology in Food Packaging and Food Safety: Barrier Materials, Antimicrobials and Sensors. *J. Colloid Interf. Sci.* **2011,** *363* (1), 1–24.
Duval, A.; Lawoko, M. A Review on Lignin-Based Polymeric, Micro- and Nanostructured Materials. *React. Funct. Polym.* **2014,** *85*, 78–96.
Ebringerová, A.; Hromádková, Z.; Heinze, T., Heinze, T., Eds. *Hemicellulose, Polysaccharides I: Structure, Characterization and Use, Advances in Polymer Science;* Springer: Berlin Heidelberg, **2005;** pp 1–67.
Emamifar, A.; Kadivar, M.; Shahedi, M.; Soleimanian-Zad, S. Effect of Nanocomposite Packaging Containing Ag and ZnO on Inactivation of *Lactobacillus plantarum* in Orange Juice. *Food Control* **2011,** *22* (3–4), 408–413.
Erich, M.; Hannes, G.; Maximilian, L. PHB—Bio Based and Biodegradable Replacement for PP: A Review. *Nov. Tech. Nutri. Food Sci.* **2018,** *2* (4), 206–209.
European Bioplastics, Nova-Institute, 2019. https://www.european-bioplastics.org/market/ (accessed Sep 17, 2020).
Farhoodi, M. Nanocomposite Materials for Food Packaging Applications: Characterization and Safety Evaluation. *Food Eng. Rev.* **2016,** *8* (1), 35–51.
Farris, S.; Introzzi, L.; Fuentes-Alventosa, J. M.; Santo, N.; Rocca, R.; Piergiovanni, L. Self-Assembled Pullulan–Silica Oxygen Barrier Hybrid Coatings for Food Packaging Applications. *J. Agric. Food Chem.* **2012,** *60*, 782–790.
Feng, Q.; Wu, J.; Chen, G.; Cui, F.; Kim, T.; Kim, J. A mechanistic Study of the Antibacterial Effect of Silver Ions on *Escherichia coli* and *Staphylococcus aureus*. *J. Biomed. Mater. Res.* **2000,** *52* (4), 662–668.
Fernández, A.; Soriano, E.; Hernández-Muñoz, P.; Gavara, R. Migration of Antimicrobial Silver from Composites of Polylactide with Silver Zeolites. *J. Food Sci.* **2010,** *75* (3), E186 E193.

Fombuena, V.; Balart, J.; Boronat, T.; Sánchez-Nácher, L.; Garcia-Sanoguera, D. Improving Mechanical Performance of Thermoplastic Adhesion Joints by Atmospheric Plasma. *Mater. Des.* **2013,** *47*, 49–56.

Forssell, P.; Lahtinen, R.; Lahelin, M.; Myllärinen, P. Oxygen Permeability of Amylose and Amylopectin Films. *Carbohydr. Polym.* **2002,** *47*, 125–129.

Fortunati, E.; Armentano, I.; Iannoni, A.; Barbale, M.; Zaccheo, S.; Scavone, M.; Visai, L.; Kenny, J. M. New Multifunctional Poly(Lactide Acid) Composites: Mechanical, Antibacterial, and Degradation Properties. *J. Appl. Polym. Sci.* **2012a,** *124* (1), 87–98.

Fortunati, E.; Peltzer, M.; Armentano, I.; Torre, L.; Jimenez, A.; Kenny, J. M. Effects of Modified Cellulose Nanocrystals on the Barrier and Migration Properties of PLA Nano-Biocomposites. *Carbohydr. Polym.* **2012b,** *90* (2), 948–956.

Fortunati, E.; Luzi, F.; Puglia, D.; Dominici, F.; Santulli, C.; Kenny, J. M.; Torre, L. Investigation of Thermo-Mechanical, Chemical and Degradative Properties of PLA-Limonene Films Reinforced with Cellulose Nanocrystals Extracted from *Phormium tenax* Leaves. *Eur. Polym. J.* **2014a,** *56* (1), 77–91.

Fortunati, E.; Rinaldi, S.; Peltzer, M.; Bloise, N.; Visai, L.; Armentano, I.; Jiménez, A.; Latterini, L.; Kenny, J. M. Nano-Biocomposite Films with Modified Cellulose Nanocrystals and Synthesized Silver Nanoparticles. *Carbohydr. Polym.* **2014b,** *101*, 1122–1133.

Gabor (Naiaretti), D.; Tita, O. Biopolymers Used in Food Packaging: A Review. *Acta Univ. Cibiniensis Ser. E: Food Technol.* **2012,** *XVI* (2), 1–19.

Ghanbarzadeh, B.; Oleyaei, S. A.; Almasi, H. Nano-Structured Materials Utilized in Biopolymer Based Plastics for Food Packaging Applications. *Crit. Rev. Food Sci. Nutr.* **2015,** *55*, 1699–1723.

Gibson, L. J. The Hierarchical Structure and Mechanics of Plant Materials. J. Royal Soc. Interf. ***2013,*** *9, 2749–2766.*

Golebiewski, J.; Rozanski, A.; Dzwonkowski, J.; Galeski, A. Low Density Polyethylene-Montmorillonite Nanocomposites for Film Blowing. *Eur. Polym. J.* **2008,** *44* (2), 270–286.

Gutierrez, T. J.; Ponce, A. G.; Alvarez, A. V. Nano-Clays from Natural and Modified Montmorillonite with and Without Added Blueberry Extract for Active and Intelligent Food Nanopackaging Materials. *Mater. Chem. Phys.* **2017,** *194*, 283–292.

Han, M.; Li, S. Study on the Medical Application of Silver Material. Micro- and Nano-technology, Materials, Processes, Packaging, and Systems IV. Melbourne, Australia, 09 December. Society of Photo-Optical Instrumentation Engineers, Washington, United States, 2008.

Härkki, O.; Karttunen, M.; Kortet, S.; Koponen, M; Harlin, A. Biaxially Oriented PLA-Montmorillonite-Nanocomposite for Barrier Film Applications. Conference on *Nanocomposites* at 2011, Paris-Est, Creteil University.

Haugaard, K.; Danielsen, B.; Bertelsen, G. Impact of Polylactate and Poly(Hydroxybutyrate) on Food Quality. *Eur. Food Res. Technol.* **2003,** *216* (3), 233–240.

Heap, B. Preface. The Environment and Human Health. *Phil. Trans. Royal Soc. Biol. Sci. Plastics* **2009,** *364*, 1971–1971.

Honarvar, Z.; Hadian, Z.; Mashayekh, M. Nanocomposites in Food Packaging Applications and Their Risk Assessment for Health. *Electron. Phys.* **2016,** *8* (6), 2531–2538.

Hooda, R.; Batra, B.; Kalra, V.; Rana, J. S.; Sharma, M. *Chitosan-Based Nanocomposites in Food Packaging: Green and Sustainable Advanced Packaging Materials*. In *Bio-based Materials for Food Packaging*; Ahmed, S., Ed.; Springer Nature: Singapore, 2018.

Hopkins, E. J.; Chang, C.; Lam, R. S. H.; Nickerson, M. T. Effects of Flaxseed Oil Concentration on the Performance of a Soy Protein Isolate-Based Emulsion-Type Film. *Food Res. Int.* **2015,** *67*, 418–425.

Huang, J. Y.; Li, X.; Zhou, W. Safety Assessment of Nanocomposite for Food Packaging Application. *Trends Food Sci. Technol.* **2015,** *45* (2), 187–199.

Introzzi, L.; Blomfeldt, T. O. J.; Trabattoni, S.; Tavazzi, S.; Santo, N.; Schiraldi, A.; Piergiovanni, L.; Farris, S. Ultrasound-Assisted Pullulan/Montmorillonite Bionanocomposite Coating with High Oxygen Barrier Properties. *Langmuir* **2012,** *28*, 11206–11214.

Jamshidian, M.; Tehrany, E. A.; Imran, M.; Akhtar, M. J.; Cleymand, F.; Desobry, S. Structural, Mechanical and Barrier Properties of Active PLA-Antioxidant Films. *J. Food Eng.* **2012,** *110* (3), 380–389.

Jensen, A.; Lim, L-T.; Barbut, S., Marcone, M. Development and Characterization of Soy Protein Films Incorporated with Cellulose Fibers Using a Hot Surface Casting Technique. *LWT Food Sci. Technol.* **2015,** *60*, 162–170.

Jiang, T.; Duan, Q.; Zhu, J.; Liu, H.; Yu, L. Starch-Based Biodegradable Materials: Challenges and Opportunities. *Adv. Ind. Eng. Polym. Res.* **2020,** *3* (1), 8–18.

Jonoobi, M.; Harun, J.; Mathew, A. P.; Oksman, K. Mechanical Properties of Cellulose Nanofiber (CNF) Reinforced Polylactic Acid (PLA) Prepared by Twin Screw Extrusion. *Compos. Sci. Technol.* **2010,** *70* (12), 1742–1747.

Kadam, D. M.; Thunga, M.; Wang, S.; Kessler, M. R.; Grewell, D.; Lamsal, B.; Yu, C. Preparation and Characterization of Whey Protein Isolate Films Reinforced with Porous Silica Coated Titania Nanoparticles. *J. Food Eng.* **2013,** *117*, 133–140.

Kang, S.; Pinault, M.; Pfefferle, L. D.; Elimelech, M. Single-Walled Carbon Nanotubes Exhibit Strong Antimicrobial Activity. *Langmuir* **2007,** *23*, 8670–8673.

Khan, A.; Khan, R. A.; Salmieri, S.; Le Tien, C.; Riedl, B.; Bouchard, J.; Lacroix, M. Mechanical and Barrier Properties of Nanocrystalline Cellulose Reinforced Chitosan-Based Nanocomposite Films. *Carbohydr. Polym.* **2012,** *90*, 1601–1608.

Kim, J. S.; Kuk, E.; Yu, K. N.; Kim, J. -H.; Park, S. J.; Lee, H. J.; Kim, S. H.; Park, Y. K.; Park, Y. H.; Hwang, C.-Y. Antimicrobial Effects of Silver Nanoparticles. *Nanomedicine* **2007,** *3* (1), 95–101.

Koh, H. C.; Park, J. S.; Jeong, M. A.; Hwang, H. Y.; Hong, Y. T.; Ha, S. Y.; et al. Preparation and Gas Permeation Properties of Biodegradable Polymer/Layered Silicate Nanocomposite Membranes. *Desalination* **2008,** *233*, *1* (3), 201–209.

Kristo, E.; Biliaderis, C. G. Physical Properites of Starch Nanocrystal Reinforced Pullulan Films. *Carbohydr. Polym*. **2007,** *68* (1), 146–158.

Kumar, P.; Sandeep, K. P.; Alavi, S.; Truong, V. D.; Gorga, R. E. Preparation and Characterization of Bio-Nanocomposite Films Based on Soy Protein Isolate and Montmorillonite Using Melt Extrusion. *J. Food Eng.* **2010,** *100*, 480–489.

Kumar, S.; Negi, Y. S. Nanoparticles Synthesis from Corn Cob (Xylan) and Their Potential Application as Colon-Specific Drug Carrier. *Macromol. Symp.* **2012,** *320*, 75–80.

Kuorwel, K. K.; Cran, M. J.; Sonneveld, K.; Miltz, J.; Bigger, S. W. Migration of Antimicrobial Agents from Starch-Based Films Into a Food Simulant. *LWT Food Sci Technol* **2013,** *50* (2), 432–438.

Langmaier, F.; Mokrejs, P.; Kolomaznik, K.; Mladek, M. Biodegradable Packing Materials from Hydrolysates of Collagen Waste Proteins. *Waste Manage*. **2008,** *28*, 549–556.

Levkane, V.; Muizniece, B. S.; Dukalska, L. Pasteurization Effect to Quality of Salad with Meat and Mayonnaise. *Foodbalt.* **2008,** 69–73.

Lin, D.; Zhao, Y. Innovations in the Development and Application of Edible Coatings for Fresh and Minimally Processed Fruits and Vegetables. *Compr. Rev. Food Sci. Food Saf.* **2007,** *6*, 3, 60–75.

Lotti, C.; Isaac, C. S.; Branciforti, M. C.; Alves, R. M. V.; Liberman, S.; Bretas, R. E. S. Rheological, Mechanical and Transport Properties of Blown Films of High Density Polyethylene Nanocomposites. *Eur. Polym. J.* **2008,** *44*, 1346–1357.

Malathi, A. N.; Santhosh, K. S.; Udaykumar, N. Recent Trends of Biodegradable Polymer: Biodegradable Films for Food Packaging and Application of Nanotechnology in Biodegradable Food Packaging. *Curr. Trends Techno. Sci.* **2019,** *3* (2), 73–79.

Man, C.; Zhang, C.; Liu, Y.; Wang, W.; Ren, W.; Jiang, L.; Reisdorffer, F.; Nguyen, T. P.; Dan, Y. Poly(Lactic Acid)/Titanium Dioxide Composites: Preparation and Performance under Ultraviolet Irradiation. *Polym. Degrad. Stab.* **2012,** *97* (6), 856–862.

Maneerat, C.; Hayata, Y. Antifungal Activity of TiO_2 Photocatalysis against Penicillium Expansum in Vitro and in Fruit Tests. *Int. J. Food Microbiol.* **2006,** *107* (2), 99–103.

Martelli, M. R.; Barros, T. T.; Assis, O. B. G. Produção de filmes plásticos apartir de polpa de frutas sobremaduras. *Revi. Bras. Prod. Agroindustr.* **2015,** *17* (3), 301–308.

McHugh, T. H.; Krochta, J. M. Milk Protein Based Edible Films and Coating. *Food Technol.* **1994,** *1*, 97–103.

McHugh, T. H.; Huxsoll, C. C.; Krochta, J. M. Permeability Properties of Fruit Puree Edible Films. *J. Food Sci.* **1996,** *61* (1), 88–91.

Moon, R. J.; Beck, S.; Rudie, A. W. Cellulosic Nanocrystals—A Material with Unique Properties and Many Potential Applications. In *Production and Applications of Cellulose Nanomaterials*, Postek, M. T., et al., Eds.; TAPPI Press: Peachtree Corners, GA, 2013; pp 9–12.

Muppalla, S. R.; Kanatt, S. R.; Chawla, S. P.; Sharma, A. Carboxymethyl Cellulose–Polyvinyl Alcohol Films with Clove Oil for Active Packaging of Ground Chicken Meat. *Food Packag. Shelf Life* **2014,** *2* (2), 51–58.

Muratore, G.; Nobile, D.; Buonocore, G. G.; Lanza, C. M.; Asmundo, N. The Influence of Using Biodegradable Packaging Films on the Quality Decay Kinetic of Plum Tomato (PomodorinoDatterino®). *J. Food Eng.* **2005,** *67* (4), 393–399.

Niaounakis, M. Introduction to Biopolymers. In *Biopolymers: Reuse, Recycling, and Disposal*; Elsevier: Amsterdam, 2013; pp 1–75.

Nile, S.; Baskar, V.; Selvaraj, D.; Nile, A.; Xiao, J.; Kai, G. Nanotechnologies in Food Science: Applications, Recent Trends, and Future Perspectives. *Nano-Micro Letters* **2020,** *12*, 1–34.

Oliveira, E. E.; Silva, A. E.; Nagashima Jr., T.; Gomes, M. C. S.; Aguiar, L. M.; Marcelino, H. R.; Arau´jo, I. B. et al. Xylan from Corn Cobs, a Promising Polymer for Drug Delivery: Production and Characterization. *Bioresour. Technol.* **2010,** *101* (14), 5402–5406.

Othman, S. H. Bio-Nanocomposite Materials for Food Packaging Applications: Types of Biopolymer and Nano-Sized Filler. *Agric. Agric. Sci. Proc.* **2014,** *2*, 296–303.

Otoni, C. G.; Avena-Bustillos, R. J.; Azeredo, H. M. C.; Lorevice, M. V.; Moura, M. R.; Mattoso, L. H. C.; McHugh, T. H. Recent Advances on Edible Films Based on Fruits and Vegetables—A Review. *Compr. Rev. Food Sci. Food Saf.* **2017. DOI.**10.1111/1541-4337.12281.

Panaitescu, D. M.; Frone, A. N.; Ghiurea, M.; Chiulan, I. Influence of Storage Conditions on Starch/PVA Films Containing Cellulose Nanofibers. *Ind. Crop. Prod.* **2015,** *70*, 170–177.

Park, S-I.; Stan, S. D.; Daeschel, M. A.; Zhao, Y. Antifungal Coatings on Fresh Strawberries (*Fragaria × ananassa*) to Control Mold Growth during Cold Storage. *J. Food Sci.* **2005,** *70*, 1–6.

Pascall, M. A.; Lin, S. J. The Application of Edible Polymeric Films and Coatings in the Food Industry. *J. Food Process Technol.* **2013,** *4* (2), 1–2.

Petersson, L.; Oksman, K. Biopolymer Based Nanocomposites: Comparing Layered Silicates and Microcrystalline Cellulose as Nanoreinforcement. *Compos. Sci. Technol.* **2006,** *66*, 2187–2196.

Plastics—The Facts 2019 An Analysis of European Plastics Production, Demand and Waste Data. https://www.plasticseurope.org/application/files/9715/7129/9584/FINAL_web_version_Plastics_the_facts2019_14102019.pdf (accessed Aug 22, 2020).

Poças, M. F. F.; Delgado, T. F.; Oliveira, F. A. R. *Smart Packaging Technologies for Fruits and Vegetables in Smart Packaging Technologies for Fast Moving Consumer Goods*; Kerry, J., Butler, P., Eds.; John Wiley & Sons: New Jersey, 2008; pp 151–166.

Pranoto, Y.; Salokhe, V. M.; Rakshit, S. K. Physical and Antibacte Rial Properties of Alginate-Based Edible Film Incorporated with Garlic Oil. *Food Res. Int.* **2005,** *38*, 267–272.

Ramanathan, T.; Abdala, A.; Stankovich, S.; Dikin, D.; Herrera-Alons, M.; Piner, R. et al. Functionalized Graphene Sheets for Polymer Nanocomposites. *Nat. Nanotechnol.* **2008,** *3* (6), 327–331.

Ramos, O. L.; Fernandes, J. C.; Silva, S. I.; Pintado, M. E.; Malcata, F. X. Edible Films and Coatings from Whey Proteins: A Review on Formulation, and on Mechanical and Bioactive Properties. *Crit. Rev. Food Sci. Nutr.* **2012,** *52* (6), 533–552.

Rasal, R. M.; Janorkar, A. V.; Hirt, D. E. Poly (Lactic Acid) Modifications. *Prog. Polym. Sci.* **2010,** *35* (3), 338–356.

Rhim, J. W.; Park, H. M.; Ha, C. S. Bio-Nanocomposites for Food Packaging Applications. *Prog. Polym. Sci.* **2013,** *38* (10), 1629–1652.

Rouf, T. B.; Kokini, J. L. *Natural Biopolymer-Based Nanocomposite Films for Packaging Applications. In Bionanocomposites for Packaging Applications*; Jawaid, M., Swain, S., Eds.; Springer: Switzerland, 2018.

Russell, A. D.; Hugo, W. B. Antimicrobial Activity and Action of Silver. In *Progress in Medicinal Chemistry*; Ellis, G. P., Luscombe, D. K., Eds.; Elsevier: Amsterdam, 1994; 31, 351–370.

Salmieri, S.; Islam, F.; Khan, R. A.; Hossain, F. M.; Ibrahim, H. M.; Miao, C.; Hamad, W. Y.; Lacroix, M. Antimicrobial Nanocomposite Films Made of Poly(Lactic Acid)-Cellulose Nanocrystals (PLA-CNC) in Food Applications—Part B: Effect of Oregano Essential Oil Release on the Inactivation of *Listeria monocytogenes* in Mixed Vegetables. *Cellulose* **2014,** *21* (6), 4271–4285.

Sanchez-Garcia, M. D.; Lagaron, J. M. On the Use of Plant Cellulose Nanowhiskers to Enhance the Barrier Properties of Polylactic Acid. *Cellulose* **2010,** *17* (5), 987–1004.

Sanchez-Garcia, M. D.; Gimenez, E.; Lagaron, J. M. Morphology and Barrier Properties of Solvent Cast Composites of Thermoplastic Biopolymers and Purified Cellulose Fibers. *Carbohydr. Polym.* **2008,** *71* (2), 235–244.

Sayanjali, S.; Ghanbarzadeh, B.; Ghiassifar, S. Evaluation of Antimicrobial and Physical Properties of Edible Film Based on Carboxymethyl Cellulose Containing Potassium Sorbate on Some Mycotoxigenic *Aspergillus* Species in Fresh Pistachios. *LWT—Food Sci. Technol.* **2011,** *44* (4), 1133–1138.

Scherer, C.; Weber, A.; Stock, F.; Vurusic, S.; Egerci, H.; Kochleus, C.; Arendt, N.; Foeldi, C.; Dierkes, G.; Wagner, M.; Brennholt, N.; Reifferscheid, G. Comparative Assessment of Microplastics in Water and Sediment of a Large European River. *Sci. Total Environ.* **2020,** *738*. DOI.org/10.1016/j.scitotenv.2020.139866.

Shahbazi, M.; Rajabzadeh, G.; Sotoodeh, S. Functional Characteristics, Wettability Properties and Cytotoxic Effect of Starch Film Incorporated with Multi-Walled And Hydroxylated Multi-Walled Carbon Nanotubes. *Int. J. Biol. Macromol.* **2017,** *104*, 597–605.

Shameli, K.; Ahmad, M. B.; Yunus, W. M. Z. W.; Ibrahim, N. A.; Rahman, R. A.; Jokar, M.; Darroudi, M. Silver/Poly(Lactic Acid) Nanocomposites: Preparation, Characterization, and Antibacterial Activity. *Int. J. Nanomed.* **2010,** *5* (1), 573–579.

Sharma, C.; Dhiman, R.; Rokana, N.; Panwar, H. Nanotechnology: An Untapped Resource for Food Packaging. *Front. Microb.*, **2017,** *8* (1735), 1–22.

Siripatrawan, U.; Harte, B. R. Physical Properties and Antioxidant Activity of an Active Film from Chitosan Incorporated with Green Tea Extract. *Food Hydrocoll.* **2010,** *24*, 770–775.

Siripatrawan, U.; Vitchayakitti, W. Improving Functional Properties of Chitosan Films as Active Food Packaging by Incorporating with Propolis. *Food Hydrocoll.* **2016,** *61*, 695–702.

Sondi, I.; Salopek-Sondi, B. Silver Nanoparticles as Antimicrobial Agent: A Case Study on *E. coli* as a Model for Gram-Negative Bacteria. *J. Colloid Interf. Sci.* **2004,** *275* (1), 177–182.

Sothornvit, R.; Rhim, J-W,; Hong, S-I. Effect of Nano-Clay Type on the Physical and Antimicrobial Properties of Whey Protein Isolate/Clay Composite Films. *J. Food Eng.* **2009,** *91*, 468–473.

Sothornvit, R.; Rodsamran, P. Effect of a Mango Film on Quality of Whole and Minimally Processed Mangoes. *Postharvest Biol. Technol.* **2008,** *47* (3), 407–415.

Sun, X.-F.; Wang, H.-H.; Jing, Z.-X.; Mohanathas, R. Hemicellulose-Based pH-Sensitive and Biodegradable Hydrogel for Controlled Drug Delivery. *Carbohydr. Polym.* **2013,** *92* (2), 1357–1366.

Sun, Y.; Yang, L.; Lu, X.; He, C. Biodegradable and Renewable Poly(Lactide)-Lignin Composites: Synthesis, Interface and Toughening Mechanism. *J. Mater. Chem. A* **2015,** 3699–3709.

Talja, R.; Helén, H.; Roos, Y; Jouppila, K. Effect of Various Polyols and Polyol Contents on Physical and Mechanical Properties of Potato Starch-Based Films. *Carbohydr. Polym.*, **2007,** *67*, 288–295.

Tang, X. Z.; Kumar, P.; Alavi, S.; Sandeep, K. P. Recent Advances in Biopolymers and Biopolymer-Based Nanocomposites for Food Packaging Materials. *Crit. Rev. Food. Sci. Nutr.* **2012,** *52* (5), 426–442.

Tee, Y. B.; Tee, L. T.; Daengprok, W.; Talib, R. A. Chemical, Physical, and Barrier Properties of Edible Film from Flaxseed Mucilage. *BioResources* **2017,** *12* (3), 6656–6664.

Tripathi, S.; Mehrotra, G. K.; Dutta, P. K. Physicochemical and Bioactivity of Cross-Linked Chitosan–PVA Film for Food Packaging Applications. *Int. J. Biol. Macromol.* **2009,** *45*, 372–376.

Tripathy, J. Polymer Nanocomposites for Biomedical and Biotechnology Applications. In *Properties and Applications of Polymer Nanocomposites*; Tripathy, D., Sahoo, B., Eds.; Springer: Berlin/Heidelberg, 2017.

Vartiainen, J.; Vähä-Nissi, M; Harlin, A. Biopolymer Films and Coatings in Packaging Applications—A Review of Recent Developments. *Mater. Sci. Appl.* **2014,** *5*, 708–718.

Weiss, J.; Takhistov, P.; McClements, D. J. Functional Materials in Food Nanotechnology. *J. Food Sci.* **2006,** *71* (9), R107–R116.

Wittaya, T. Protein-Based Edible Films: Characteristics and Improvement of Properties, Structure and Function of Food Engineering, Ayman Amer Eissa, IntechOpen, (2012). https://www.intechopen.com/books/structure-and-function-of-food-engineering/protein-based-edible-films-characteristics-and-improvement-of-properties

Wu, J.; Ge, S.; Liu, H.; Wang, S.; Chen, S.; Wang, J.; Li, J.; Zhang, Q. Properties and Antimicrobial Activity of Silver Carp (*Hypophthalmichthys molitrix*) Skin Gelatin-Chitosan Films Incorporated with Oregano Essential Oil for Fish Preservation. *Food Pack. Shelf Life* **2014,** *2*, 7–16.

Xiao-e, L., Green, A. N. M., Haque, S. A., Mills, A., Durrant, J. R. Light-Driven Oxygen Scavenging by Titania/Polymer Nanocomposite Films. *J. Photochem. Photobiol. A,* **2004,** *162* (2), 253–259.

Xiong, X.; Zhang, K.; Chen, X.; Shi, H.; Luo, Z.; Wu, C. Sources and Distribution of Microplastics in China's Largest Inland Lake—Qinghai Lake. *Environ. Pollut.* **2018,** *235*, 899–906.

Yadav, S. K. Tissue Science and Engineering Realizing the Potential of Nanotechnology for Agriculture and Food Technology. *J. Tissue Sci. Eng.* **2017,** *8*, 8–11.

Yu, D-G.; Lin, W.-C.; Yang, M.-C. Surface Modification of Poly(L-lactic acid) Membrane via Layer-by-Layer Assembly of Silver Nanoparticle-Embedded Polyelectrolyte Multilayer. *Bioconjug. Chem.* **2007,** *18* (5), 1521–1529.

Zhao, R. X.; Torley, P.; Hally, P. J. Emerging Biodegradable Materials; Starch and Protein-Based Bionanocomposites. *J. Mater. Sci.* **2008,** *43*, 3058–3071.

Zhu, Y.; Buonocore, G. G.; Lavorgna, M.; Ambrosio, L. Poly(Lactic Acid)/Titanium Dioxide Nanocomposite Films: Influence of Processing Procedure on Dispersion of Titanium Dioxide and Photocatalytic Activity. *Polym. Compos.* **2011,** *32* (4), 519–528.

CHAPTER 12

Fabrication and Applications of Nanocomposites

SALINEY THOMAS[1], JIJI ABRAHAM[2], SONEY C. GEORGE[3*], SABU THOMAS[4], and SHAJI P. THOMAS[5]

[1]School of Chemical Sciences, Mahatma Gandhi University, Kottayam 686560, Kerala, India

[2]Vimala College (Autonomous) Thrissur 680009, Kerala, India

[3]Centre for Nano Science & Technology, Amal Jyothi Engineering College, Kanjirappally 686518 Kerala, India

[4]International and Inter University Centre for Nano Science & Technology, Mahatma Gundhi University, Kottayam 686560 Kerala, India

[5]Bishop Abraham Memorial College, Thuruthicad 689597 Kerala, India

[*]Corresponding author. E-mail: soneygeo@gmail.com

ABSTRACT

The topic "nanocomposites" remains a hot and a fast-developing area of research. Scientists across the world have made significant attempts to obtain control of the nanostructures by novel methods. The properties of nanocomposite depend on the properties of their individual components, and also on their morphology and interfacial characteristics. Efforts have been made in this review to offer proper definition and vivid explanation for the nanocomposites. Different techniques, such as solution blending, latex compounding melt intercalation, and in situ polymerization are used for the preparation of nanocomposites. Different types of nanocomposites, such as

The Science of Nanomaterials: Basics and Applications. Suresh C. Ameta & Rakshit Ameta (Eds.)

nanoparticle-reinforced, nanoplatelets-reinforced, and nanofibers-reinforced are illustrated. Various elastomer nanocomposites including other polymer nanocomposites are discussed with special emphasis to their physicomechanical properties. Analytical results, such as morphological, spectroscopic, crystallographic, and so on are presented. Nanotechnology offers promising applications in the various fields, such as electronics, energy, medicine, diagnosis, environment and so on. These promising applications of nanotechnology area are also detailed with appropriate examples.

12.1 INTRODUCTION

Nanotechnology can be broadly defined as, "The development of multifunctional materials and systems (1–100 nm) that exhibits unique features owing to its nano size" (Ono-Ogasawara et al., 2009).

The contemporary interests in nanotechnology include nanobiotechnology, nanomedicines, nanoelectronics, and nanostructured materials, whose key parts are nanocomposites.

Nanocomposites contain a wide variety of systems, such as one-dimensional, two-dimensional, three-dimensional, and amorphous materials, made of dissimilar components and mixed at the nanometer scale. The combination of diverse properties, which are not offered by a single material, such as flexible mechanical properties and superconducting properties is possible in nanocomposites. Another interesting and meritorious issue is the possibility of creating heterostructures composed of different kinds of inorganic layers, which could lead to novel multifunctional materials (Jordan et al., 2005).

Polymer nanocomposites are the materials in which nano-dimensional particles are suitably dispersed in polymer matrix with a view to improve the performance properties of the polymer dramatically. By virtue of nanometer scale, which can minimize the scattering of light, nanocomposites are usually transparent. Polymer nanocomposites are new alternatives to polymer microcomposites. Due to the nanometer size, fine and uniform filler dispersion, nanocomposites exhibit markedly improved properties including increased modulus strength, outstanding barrier properties, improved solvent heat resistance and high flame retardance as compared with the neat polymers and their traditional microcomposites.

The research in nanotechnology is a fast-developing area of research. Scientists across the world have made significant attempts to obtain control of the nanoscale structures by novel synthetic methodologies. The properties of nanocomposite depend on the properties of their individual components, and also on their morphology and interfacial characteristics. The reports published by Kelly (1985) have paved the path for exciting developments in multiple aspects of composites research and technology during these decades. Nanotechnology is aiming at the development of nanostructured materials using a bottom-up approach. Technologists working in the field of fiber-reinforced composites have practiced the bottom-up approach (Whitesides and Grzybowski, 2002; Balzani et al., 2008). The properties of materials are fine-tuned in length scales to obtain a best-performing composite.

The technologists concentrate their studies in search to make practical materials and devices using nanostructures. A clear understanding of the material behavior across length scales from the nanoscopic to macroscopic levels is a basic requirement. Understanding of the influence of nanoscale structure on bulk properties will support the design of the nanostructure to create multifunctional composites. This review encompasses the salient features of polymer nanocomposites in detail.

12.2 GENERAL STRATEGIES OF SYNTHESIS OF NANOPARTICLES

12.2.1 GAS PHASE AND SOL–GEL PROCESS

Two methods are generally used to produce nanomaterials. In the first method, bulk material is broken down into smaller particles using various forms of energy (top-down). In the bottom-up method, the nanomaterial is synthesized from atomic or molecular species through chemical reactions, allowing the precursor particles to grow in size. Both methods can be done in suitable medium. The technologists are usually interested in the ability to control (1) particle size, (2) particle shape, (3) size distribution, (4) particle composition, and (5) degree of particle agglomeration. Sol–gel technique is a traditional process to generate colloidal nanoparticles from liquid phase. Sol–gel processes are better suited for the synthesis of oxide nanoparticles and nanopowders. The notable advantages of sol–gel techniques are low temperature of processing, cost-effectiveness, and ease of preparation. Alkoxides remain the most suitable oxide precursors owing to their accessibility and susceptibility of M-OR bond allowing in situ facile tailoring during processing. Sol–gel process is a wet chemical synthetic approach, which can be used to produce nanoparticles (Kung and Ko, 1996). Size distribution of nanoparticles can be estimated either by the introduction of dopant (Kyprianidou-Leodidou et al., 1994) or by heat treatment (Wang et al., 1997). Monitoring of size and stability of nanoparticles is carried out through inverted micelles (Gacoin et al., 1997), block copolymers (Sankaran et al., 1993), or polymer blends (Justus et al., 1992), porous glasses, and particle-capping methods (Majetich and Carter, 1993; Olshavsky and Allcock, 1997).

12.2.2 OTHER STRATEGIES

Other techniques which can be adopted for nanoparticle synthesis are sonochemical reactions, microemulsion processing, and high-energy ball milling (Nalwa, 2004). In sonochemical method, the collapse of bubbles leads to the formation of transient localized hot spots with extremely high temperature and pressure (Suslick et al., 1999, 1999b). The resulting heating and cooling effects can lead to chemical and physical changes.

In hydrodynamic cavitation process, a large pressure drop is created in a cavitation chamber using high-pressure fluid system. The pressure drop generates bubbles and they collapse causing localized heating effect controlling the growth kinetics of the nanoparticles in the sol–gel. The particle size relies on the pressure and retention time of the solution in the cavitation chamber.

Microemulsions are nowadays used for the synthesis of metal (Kortan et al., 1992) semiconductor, silica (Pileni et al., 1992), barium sulfate (Arriagada and Osseo-asare, 1995), magnetic (Hopwood and Mann, 1997), and superconductor (Pillai et al., 1995) nanoparticles. When microemulsions containing reactants are mixed together, exchange of reactants will take place during the collision of water droplets in the microemulsion. Since the exchange of reactant is too fast, precipitation reaction will occur in the nanodroplets, followed by nucleation, growth, and coagulation of primary particles, resulting in the formation of the nanoparticles stabilized by water and surfactants. Simple and inexpensive devices were used for the large-scale production of nanoparticles.

Magnetic, catalytic, and structural nanoparticles have been produced by high energy ball milling (De CArvalho et al., 2013; Ying and Sun, 1991). In this mechanical alloying, the powder mixture undergone high-energy collision resulting in fine and uniform dispersions of oxide particles. The main drawback of the technique is the chance of contamination, which is overcome by using tungsten carbide components under inert medium or in high vacuum.

12.2.3 PREPARATION OF NANOCOMPOSITES

For achieving different performance parameters as required by the corresponding finished goods, polymers are usually reinforced by suitable fillers, such as carbon black, clays, silicates, calcium carbonate, zinc oxide, etc. as discussed in detail by various authors (Weimer et al., 1999; Chen et al., 2000; Lim and Park, 2000; Fu and Qutubuddin, 2001; Tseng et al., 2001; Wu et al., 2001). Such polymer nanocomposites are conveniently prepared by these techniques:

- Solution blending
- Latex compounding
- Melt intercalation and
- In situ polymerization

12.2.3.1 SOLUTION BLENDING

In solution blending technique, the elastomer is dissolved in a suitable solvent and then followed by the dispersion of the filler obtained using the same solvent. The elastomer chains, which become well separated in the solution, can now easily be exfoliated or intercalate nanofillers or can interact with the particulate nanofillers. The solvent is evaporated off in vacuum after preparing the elastomer–filler solution. Other compounding ingredients can be added during the preparation of the dispersion (Zhu et al., 2001), but they are usually compounded with the intercalated material after partial or complete solvent removal (Zeng et al., 2002; Beyer et al., 2002; Gilman et al., 2002) and then vulcanized.

Nanocomposites of natural rubber (NR), epoxidized natural rubber (ENR), nitrile butadiene rubber (NBR), etc. have been synthesized by this method (Xiao et al., 2001; Pramanik et al., 2003; Gelfer et al., 2003). Carbon-based nanofillers have been also dispersed by this method, even though often such a mixing process is preceded by certain preprocessing operations, such as dispersion of individual nanofillers by deagglomeration, removal of non-nanotube material, and functionalization of the fillers by chemical modification.

12.2.3.2 LATEX COMPOUNDING

The elastomers, most of which are available in the form of latex, are aqueous dispersion of elastomer particles in the submicron–micron range.

Clay nanoparticles or their aqueous dispersion can be added directly to the elastomer latex. The hydrophilic nature of clay helps them to adsorb water molecules followed by an expansion of their intergallery spacing. After mixing, they are co-precipitated and dried. By incorporating suitable curatives in the clay–elastomer latex, vulcanized elastomer composites are obtained. Nanocomposites of NR, styrene-butadiene rubber (SBR), NBR, and so on have been prepared by this method (Ogata et al., 1997, Gao et al., 1999; Xia et al., 2003). Latex compounding has been effectively used for dispersing nano-calcium carbonate in NR latex (Strawhecker and Manias, 2000).

12.2.3.3 MELT INTERCALATION

By far, melt mixing remains the most economic and environment-friendly technique, which is widely used for preparing elastomer/organoclay nanocomposites. In this technique, internal mixers and open two-roll mills are used for preparing the nanocomposites. Various factors, such as mixing temperature and shear rate also influence the intercalation/exfoliation phenomena in addition to the elastomer modification, compounding, and curing chemistry (Komori et al., 1999).

12.2.3.4 IN SITU POLYMERIZATION

In this method, monomer dispersions are mixed with the filler dispersions, during which exfoliation/intercalation of layered nanofillers will occur. By adding suitable initiator, the monomer/filler composite is allowed to polymerize to get in situ nanocomposite. The period of time required is determined by the factors, such as the monomer polarity, surface treatment of the filler and swelling temperature after the initiation of the reaction. Acrylate, methacrylate, or similar monomers are mostly used as the organic monomer components (Zhao et al., 1989; Fournaris et al., 1999; Nisha et al., 2000).

12.2.4 SYNTHESIS ISSUES AND NANOSTRUCTURES

12.2.4.1 MONODISPERSITY

Production of monodispersed nanoparticles with size uniformity remains as one of the most challenging issues in nanoparticle synthesis. Among the various synthetic methodologies adopted, till today gas-phase synthesis remains one of the best techniques to achieve size monodispersity. This is obtained by the combination of selective nucleation–condensation growth control and the elimination of coagulation by diffusion and turbulence. Through sintering, stability of the nanoparticles against agglomeration can be offered, and the changes in constituents can be guaranteed by procuring the nanoparticles in liquid suspension. The stable semiconducting particles are prepared by the addition of polar solvent in the liquid suspension. The addition of surfactant molecules in the liquid suspension of metallic nanoparticles offers stability (Murray et al., 1993). Alternatively, metallic nanoparticles can be safely encapsulated using inert silica and it is also an effective technique (Andrews et al., 1999).

Scientists have been experimenting to develop novel techniques for the production of monodispersed nanoparticles that undemands the use of a size classification procedure. In Japan, gold colloidal nanoparticles (1 nm) have produced in the presence of dendrimers by reducing its salt with UV rays (Esumi et al., 1998). In spite of lack of scientific explanations for the specific role of dendrimers in the formation of monodispersed nanoparticles, the better monodispersity is believed to occur due to the complex reaction followed by the decomposition of dendrimers, leading to the conversion of solution ions to gold nanoparticles.

12.2.4.2 SCALE-UP

Scaling up is one of the major interesting issues in nanoparticle synthesis. The high energy ball milling, being a commercial high-volume process, has been successfully used in generating magnetic, structural, and catalytic nanoparticles. Since polydispersed amorphous powder is obtained by this process, subsequent partial recrystallization is demanded prior to the consolidation into nanostructured materials. Continuous processing techniques based on batch processing are to be addressed in the case of sol–gel process for economic scale-up of nanoparticles. Other major sol–gel issues to be addressed are the cost of precursors and the recycling of solvent. In short, sol–gel processing is attractive for commercial scale-up production.

12.2.4.3 NANOPARTICLE CHAINS OR WIRES

There is a paradigm shift toward the development of assembly of molecular or quantum wires for optoelectronics and computational devices (Chidsey and Murray, 1986). Many polymeric molecules have been used as nano-building blocks for the preparation of biocompatible materials, nanoporous molecular optical switching sieves, and other nonlinear optical components. The chain aggregates of nanoparticles are considered as quantum wires, many of which have been extensively investigated as magnetic materials (Zhang and Manthiram, 1997), as reinforced elastomers (Pu et al., 1996), and as concrete additives (Sabir, 1997). These chain aggregates are shown to exhibit chemical and mechanical properties, which are different from those of individual primary particles (Friedlander et al., 1998). The bonding force from weak van der Waals to strong chemical/and very strong magnetic dipolar force is found responsible for holding the aggregates together. Agglomeration of nanoparticles generated by different synthetic techniques discussed earlier can result in chain aggregates with the exception that high-energy ball milling generates particles with low surface area and high anisotropic morphologies, both of which are detrimental to the formation of chain aggregates. The fractal dimension of aggregates can be controlled by using an external magnetic or electric field by depending on the magnetic and electric charging properties the nanoparticles. Chain aggregates of with lower fractal dimensions are apt candidates for optical applications.

12.2.4.4 NANOMETER FIBERS OR TUBES

Recent trends in the designing of nanofibers/tubes have contributed a novel form of building units for nanostructured materials. Membrane-template technique has been proved as an effective and suitable method to produce nanometer fibers/tubes (Martin, 1994). Nanochannels created by fission-fragment tracks or by electrochemical etching of aluminum metal are suitably used as templates for chemical or electrochemical deposition of conductive polymers (Parthasarathy and Martin, 1994). Since the nanochannels on membranes are uniform in size, the nanodimension and the aspect ratio of the nanofibers/tubes obtained by the membrane-template technique can be precisely controlled. Scientists have practiced catalytic development of single-crystal semiconductor nanofibers by metal–organic vapor-phase epitaxy and laser ablation vapor–liquid–solid techniques (Hiruma et al., 1995). The success in the production of these one-dimensional structures with diameters in the range of 3–15 nm is a notable technological landmark for optoelectronic device applications, such as the p–n junctions for light emission. The success in the synthesis of carbon-based nanotubes has created a novel way to fabricate nanosize fibers and tubes. The success in the synthesis of carbon-based nanotubes has paved another way to fabricate nanometer fibers and tubes. The scientists and technologists have begun to use carbon nanotubes as templates for the fabrication of carbide and oxide nanotubes (Dai et al., 1995; Kasuga et al., 1998). Reports on the synthesis of nanotubes based on boron nitrides and carbides are found in the literature (Chopra et al., 1995). These nanotubes are reported to be potential in possessing large third-order optical nonlinearity and other unusual properties. Metallic nanofibers produced by carbon nanotube (CNT)-template techniques have been shown to be potential infrared absorption materials. Scientists have also become successful in producing carbon nanotubes by catalytic method in bulk quantities for using as reinforcing fillers in nanostructural composite materials and concrete (Peigney et al., 1997).

12.3 TYPES OF NANOCOMPOSITES

12.3.1 NANOPARTICLE-FILLED COMPOSITES

In the day-to-day life, plenty of composites containing nanoparticles are widely and potentially used (Niihara, 1991; Thostenson et al., 2002). Nanoparticles are embedded in the matrix with an objective to improve the elastic modulus and yield strength. It has been shown to be practical to achieve novel material properties by scaling the particle size down to the nanometer scale. Micron-scale particles by virtue of their ability to scatter light make otherwise transparent matrix materials to appear opaque. In their study of SiO_2/epoxy composites, Naganuma and Kagawa (2002) have shown that decreasing the particle size resulted in significant improvement in transmittance of visible light. Singh et al. (2002) investigated the fracture toughness of polyester resin containing differently aluminum particles. It was noticed that fracture toughness depends on the volume fraction of aluminum particles at a constant particle dispersion and deagglomeration for a specific size.

Guo et al. (2007) reported the synthesis of nano-cupric oxide (modified with a bifunctional coupling agent methacryloxypropyltrimethoxysilane or MPS)-vinyl ester resin nanocomposites. They noted the enhanced thermal stability and improved mechanical properties for the composites filled with functionalized nanoparticles, which were explained by the better dispersion of nanoparticle in the polymer matrix and also by the newly established chemical bonding between the nanoparticles and the polymer matrix. The presence of MPS prevented the intimate contact of the CuO nanoparticle with the vinyl ester resin by making the particle surface passive. Shao et al. (2002) prepared SiO_2-nanoparticle-filled polyphthalazine ether sulfone ketone (PPESK) composites at various filler volume fractions by compression molding. They investigated the tribological behavior of the PPESK composites using a block-on-ring test by sliding PPESK-based composite blocks against a mild carbon steel ring. They reported that SiO_2 nanoparticle-filled PPESK composites exhibited good wear resistance and friction-reduction behavior. Lin et al. (2006) prepared epoxy filled with cloisite 30B and titanium dioxide. Jose and Thomas (2014a) made a systematic analysis on the alumina–clay nanoscale hybrid filler assembly in cross-linked polyethylene (XLPE) and their effect on the overall performance of composites. These ternary hybrid nanocomposites displayed superior mechanical performance, thermal barrier, and dielectric behavior because of the proper interaction between clay and alumina, which gave rise to inimitable filler architecture (Figure 12.1) (Jose and Thomas, 2014a, 2014b).

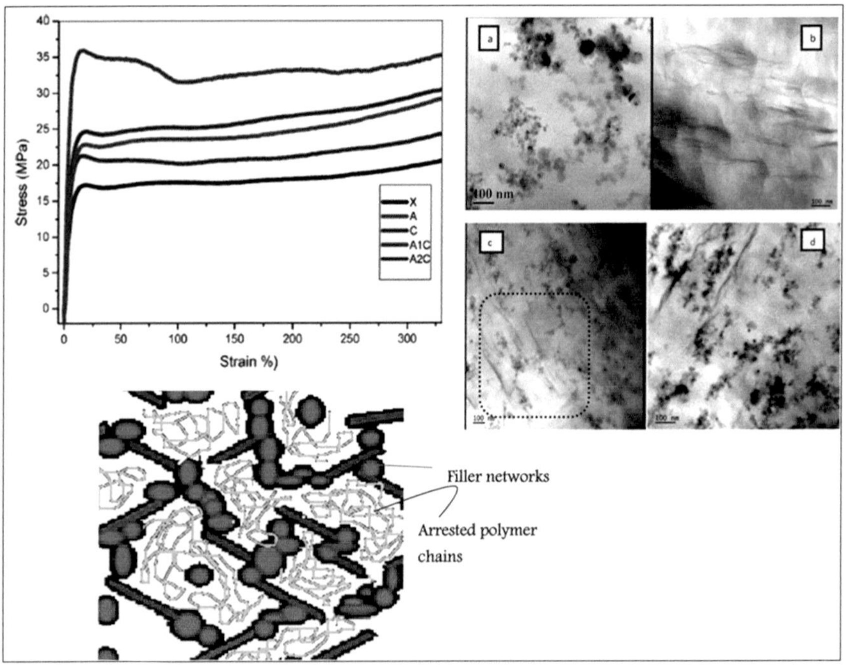

FIGURE 12.1 Unique hybrid filler networks in XLPE polymer matrix and the effect of filler networks on mechanics, Stress–strain behavior of XLPE (X), XLPE/Al_2O_3 (A), XLPE/clay (C), XLPE/Al_2O_3: clay = 1: 1 (A1C), XLPE/Al_2O_3: clay = 2: 1 (A_2C).

Source: Reprinted with permission from Jose and Thomas (2014b). © 2014 Royal Society of Chemistry.

Many researchers investigated the non-isothermal crystallization behavior of polymers filled with nanoparticle. Kuo et al. (2006) prepared nanoalumina-filled poly (ether ether ketone) (PEEK) nanocomposite and studied the nonisothermal crystallization behavior. They observed that the incorporation of the nanosized alumina particles could reduce the overall crystallization time of the PEEK composite as a result of the combined effects of rapid heterogeneous nucleation and the shorter time of growth to the small final crystal grain size.

Dielectric properties of nanoparticle-filled polymer nanocomposites are well explored by many researchers. The incorporation of high dielectric constant nanoparticles into an insulating polymer is an imperative method to make high dielectric constant nanocomposites with numerous applications (Hu et al, 2017). Various parameters, such as dielectric constant, dielectric loss, breakdown strength, and energy density have been investigated to study their applications in energy storage devices. Creation of large interfacial area to trap electron is very important for making the dielectric materials. Interfacial coupling enhances permittivity and high breakdown strength and along with high electrical energy storage density. Many dielectric fillers, such as alumina (Al_2O_3) (Mallakpour and Khadem, 2015), silica (SiO_2) (Virtanen et al., 2014), zinc oxide (ZnO) (Matei et al., 2008), titanium dioxide (TiO_2) (Hamming et al., 2009), $BaTiO_3$(Lott et al., 2008), metal nanoparticles (Sambhy et al., 2006; Li et al., 2015) core shell nanoparticles (Li et al., 2008; Yang et al., 2015), doped nanoparticles (Wang et al., 2018), and hybrid nanoparticles (Zhang et al., 2016) have been used to fabricate electric materials for power-energy storage and dielectric applications.

Several nanoparticles filled polymer nanocomposites have been used in many biomedical fields, such as tissue engineering scaffolds (Ravichandran et al., 2012), drug and gene delivery, chemotherapy, orthopedic applications, dental materials, biomedical devices and implants (Das et al., 2013), biosensors, smart antibacterial, and antimicrobial materials. Success of any tissue engineering scaffolds depends on cell adhesion and cell migration. Most of the synthetic materials are flammable in nature. Much attention must be given to the flammability of fabricated polymer-based nanocomposites. Now researchers are exploring many flame-retardant materials as reinforcing agent for polymers.

12.3.2 NANOPLATELET-REINFORCED COMPOSITES

Graphite and clay-filled composites represent this category. In the bulk state, both clay and graphite exist as layered materials. To show maximum efficiency, the layers must be separated and dispersed throughout the matrix phase. In the conventional miscible state, the interlayer spacing in a clay particle is at its minimum. When polymer resin gets embedded into the gallery between the adjacent layers, the spacing expands to the intercalated state. When the layers are completely separated, the clay is said to be in exfoliated state. In the transmission electron microscopy (TEM) image of montmorillonite poly (L-lactic acid) (PLLA) matrix nanocomposite (Figure 12.2), intercalated and exfoliated clay layers are illustrated (Krikorian and Pochan, 2003).

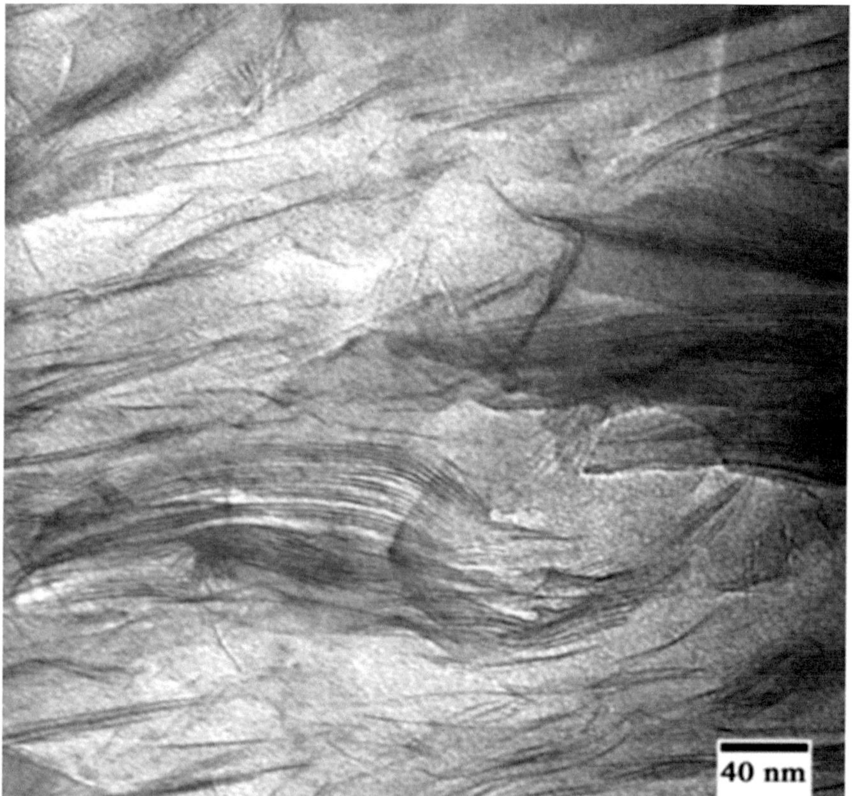

FIGURE 12.2 TEM micrograph of a montmorillonite poly (L-lactic acid) nanocomposite, showing both intercalated and exfoliated states.

Source: Reprinted with permission from Krikorian and Pochan (2003). © 2003, American Chemical Society.

Fogelström et al. (2010) prepared unmodified montmorillonite (MMT)-filled hyperbranched polymer resin and investigated the mechanical, electrical, thermal, and barrier properties. They reported that the composite films obtained had exfoliated structure. Thermogravimetric analysis showed an upward shift of decomposition temperature in oxygen atmosphere, which could be attributed to the improved barrier properties caused by the clay. Chen et al. (2003) investigated the mechanical and fracture properties of maleic anhydride-modified polypropylene composites (MAPP) filled with surface-modified MMT. They observed two-fold increase in yield strength when compared with that of un-reinforced MAPP, whereas the tensile modulus was sixfold than that of un-reinforced MAPP at 40 wt.% clay, and the observed improvement in modulus on increasing the clay content was due to the hindrance to the mobility of polymer chains.

Mohan and Kanny (2011) conducted studies on the effects of synthetic procedures (extrusion and casting) on the dispersion characteristics of nano-layered silicate clay particles in the polypropylene (PP) polymer matrix. Three different molecular weight PP, such as low molecular weight PP (LMPP), medium molecular weight PP (MMPP), and high molecular weight PP (HMPP) were used for preparing the composites using 1, and 3 wt.% organically modified nanoclay (cloisite 15A). The rate of increase in viscosity was higher for HM-PP-clay series than other series. Among the nanocomposite series, LMPP series filled with nanoclay showed higher rate of crystallization. Since the molecular weight of LMPP is lower, the crystal formation is relatively easier than that in higher molecular weight PP nanoclay composites as shown in Table 12.1. The higher crystallization rate and % crystallinity of nanoclay-filled PP composites indicate that the nanoclay acts as a nucleating agent.

TABLE 12.1 Crystallization Behavior PP-Clay Series.

	LMPP series			MMPP series			HMPP series		
Time (min)	**0% clay**	**1% clay**	**3% clay**	**0% clay**	**1% clay**	**3% clay**	**0% clay**	**1% clay**	**3% clay**
0	0	0	0	0	0	0	0	0	0
1	17	24	26	13	16	17	12	14	16
2	36	41	43	30	36	38	28	32	34
3	47	61	66	41	46	47	38	43	45
5	62	71	74	56	62	64	54	59	61

Source: Reprinted with permission from Mohan and Kanny (2011). https://creativecommons.org/licenses/by/4.0/

Rajasekar et al. (2009a) prepared ENR–compatibilized cloisite-filled NBR composites and investigated morphological and mechanical properties. The dispersion state of clay in polymer matrix was investigated using

X-ray diffraction technique (XRD). The nanoclay (Cloisite 20A) showed an intense peak around 3.144° (d- 2.82 nm) (Figure 12.3A). The ENR–cloisite (EC) composite exhibited a shifting of this peak to a lower angle 2.31° corresponding to basal spacing 3.83 nm, indicating intercalated structure. The peak corresponding to 1.9 nm is due to the re-aggregation of clay layers. But, when the ENR–cloisite composite is incorporated into the SBR matrix, exfoliation occurs as it is evident from the disappearance of the peak. The secondary peak is due to the re-aggregated clay platelets, and it is confirmed using TEM images (Figure 12.3). The dark hair-like structures in Figure 12.3B show intercalated clay, whereas the disorderly oriented clay platelets in the Figure 12.3C and D indicate exfoliation.

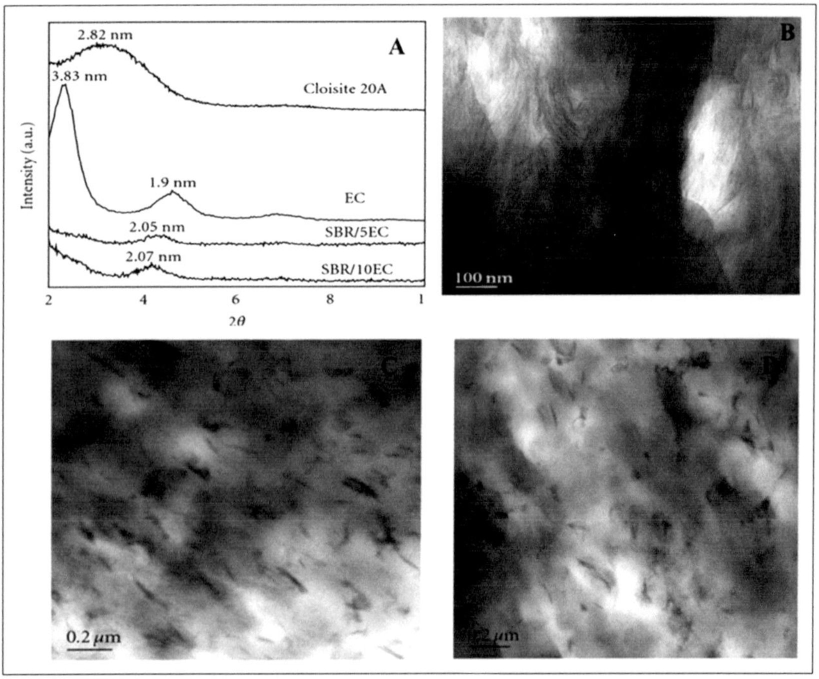

FIGURE 12.3 (A) XRD of cloisite, EC, SBR/5EC and SBR/10EC TEM images of (B) ENR–cloisite composite (EC), (C) SBR-5EC, and (D) SBR-10EC.

Source: Reprinted with permission from Rajasekar et al. (2009b). https://creativecommons.org/licenses/by/3.0/

Sridhar and Tripathy (2006) prepared chlorobutyl–nanoclay composites and investigated the barrier properties. These exfoliated nanofillers are able to form individual platelets, when dispersed in a polymer matrix. The nanoplatelets create tortuous path and it restricts the diffusion of gases through them and thus acts as a barrier for gases. They observed that the concentration-dependent diffusion coefficient D was one-order less in lower concentration region than that in higher concentrations. Follain et al. (2016) investigated the effect of clay structure and mixing protocol on the barrier properties of polyamide 12/montmorillonite nanocomposites. Changes in mixing shear rate control the fragmentation level and exfoliation of clay. Barrier effect is improved, when clay platelets are exfoliated and a reduction is observed when these are aggregated.

Compatibilizing action and localization of clay in a polypropylene/natural rubber (PP/NR) blend was explored by Chandran et al. (2015). In their study, the blend composition was 70/30 (PP/NR). Domain size of dispersed NR was decreased by the addition of nanoclay upto 5 wt.% and after that a leveling effect is observed. Here, nanoclay decreased the interfacial tension between NR and PP by the preferential localization of clay at the interface which point toward the role of clay as a compatibilizer also. In another study, Ock et al. (2016) used small-amplitude oscillatory shear (SAOS) and large-amplitude oscillatory shear (LAOS) tests to get an idea about the compatibilizing effect of organoclay in immiscible poly(lactic acid) (PLA) and natural rubber (NR) blends. Two parameters, such as a quantified nonlinear parameter ($I_{3/1}$) from FT-rheology and complex moduli ($|G^*|$) were used to investigate the structural modifications of the PLA/NR blends by the addition of nanoclay (Figure 12.4).

The nonlinear–linear viscoelastic ratio (NLR = Normalized nonlinear viscoelastic property/normalized linear viscoelastic property) is used for a quantitative estimation of compatibilizing effect of organoclay in immiscible polymer blends.

In another study by Rajasekar et al. (2009b), epoxidized natural rubber (ENR)–compatibilized cloisite-filled nitrile rubber (NBR) composites were prepared by separately preparing ENR–cloisite composite (EC) by solution mixing and then incorporating it into NBR. They investigated the morphology which indicated the intercalation of nanoclay and exfoliation of ENR–clay composite in ENR and NBR matrices, respectively. Cure study showed a reduction in scorch time and cure time with an increase in maximum torque for the ENR incorporated NBR compounds than NBR. ENR-incorporated NBR matrix exhibits rise in storage modulus and decrease in damping characteristics. Owing to the better reinforcement of nanoclay in NBR, the composites exhibit enhanced mechanical properties and oil and solvent resistance along with decrease in compression set. Stress relaxation behavior of organically modified montmorillonite-filled natural rubber/nitrile rubber nanocomposites was investigated by Maria et al. (2014). The effects of filler loading, blend composition, filler polarity, and temperature on the stress relaxation blend nanocomposites has been studied. Stress relaxation test is useful in predicting long-term behavior of polymer nanocomposites.

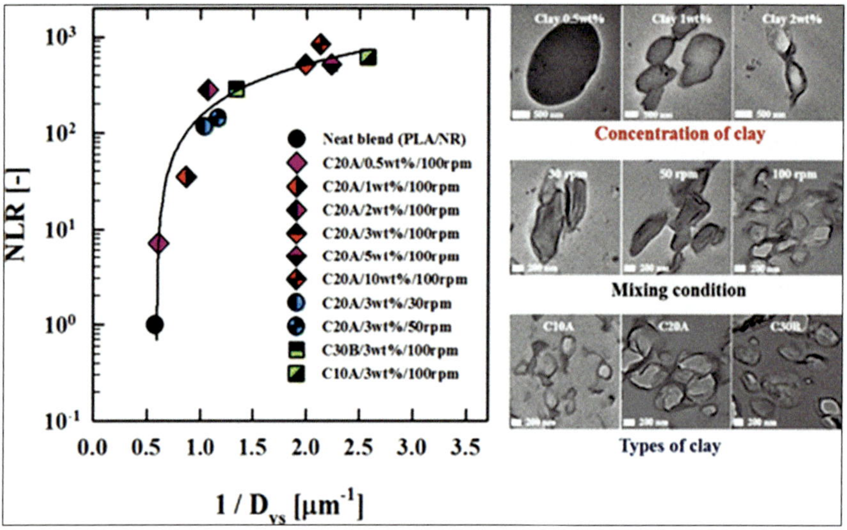

FIGURE 12.4 Quantified nonlinear parameter ($I_{3/1}$) from FT-rheology and complex moduli.
Source: Reprinted with permission from Ock et al. (2016). © 2016 American Chemical Society.

Maiti et al. (2004) prepared nanocomposite of brominated poly (isobutylene-*co*-paramethylstyrene) (BIMS) rubber and octadecyl amine-modified montmorillonite nanoclay. They studied the effect of carbon black loading in these composites on physicomechanical properties of the resultant composites. It was found that addition of 4 phr nanoclay to 20 phr N550 carbon black-filled samples could increase the tensile strength by 53%. Ali et al. (2010) investigated the influence of nanoclay in HNBR/NR blends on various properties using electrical conductance measurement during the mixing process. It was observed that the electrical conductance obtained through online depends on factors, such as nanoclay localization and blend morphology. This is because of the interaction as well as the compatibilization of nanoclay with respective blend phases. It was evident that the presence of clay influenced the morphology of the blends significantly in such a way that the cocontinuous phase morphology has been transformed to island–matrix morphology. In short, online conductance measurement is an effective tool to study the dispersion processes of nanoclay in the polymer matrix and to monitor its quality.

Praveen et al. (2009) investigated the effects of carbon black (CB) and organo-modified nanoclay (NC) on styrene-butadiene rubber (SBR). SBR composite with 20 phr CB and10 phr NC exhibited synergic effect as it is evident from the tremendous improvement in tensile strength, elongation at break, and stress improvement at 100% strain as 153%, 157%, 144%, respectively. It was found that one additional peak at d = 2.52 nm appeared in the XRD spectrum of SBR/NC nanocomposites containing CB, which was not in SBR/NC nanocomposites without containing CB. In this case, the CB particles increased the gap among the clay platelets, resulting in an increase in the extent of intercalation of SBR macromolecules through the clay platelets.

Bakhtiar et al. (2016) reported the preparation of hybrid filler composed of CNT grown directly on muscovite particles by chemical vapor deposition using methane as the carbon precursor. They incorporated CNT–muscovite hybrid compound into the epoxy matrix at various filler loadings (1–5 wt.%) and compared with physically mixed CNT–muscovite. The tensile strength, tensile modulus, and microhardness of epoxy/CNT–MUS HYB nanocomposites were found to exhibit an improvement up to 86.58, 134.59, and 14%, respectively, compared with the neat epoxy composites, which is attributed to the better dispersion of CNT–muscovite hybrid in the epoxy composites. Synergistic effect of clay and carbon nanotubes in wide variety of polymers has been investigated by Terzopoulou et al. (2016). Here, nanotubes are absorbed and intercalated on the surface and the gap between clay particles. This has significant effect on the mechanical and thermal crystallization behavior of composites. In some cases, incorporation of clay can reduce the electrical percolation threshold of CNT polymer nanocomposites, thereby reducing the amount of CNT in composites (Figure 12.5). It can be guessed that the introduction of nanoclay enhanced the conductivity by reducing the shear stress or polymer crystallinity (Al-Saleh, 2017). The macrodispersion analysis did not differentiate the dispersion of CNT in different CNT-based composites. There is a drop in torque and energy during mixing due to OMC. The drop in torque or energy due to mixing maintains the CNT aspect ratio. Moreover, the DSC analysis exhibited a drop in composite crystallinity owing to the presence of OMC.

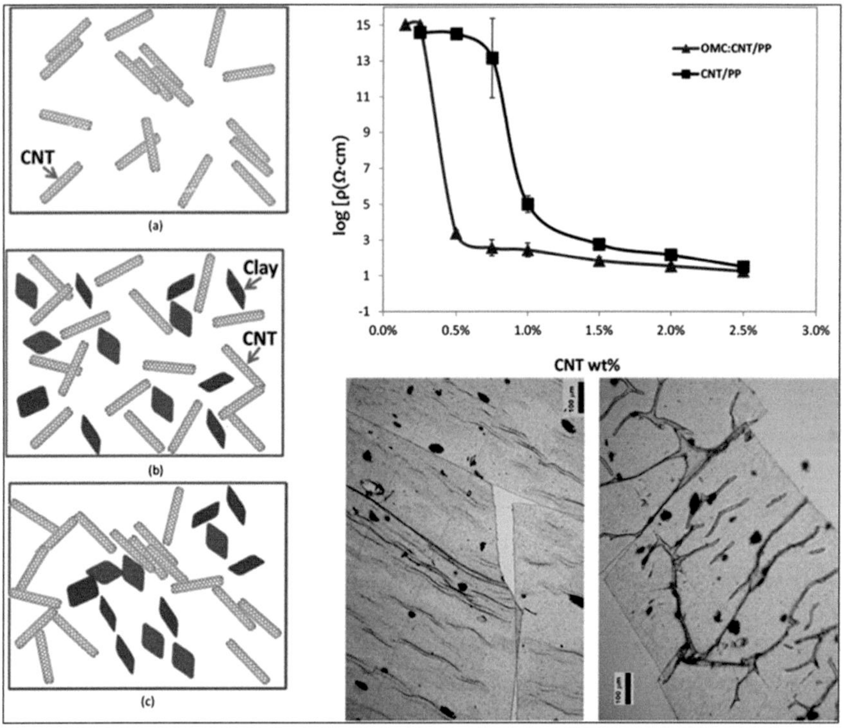

FIGURE 12.5 Dispersion of CNT particles. (a) Before addition of clay. (b) Clay improving the dispersion of CNT. (c) Clay creating CNT-free zones, electrical percolation behaviors of CNT/PP composite with and without OMC, Optical micrographs of (a) 1 wt.% CNT/PP composite and (b) 2 wt.% CNT:OMC/PP composite.

Source: Reprinted with permission from Al-Saleh (2017). © 2017 Elsevier.

Sometimes, polyhedral oligomeric silsesquioxane (POSS) can be used as interfacial modifier for clay. Here, cations present in the clay are replaced by POSS molecules. Final properties of the hybrid filler-reinforced composites depend on the arrangement of POSS molecules and not on their concentrations. Here, clay molecules have an exfoliated structure in the polymer matrix due the presence of POSS molecules in the gallery space. Prepared composites have improved mechanical and thermal properties (Kuo and Chang, 2011; Ayandele et al., 2012).

The versatile properties of graphitic fillers gave rise to profound attention in the fabrication of polymer nanocomposites. Interfacial interaction is improved by the addition of these kinds of fillers, which in turn leads to multifunctional polymer composites for potential applications. Natural graphite flakes (NFG) are polycrystalline forms of carbon containing layered planes form an atomically flat-stacked material in three dimensions. Exfoliation graphite layers resulted in the production of single to many layer graphene. Graphene fillers have high surface area, high modulus, tensile strength, fracture strength, thermal conductivity, and gas impermeability, which make them an attractive candidate for making multifunctional composites. Figure 12.6 represents a schematic diagram for the fabrication of polymer nanocomposites containing different types of graphitic fillers (Sadasivuni et al., 2014).

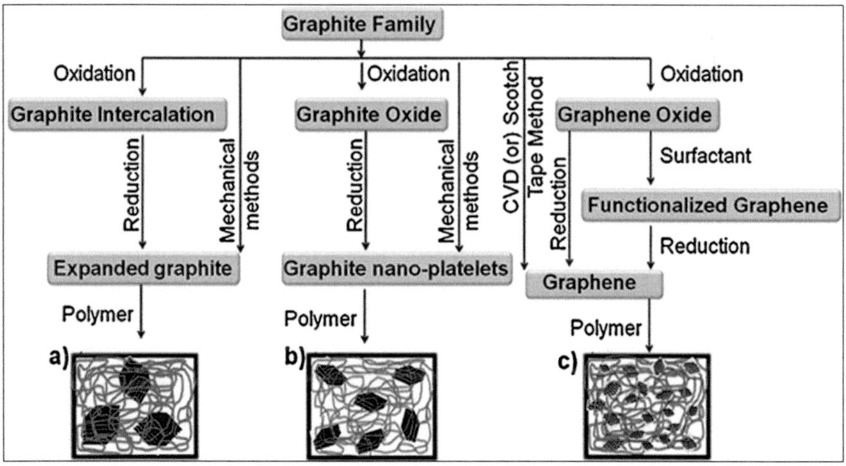

FIGURE 12.6 Schematic representation of graphitic-based filler systems and elastomer nanocomposite containing (a) expanded graphite, (b) graphite nanoplatelets, and (c) graphene.
Source: Adapted with permission from Sadasivuni et al. (2014).

The excellent mechanical properties of graphitic fillers paved the way for developing cost-effective and conducive composites (Kuilla et al., 2010) Several studies are reported on the role of graphene fillers as reinforcement in polymers. Primary or secondary interaction between graphene and polymers leads to the restriction in chain mobility. On the incorporation of reduced graphene oxide in natural rubber, a dramatic improvement in modulus was observed because of the better interfacial interaction between filler and rubber, that is, π–π interactions. Kraus plot studies clearly revealed that thermally reduced graphene (TRG) offers high reinforcement in the NR matrix (Yaragalla et al., 2015a). Various properties of fluoroelastomer are improved by the addition of microwave-reduced graphene oxide. The better electrostatic interaction between RGO and fluoroelastomer is evident from the marvellous dielectric performance (Moni et al., 2018a). Same group also reported the improved solvent resistant characteristics of flouroelastomer reinforced with microwave-reduced graphene oxide (Moni et al., 2018b). Kang et al. (2014) fabricated graphene oxide/carboxylated acrylonitrile butadiene rubber (GO/XNBR) nanocomposites by eco-friendly latex co-coagulation method. They reported 357 and 117% increase in tensile strength and tear resistance respectively with 1.9 vol.% of GO, which was attributed to the increased GO–polymer interaction through hydrogen bonding as supported by FTIR and DSC techniques. Highly exfoliated and uniform dispersion of GO sheets in the matrix was proved using HRTEM and XRD.

Electrical and dielectric properties of polymers are altered by the incorporation of graphene-based fillers. High dielectric permittivity of graphene polymer nanocomposites is explained on the basis of Maxwell Wagner Sillar effect. When exfoliated graphene (percolation threshold –0.52 vol.%) is added to the epoxy matrix, electrical conductivity of $\sim 10^{-2}$ S/m @ 3 wt.% graphene nanoplates (GNP) is obtained. Variation of maximum conductivity with aspect ratio (and lateral dimension of GNP) can be explained by the reduction of density of defects in the graphitic structure. The higher aspect ratio means a larger lateral dimension and/or a lower thickness of GNP. Density defects are reduced considerably in the graphitic structures due to the better dispersion of GNP. Sometimes, graphene mixed with CNTs is used as hybrid fillers to improve the electrical conductivity of polymer composites. Presence of graphene improves the network formation of CNT and insulator-to-conductor transition was observed at low filler loading (Yaragalla et al., 2015b; Hung et al., 2018).

Gas barrier performance of graphene is significant because of the layered structure of graphene, which offers restriction for the transport of gases and organic vapors. Gaphene fillers have sufficient aspect ratio to alter the diffusion path of gas-penetrant molecules. Important parameters which are influencing the gas barrier properties of nanocomposites include the size, stacking orientation, and exfoliation of graphene nanosheets in the polymer matrix. The effect of graphene reinforcement on the gas barrier properties of epoxidized natural rubber nanocomposites has been studied by Yaragalla et al. (2015c). Here, the homogeneous distribution of fillers as well as hydrogen bonds between fillers and polymers lead to the formation of constrained regions and it may create a tortuous pathway throughout the matrix, which hinders the diffusion of gas molecules throughout the composite. Moreover, high aspect ratio of graphene fillers offers some extent of impermeability toward oxygen molecules.

12.3.3 NANOFIBER-REINFORCED COMPOSITES

Vapor-grown carbon nanofibers (CNFs) have been used to reinforce a number of polymers, such as polypropylene, polycarbonate, nylon, poly(ether sulfone), poly(ethylene terephthalate), poly(phenylene sulfide), acrylonitrile-butadiene-styrene (ABS), and epoxy polymers. Reports show that CNFs have wide-ranging morphologies (Rath, 2010) from disordered bamboo-like structure to highly graphitized "cup-stacked" structures (Endo et al., 2002), where the conical shells of the nanofiber are nested within each other. Carbon nanofibers are reported to have diameters of 50–200 nm. Yang et al. (2005a) conducted studies on EMI shielding properties of CNF and multiwalled carbon nanotube (MWCNT)-filled polystyrene composites in the frequency range of 8.2–12.4 GHz (X-band). The composites exhibited frequency-independent shielding property, but it increased with the increase of MWCNT or CNF loading. MWCNT-filled polystyrene composites showed higher shielding capacity compared with the CNF-filled samples at the same filler loading. They showed that the shielding capacity of the composite containing 7 wt.% CNTs reached more than 26 dB level denoting the potential use of composite as EMI shielding material. In another study, Yang et al. (2005b) reported a novel foam structure of CNF-filled polymer composite with improved electrical conductivity and percolation behavior. Owing to light weight, electrical conduction, and easy processability, this foam structure is a promising candidate for use as an effective electromagnetic interference shielding material.

Investigations on large deformation behavior of flexible nanofiber (with high aspect ratio)-reinforced polymer composites were conducted by Dalmas et al. (2006). They used two nanofibers, viz., cellulose nanofibrils and CNF for studying the role of entanglements and possible interactions between fibers of the nanofiber networks in the composite material, on its properties. Strong hydrogen bonds between cellulose nanofibers contributed to a remarkable reinforcement effect and a reduction in the composite ultimate strain and an irreversible damage of composite properties after first deformation.

Nowadays, carbon nanotubes have got tremendous attention because of their excellent properties, such as high tensile strength, tensile modulus, good electrical, and thermal conductivity. Carbon nanotubes can be considered as an ideal reinforcing agent in many polymers (Coleman et al., 2006; Abraham et al., 2015). The incorporation of CNTs into polymer matrices is hindered by its poor dispersion and weak interfacial bonding. Due to the possibility of formation of agglomerates, these are functionalized by various routes (Ma et al., 2010; Abraham et al., 2018). Carbon nanotube–polymer composites offer so many applications (Figure 12.7) (Chen et al., 2018).

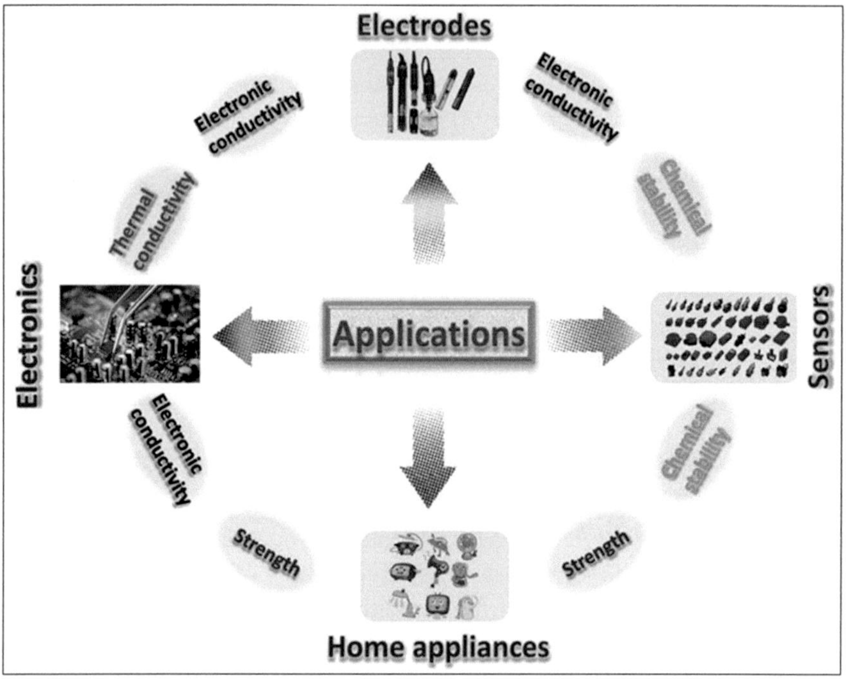

FIGURE 12.7 Applications of MWCNTs.

Source: Reprinted with permission from Chen et al. (2018). https://creativecommons.org/licenses/by-nc/3.0/

Ning et al. (2015) prepared elastomeric composites with both high elasticity and conductivity for the first-time using carbon nanotubes array (CNTA) as nanosprings. A special CNTA bundle was prepared by electrostatic self-assembly and it was blended with the polymethylvinylsiloxane (PMVS) matrix. The results revealed that these CNTAs were well dissociated into many single CNTs without almost any change in the length of the PMVS matrix with low viscosity during mechanical shearing, and exhibited a nanospring effect. The CNTA composites exhibited higher elasticity than the conventionally used conductive carbon black composites at the same filler level in addition to higher conductivity and electromagnetic shielding effectiveness. Abraham et al. (2016) prepared conductive flexible styrene-butadiene rubber nanocomposites using ionic liquid-modified MWCNT. Here, ionic liquid acts as a dispersing agent for CNT, which improves the chance of formation of networks of the CNT. There is a synergistic effect of electrical conductivity of CNT and ionic conductivity of ionic liquid. Strain sensing properties of polymer/carbon nanotube composites were investigated by Georgousis et al. (2015). The improvement in composite properties in relation to its mechanical and electrical behaviors makes the possibility of using these materials as various sensors (Figure 12.8). The strain sensing capabilities of the conducting samples were studied by in situ conductivity measurements upon tensile loading. Here, nanocomposite with high filler content, much above the percolation threshold like 8 wt.% in the present study, is more suitable, because it can withstand higher strain levels without breaking. However, for applications where relatively low strain levels are expected, a filler content of 1.5 wt.% is suitable for strain sensing.

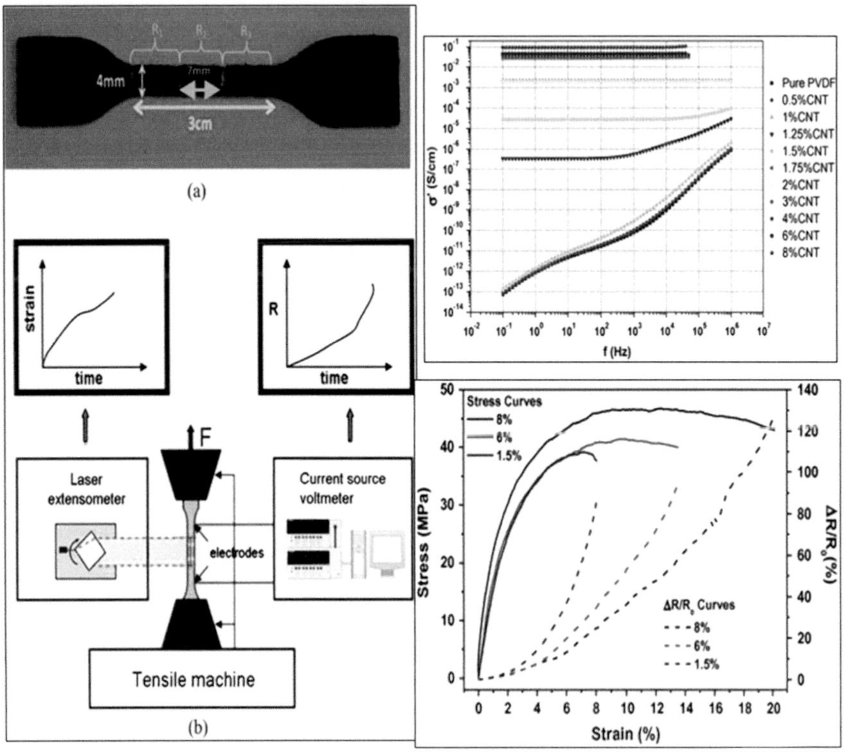

FIGURE 12.8 (a) Sample structure for measuring resistance, and (b) experimental setup, real part of conductivity σ' against frequency measured at room temperature, Variation of stress and relative resistance with strain as a function of filler concentration.

Source: Reprinted with permission from Georgousis et al. (2015). © Elsevier.

12.4 APPLICATIONS OF NANOTECHNOLOGY

12.4.1 MEDICAL APPLICATIONS

Nanotechnology finds a lot of applications in medical field, such as in drugs, diagnosis of diseases and abnormal body conditions, surgery, medical robotics, and all allied areas related to body of human beings.

The traditional drug delivery systems face several challenges including poor bioavailability, in vivo stability, solubility, and side effects. Nanotechnology promises a novel strategy to overcome these challenges by offering the techniques for development and fabrication of nanostructures at nanoscale, which are mainly polymeric in nature (Prabha and Raj, 2016). Nanostructures are shown to possess the capacity to protect the encapsulated drugs from hydrolytic and enzymatic degradation in the gastrointestinal tract. They also possess the ability of controlled delivery of a wide range of drugs at various sites of the body, making them capable of delivering drugs, proteins, and genes by oral administration (Jung et al., 2000; Soppimath et al., 2001; Nimesh et al., 2006). Controlled releases of encapsulated drug are made possible by manipulating and tailor making the nature of polymers to reach an optimum therapeutic dosage during the period. For the controlled release and targeted delivery of the drugs at the desired site of action, nanostructures can be conjugated with targeting moieties with an objective of manipulating the bonding between the polymer and the active substance. By the proper incorporation of spacer molecules, such as amino acids, lipids, peptides, or small chains, the linkage between the polymer and active substance may be established (Peppas, 2004). Such drug targeting is crucial in chemotherapy, where a drug delivery system should target only the malignant portions without detrimental to the healthy cells. The application of polymeric nanostructures in the treatment of diseases, such as cerebrovascular and inflammatory diseases has wider scope (Koo et al., 2005).

Using nanotechnology, relevant nano-based devices/robots can be made to conduct microsurgeries on body. Without detrimental to the healthy parts of the body, these instruments can precisely and accurately target the surgical site. Computers can replace surgeon to control the nanosized surgical instruments. "Nanocameras" are useful to visualize the surgery and they enable to eliminate faults. Tissue, genetic, and cellular levels surgery can be successfully done. Nanorobotics finds a variety of potential applications, such as premature detection of cancer and its inactivation by focused drug delivery, pharmacokinetics, monitoring of diabetes, and health care uses in medical fields. Nanotechnology is a powerful tool to understand the working of cells, bacteria, viruses, etc. It also helps to understand the causes of present-day new diseases so as to enable us to take preventive measures. Genome sequencing can be made much easier using nanotechnology. Moreover, biological causes of mental disorders can be monitored and identified.

Development in tissue engineering and allied areas brought appropriate systems to address issues in replacement (Furth et al., 2007) Biodegradability, high mechanical strength, cell adhesion, biocompatibility, and cell proliferation are required for tissue engineering materials. Polymer nanocomposite scaffolds are widely used for tissue engineering, which include bone tissue engineering (Venkatesan et al., 2015), cartilage tissue engineering (Lee et al., 2017), wound healing (Alberti et al., 2017), etc. Nanomaterials are highly useful in biotechnology. Nanomaterial-based scaffolds and growth factors are in use for artificially stimulated cell proliferation. Since our tissues are built up of proteins in nanoscale dimension and cells can directly interact with nanostructured extracellular matrices (ECM), lot of nano-biomaterials, such as nanofibers, nanotubes, nanoparticles, and other nanofabricated devices within nanoscale can contribute to cell growth and tissue regeneration (Yu et al., 2011).

Many polymer nanocomposites have antibacterial and antimicrobial properties (Palza, 2015). The development of polymer nanocomposites with antimicrobial properties has been a key factor for controlling or inhibiting the growth of microorganisms and preventing foodborne diseases and nosocomial infections. Antibacterial surfaces are very important in the field of food packaging, medical devices, textiles and pharmaceuticals, and water treatment. Similarly, polymer nanocomposites have potential applications as antimicrobial films. Nanocomposites are multifunctional in nature and are widely found in variety of applications. Use of coatings of conducting polymers for corrosion protection has got special attention in industrial field. Polymers possess long-chain carbon linkages and therefore, upon adsorption, these are able to block large areas of the corroding metal surfaces. The thin films adsorbed on the metal substrate provide a barrier effect between the metal and its environment (Ates, 2016). Many conducting polymer nanocomposites can be used as electrostatic charge dissipating materials (Pramanik and Karak, 2017).

12.4.2 ENVIRONMENTAL APPLICATIONS

Polymer nanocomposites (PNCs) have been used as candidate for selective removal of pollutants, such as organic dyes (Wang et al., 2012), oil contaminants, and heavy metals from water and also for water treatment (Singh and

Susan, 2018; Lofrano et al., 2016). Polymer–clay composites are considered as fruitful adsorbents for water remediation. CPNs efficiently remove all kinds of micropollutants from aqueous solutions (Unuabonah and Taubert, 2014). Polymer nanocomposites are also used to detect pollutants of all kinds and are removed by catalytic degradation and adsorption. Also, PNCs are suitable for sensing and detecting pollutants particularly at trace levels, as well as for green chemistry to minimize the discharge of pollutants into the receiving environment (Qu et al., 2012). Polymer nanocomposite membranes are widely used for desalination (Al Aani et al., 2017). Pervaporation is a membrane-based technique to remove organic to separate organic mixtures or to remove organic solvents from water. Usually, polymer nanocomposites are used for pervaporation technique (Jose et al., 2014c).

12.4.3 ENERGY STORAGE AND CONDUCTING APPLICATIONS

Polymer nanocomposites with high dielectric permittivity, high energy density, and low dielectric loss serve as ideal materials for energy storage applications. Addition of dielectric filler or conducting nanomaterials improved the dielectric permittivity of polymers (Abraham et al., 2017). Conductive polymer nanocomposites can be used as various types of sensors, such as electromechanical sensor (Boland et al., 2016), strain sensor (Amjadi et al., 2016), biosensors (Shrivastava et al., 2016), gas humidity sensors (Zhang et al., 2014), organic vapor sensor (Li et al., 2016; Liu et al. 2016), pressure sensor (Tung et al., 2016), etc. Electromagnetic interference shielding applications of polymer nanocomposites have got tremendous attention. Owing to their appropriate properties, conductive polymer composites are attractive materials for electromagnetic interference (EMI) shields (Sharma et al., 2014; Poothanari et al., 2018). Various conducting and dielectric fillers are used to fabricate materials for EMI shielding. Incorporation of magnetic nanoparticles in addition to conducting nanoparticles further shields the passage of electromagnetic radiations. Socarbon/ferromagnetic composites exhibit a good balance between dielectric permittivity and magnetic permeability (Pawar et al., 2016).

Nanostructured oxide/conjugated polymer composite photovoltaic were made and these nanostructures were fabricated by first growing ZnO fibers directly onto a transparent conducting oxide substrate using a chemical solution growth technique (Shaheen et al., 2005). The fibers were then intercalated with a hole-conducting organic semiconductor into the pores of the oxide. These composites have the advantage of the light-absorbing and hole-transporting properties of the organic compounds combined with the high electron mobility of the oxide nanofibers, when used as active layer in solar cells. Tang et al. (2010) conducted studies to establish the relationship between the filler aspect ratio in nanocomposites and energy storage capacity. They prepared nanocomposites using lead zirconatetitanate (PZT) with fillers in polyvinylidene fluoride (PVDF) matrix. Higher permittivity constants were observed with nanowires (NWs) than that of nanorods (NRs) at the same inclusion volume fraction. Samples with PZT nanowires exhibited high frequency loss tangent than that of nanorods revealing their high electrical energy storage efficiency. Jamail et al. (2014) prepared (linear low-density polyethylene) LLDPE/NR compounds (80:20) filled with nano-SiO_2/TiO_2 and conducted polarization–depolarization current (PDC) measurements to determine the conductivity levels. They observed less conductivity LLDPE/NR/ SiO_2 nanocomposite compared with LLDPE/NR/TiO_2 nanocomposite enabling to use as anapt candidate for high-voltage applications (Chen et al., 2016). Prepared polyurethane/TiO_2-functionalized graphene composites demonstrated good electric stimulus responses and strain inducing on the application of electric field. The superior properties achieved, such as higher dielectric constant, less loss modulus, elevated breakdown strength, and better electric stimulation responses were attributed to the enhanced filler–matrix interaction owing to the hydrophilicity and wettability of graphene on the decoration of TiO_2 on to graphene surface.

12.4.4 NANOELECTRONICS

Studies are on the one-step direct nanoimprinting of metal nanoparticles to develop nano-/microscale metallic structures including nanodot and nanowire arrays. The study was done at low temperatures and pressures in order to take advantage of the low melting temperature and viscosity of metal nanoparticle solutions. They became successful in demonstrating nanoscale electronic devices, including nanowire resistors and nanochannel organic field effect transistors with an air-stable semiconducting polymer (Ko et al., 2007).

12.4.5 DEFENSE/MILITARY APPLICATIONS

Polymer nanocomposites have promising applications in defense filed. These include aerodynamics, mobility, stealth, sensing, power generation and management, smart structures materials, resilience, and robustness. Nanocomposites can be used in the area of high performance fiber/fabrics, ballistic protection, microwave absorbers, refractive index tuning, solid lubricants, electrostatic charge dissipation in space environment, ultraviolet irradiation resistance, fire retardation, corrosion protection, signature reduction, diffusion barriers, and so on (Kurahatti et al., 2010).

12.4.6 OTHER APPLICATIONS

Polymer nanocomposites are used in the field of photovoltaic (PV) cells and photodiodes, supercapacitors, printable conductors, light emitting diodes (LEDs), field effect transistors, solar cells, shape memory materials, self-healing materials, biocompatible/degradable/renewable materials, packaging/films, gaskets, etc.

12.5 CONCLUSION

Scientists and technologists resort to different methods, such as solution blending, latex compounding, melt intercalation, in situ polymerization for preparing polymer nanocomposites, among which melt intercalation remains more economic and environment-friendly. One of the challenges in the field of nanocomposite preparation is achieving the monodispersity in size of the nanoparticles. Among all the synthetic methodologies, gas-phase method is apt technique to achieve size monodispersity, which is feasible by control of nucleation–condensation growth, avoiding coagulation by diffusion and turbulence and also by the effective collection of nanoparticles. Enhanced thermal stability and improved mechanical properties are exhibited by the composites filled with functionalized nanoparticles, which is attributed to better nanoparticle dispersion in the polymer matrix and also to the newly introduced bonding between the nanoparticles and the polymer matrix. Microscopic imaging techniques, such as FESEM and TEM are effective tools for understanding the morphological features of nanocomposites. Usage of hybrid fillers is a suitable technique to improve mechanical properties of nanocomposites. Nanoscience and technology offer promising applications in the field of medicine, diagnosis, and energy.

ACKNOWLEDGMENT

The author (Soney C. George) is thankful to Department of Science and Technology for the financial assistance under FIST Scheme (Order No: SR/FST/COLLEGE-346/2018 dated 20th December 2018).

KEYWORDS

- **nanoparticles**
- **nanocomposites**
- **morphology**
- **mechanical properties**
- **aspect ratio**

REFERENCES

Abraham, J.; Kailas, L.; Kalarikkal, N.; George, S. C.; Thomas, S. Developing Highly Conducting and Mechanically Durable Styrene Butadiene Rubber Composites with Tailored Microstructural Properties by a Green Approach Using Ionic Liquid Modified MWCNTs. *RSC Adv*. **2016,** *6*, 32493–32504.

Abraham, J.; Maria, H. J.; George, S. C.; Kalarikkal, N.; Thomas, S. Transport Characteristics of Organic Solvents through Carbon Nanotube Filled Styrene Butadiene Rubber Nanocomposites: The Influence of Rubber–Filler Interaction, the Degree of Reinforcement and Morphology. *Phys. Chem. Chem. Phys*. **2015,** *17*, 11217–11228.

Abraham, J.; Thomas, J.; Kalarikkal, N.; George, S. C.; Thomas, S. Static and Dynamic Mechanical Characteristics of Ionic Liquid Modified MWCNT-SBR Composites: Theoretical Perspectives for the Nanoscale Reinforcement Mechanism. *J. Phys. Chem. B*. **2018,** *122*, 1525–1536.

Abraham, J.; Xavier, P.; Bose, S.; George, S. C.; Kalarikkal, N.; Thomas, S. Investigation into Dielectric Behaviour and Electromagnetic Interference Shielding Effectiveness of Conducting Styrene Butadiene Rubber Composites Containing Ionic Liquid Modified MWCNT. *Polymer*. **2017,** *112*, 102–115.

Al Aani, S.; Wright, C. J.; Atieh, M. A.; Hilal, N. Engineering Nanocomposite Membranes: Addressing Current Challenges and Future Opportunities. *Desalination*. **2017,** *401*, 1–5. doi.org/10.1016/j.desal.2016.08.001.

Alberti, S.; Coelho, D.; Voytena, A.; Pitz, H.; de Pra, M.; Mazzarino, L. et al. Nanotechnology: A Promising Tool Towards Wound Healing. *Curr. Pharm. Des*. **2017,** *23*, 3515–3528.

Ali, Z.; Le, H. H.; Ilisch, S.; Thurn-Albrecht, T.; Radusch, H. J. Morphology Development and Compatibilization Effect in Nanoclay Filled Rubber Blends. *Polymer*. **2010,** *51*, 4580–4588.

Al-Saleh, M. H. Clay/Carbon Nanotube Hybrid Mixture to Reduce the Electrical Percolation Threshold of Polymer Nanocomposites. *Compos. Sci. Technol.* **2017,** *149*, 34–40.

Amjadi, M.; Kyung, K. U.; Park, I.; Sitti, M. Stretchable, Skin-Mountable, and Wearable Strain Sensors and Their Potential Applications: A Review. *Adv. Funct. Mater.* **2016,** *26*, 1678–1698.

Andrews, R.; Jacques, D.; Rao, A. M.; Rantell, T.; Derbyshire, F.; Chen, Y.; Chen, J.; Haddon, R. C. Nanotube Composite Carbon Fibers. *App. Phys. Lett.* **1999,** *75*, 1329–1331.

Arriagada, F. J.; Osseo-Asare, K. Synthesis of Nanosize Silica in Aerosol OT Reverse Microemulsions. *J.Colloid Interface Sci.* **1995,** *170*, 8–17.

Ates, M. A Review on Conducting Polymer Coatings for Corrosion Protection. *J. Adhesion Sci. Technol.* **2016,** *30* (14), 1510–1536.

Ayandele, E.; Sarkar, B.; Alexandridis, P. Polyhedral Oligomeric silsesquioxane (POSS)-Containing Polymer Nanocomposites. *Nanomaterials*. **2012,** *2*, 445–475.

Bakhtiar, N. S.; Akil, H. M.; Zakaria, M. R.; Kudus, M. H.; Othman, M. B. New Generation of Hybrid Filler for Producing Epoxy Nanocomposites with Improved Mechanical Properties. *Mater. Design*. **2016,** *91*, 46–52.

Balzani, V.; Credi, A.; Venturi, M. *Molecular Devices and Machines: Concepts and Perspectives for the Nanoworld*; John Wiley & Sons: Weinheim, 2008.

Beyer, F. L.; Beck Tan, N. C.; Dasgupta, A.; Galvin, M. E. Polymer–Layered Silicate Nanocomposites from Model Surfactants. *Chem. Mater.* **2002,** *14*, 2983–2988.

Boland, C. S.; Khan, U.; Ryan, G.; Barwich, S.; Charifou, R.; Harvey, A. et al. Sensitive Electromechanical Sensors Using Viscoelastic Graphene-Polymer Nanocomposites. *Science* **2016,** *354*, 1257–1260.

Chandran, N.; Chandran, S.; Maria, H. J.; Thomas, S. Compatibilizing Action and Localization of Clay in a Polypropylene/Natural Rubber (PP/NR) Blend. *RSC Adv.* **2015,** *5*, 86265–86273.

Chen, G.; Liu, S.; Zhang, S.; Qi, Z. Self-Assembly in a Polystyrene/Montmorillonite Nanocomposite. *Macromo. Rapid Commun.* **2000,** *21*, 746–749.

Chen, J.; Liu, B.; Gao, X.; Xu, D. A Review of the Interfacial Characteristics of Polymer Nanocomposites Containing Carbon Nanotubes. *RSC Adv*. **2018,** *8*, 28048–28085.

Chen, L.; Wong, S. C.; Pisharath, S. Fracture Properties of Nanoclay-Filled Polypropylene. *J. Appl. Polym. Sci.* **2003,** *88*, 3298–3305.

Chen, T.; Qiu, J.; Zhu, K.; Li, J. Electro-Mechanical Performance of Polyurethane Dielectric Elastomer Flexible Micro-Actuator Composite Modified with Titanium Dioxide-Graphene Hybrid Fillers. *Mater. Design*. **2016,** *90*, 1069–1076.

Chidsey, C. E.; Murray, R. W. Electroactive Polymers and Macromolecular Electronics. *Science* **1986,** *231*, 25–31.

Chopra, N. G.; Luyken, R. J.; Cherrey, K.; Crespi, V. H.; Cohen, M. L.; Louie, S. G.; Zettl, A. Boron Nitride Nanotubes. *Science* **1995,** *269*, 966–967.

Coleman, J. N.; Khan, U.; Gun'ko, Y. K. Mechanical Reinforcement of Polymers Using Carbon Nanotubes. *Adv. Mater*. **2006,** *18*, 689–706.

Dai, H.; Wong, E. W.; Lu, Y. Z.; Fan, S.; Lieber, C. M. Synthesis and Characterization of Carbide Nanorods. *Nature* **1995,** *375*, 769–772.

Dalmas, F.; Chazeau, L.; Gauthier, C.; Cavaillé, J. Y.; Dendievel, R. Large Deformation Mechanical Behavior of Flexible Nanofiber Filled Polymer Nanocomposites. *Polymer* **2006,** *47* 2802–2812.

Das, B.; Mandal, M.; Upadhyay, A.; Chattopadhyay, P.; Karak, N. Bio-based Hyperbranched Polyurethane/Fe3O4 Nanocomposites: Smart Antibacterial Biomaterials for Biomedical Devices and Implants. *Biomed. Mater*. **2013**. DOI: 10.1088/1748-6041/8/3/035003.

De Carvalho, J. F.; De Medeiros, S. N.; Morales, M. A.; Dantas, A. L.; Carriço, A. S. Synthesis of Magnetite Nanoparticles by High Energy Ball Milling. *Appl. Surf. Sci.* **2013,** *275*, 84–87.

Endo, M.; Kim, Y. A.; Hayashi, T.; Fukai, Y.; Oshida, K.; Terrones, M.; Yanagisawa, T.; Higaki, S.; Dresselhaus, M. S. Structural Characterization of Cup-Stacked-Type Nanofibers with an Entirely Hollow Core. *Appl. Phy. Lett.* **2002,** *80*, 1267–1269.

Esumi, K.; Suzuki, A.; Aihara, N.; Usui, K.; Torigoe, K. Preparation of Gold Colloids with UV Irradiation Using Dendrimers as Stabilizer. *Langmuir* **1998,** *14*, 3157–3159.

Fogelström, L.; Malmström, E.; Johansson, M.; Hult, A. Hard and Flexible Nanocomposite Coatings Using Nanoclay-Filled Hyperbranched Polymers. *ACS Appl. Mater. Interf.* **2010,** *2*, 1679–1684.

Follain, N.; Alexandre, B.; Chappey, C.; Colasse, L.; Médéric, P.; Marais, S. Barrier Properties of Polyamide 12/Montmorillonite Nanocomposites: Effect of Clay Structure and Mixing Conditions. *Compos. Sci. Technol.* **2016,** *136*, 18–28.

Fournaris, K. G.; Karakassides, M. A.; Petridis, D.; Yiannakopoulou, K. Clay− Polyvinylpyridine Nanocomposites. *Chem. Mater.* **1999,** *11*, 2372–2381.

Friedlander, S. K.; Jang, H. D.; Ryu, K. H. Elastic Behavior of Nanoparticle Chain Aggregates. *Appl. Phys. Lett.* **1998,** *72*, 173–175.

Fu, X.; Qutubuddin, S. Polymer–Clay Nanocomposites: Exfoliation of Organophilic montmorillonite nanolayers in Polystyrene. *Polymer* **2001,** *42*, 807–813.

Furth, M. E.; Atala, A.; Van Dyke, M. E. Smart Biomaterials Design for Tissue Engineering and Regenerative Medicine. *Biomaterials* **2007,** *28*, 5068–5073.

Gacoin, T.; Malier, L.; Boilot, J. P. New Transparent Chalcogenide Materials Using a Sol−Gel Process. *Chem. Mater.* **1997,** *9*, 1502–1504.

Gao, D.; Heimann, R. B.; Williams, M. C.; Wardhaugh, L. T.; Muhammad, M. Rheological Properties of Poly Acrylamide)-Bentonite Composite Hydrogels. *J. Mater. Sci.* **1999,** *34*, 1543–1552.

Gelfer, M. Y.; Song, H. H.; Liu, L.; Hsiao, B. S.; Chu, B.; Rafailovich, M.; Si, M.; Zaitsev, V. Effects of Organoclays on Morphology and Thermal and Rheological Properties of Polystyrene and Poly (Methyl Methacrylate) Blends. *J. Polym. Sci. Part B: Polym. Phys.* **2003,** *41*, 44–54.

Georgousis, G.; Pandis, C.; Kalamiotis, A.; Georgiopoulos, P.; Kyritsis, A.; Kontou, E. et al. Strain Sensing in Polymer/Carbon Nanotube Composites by Electrical Resistance Measurement. *Compos. Part B: Eng.* **2015,** *68*, 162–169.

Gilman, J. W.; Awad, W. H.; Davis, R. D.; Shields, J.; Harris, R. H.; Davis, C.; Morgan, A. B.; Sutto, T. E.; Callahan, J.; Trulove, P. C.; De Long, H. C. Polymer/Layered Silicate Nanocomposites from Thermally Stable Trialkylimidazolium-Treated Montmorillonite. *Chem. Mater.* **2002,** *14*, 3776–3785.

Guo, Z.; Liang, X.; Pereira, T.; Scaffaro, R.; Hahn, H. Curing Mechanism Study of Cupric Oxide Nanoparticle Filled Polymeric Nanocomposite and its Characterization. *Compos. Sci. Techn.* **2007,** *67*, 2036–2044.

Hamming, L. M.; Qiao, R.; Messersmith, P. B.; Brinson, L. C. Effects of Dispersion and Interfacial Modification on the Macroscale Properties of TiO_2 Polymer–Matrix Nanocomposites. *Compos. Sci. Technol.* **2009,** *69*, 1880–1886.

Hiruma, K.; Yazawa, M.; Katsuyama, T.; Ogawa, K.; Haraguchi, K.; Koguchi, M.; Kakibayashi, H. Growth and Optical Properties of Nanometer-Scale GaAs and InAs Whiskers. *J. Appl. Phys.* **1995,** *77*, 447–462.

Hopwood, J. D.; Mann, S. Synthesis of Barium Sulfate Nanoparticles and Nanofilaments in Reverse Micelles and Microemulsions. *Chem. Mater.* **1997,** *9*, 1819–1828.

Hu, J.; Zhang, L.; Dang, Z. M.; Wang, D. Improved Dielectric Properties of Polypropylene-Based Nanocomposites via Co-filling with Zinc Oxide and Barium Titanate. *Compos. Sci. Technol.* **2017,** *148*, 20–26.

Hung, P. Y.; Lau, K. T.; Fox, B.; Hameed, N.; Lee, J. H.; Hui, D. Surface Modification of Carbon Fibre Using Graphene–Related Materials for Multifunctional Composites. *Compos. Part B: Eng.* **2018,** *133*, 240–257.

Jamail, N. A.; Piah, M. A.; Muhamad, N. A.; Salam, Z.; Kasri, N. F.; Zainir, R. A.; Kamarudin, Q. E. Effect of Nanofillers on the Polarization and Depolarization Current Characteristics of New LLDPE-NR Compound for High Voltage Application. *Adv. Mater. Sci. Eng.* **2014**. DOI: 10.1155/2014/416420

Jordan, J.; Jacob, K. I.; Tannenbaum, R.; Sharaf, M. A; Jasiuk, I. Experimental Trends in Polymer Nanocomposites—A Review. *Mater. Sci. Eng.: A* **2005,** *393*, 1–11.

Jose, J. P.; Thomas, S. Alumina–Clay Nanoscale Hybrid Filler Assembling in Cross-Linked Polyethylene Based Nanocomposites: Mechanics and Thermal Properties. *Phys. Chem. Chem. Phys*. **2014a,** *16*, 14730–14740.

Jose, J. P.; Thomas, S. XLPE Based Al_2O_3–Clay Binary and Ternary Hybrid Nanocomposites: Self-Assembly of Nanoscale Hybrid Fillers, Polymer Chain Confinement and Transport Characteristics. *Phys. Chem. Chem. Phys*. **2014b,** *16*, 20190–20201.

Jose, T.; George, S. C.; Maria, H. J.; Wilson, R.; Thomas, S.Effect of Bentonite Clay on the Mechanical, Thermal, and Pervaporation Performance of the Poly(Vinyl Alcohol) Nanocomposite Membranes. *Industr. Eng. Chem. Res*. **2014c,** *53*, 16820–16831.

Jung, T.; Kamm, W.; Breitenbach, A.; Kaiserling, E.; Xiao, J. X.; Kissel, T. Biodegradable Nanoparticles for Oral Delivery of Peptides: Is There a Role for Polymers to Affect Mucosal Uptake? *Eur. J. Pharm. Biopharm*. **2000,** *50*, 147–160.

Justus, B. L.; Tonucci, R. J.; Berry, A. D. Nonlinear Optical Properties of Quantum-Confined GaAs Nanocrystals in Vycor Glass. *Appl. Phys. Lett.* **1992,** *61*, 3151–3153.

Kang, H.; Zuo, K.; Wang, Z.; Zhang, L.; Liu, L.; Guo, B. Using a Green Method to Develop Graphene Oxide/Elastomers Nanocomposites with Combination of High Barrier and Mechanical Performance. *Compos. Sci. Technol*. **2014,** *92*, 1–8.

Kasuga, T.; Hiramatsu, M.; Hoson, A.; Sekino, T.; Niihara, K. Formation of Titanium Oxide Nanotube. *Langmuir* **1998,** *14*, 3160–3163.

Kelly, A. Composites in Context. *Compos. Sci. Technol*. **1985,** *23* (3), 171–199.

Ko, S. H.; Park, I.; Pan, H.; Grigoropoulos, C. P.; Pisano, A. P.; Luscombe, C. K.; Fréchet, J. M. Direct Nanoimprinting of Metal Nanoparticles for Nanoscale Electronics Fabrication. *Nano Lett.* **2007,** *77*, 1869–19877.

Komori, Y.; Sugahara, Y.; Kuroda, K. Direct Intercalation of Poly (Vinylpyrrolidone) into Kaolinite by a Refined Guest Displacement Method. *Chem. Mater.* **1999,** *11*, 3–6.

Koo, O. M.; Rubinstein, I.; Onyuksel, H. Role of Nanotechnology in Targeted Drug Delivery and Imaging: A Concise Review. *Nanomed. NBM* **2005,** *1*, 193–212.

Kortan, A. R.; Kopylov, N.; Glarum, S.; Gyorgy, E. M.; Ramirez, A. P.; Fleming, R. M.; Zhou, O.; Thiel, F. A.; Trevor, P. L.; Haddon, R. C. Superconductivity in Barium Fulleride. *Nature* **1992,** *360*, 566.

Krikorian, V.; Pochan, D. J. Poly (L-lactic acid)/Layered Silicate Nanocomposite: Fabrication, Characterization, and Properties. *Chem. Mater*. **2003,** *15*, 4317–4324.

Kuilla, T.; Bhadra, S.; Yao, D.; Kim, N. H.; Bose, S.; Lee, J. H. Recent Advances in Graphene Based Polymer Composites. *Pro. Poly. Sci.* **2010,** *35*, 1350–1375.

Kung, H. H.; Ko, E. I. Preparation of Oxide Catalysts and Catalyst Supports—A Review of Recent Advances. *Chem. Eng. J. Biochem. Eng. J.* **1996,** *64*, 203–214.

Kuo, M. C.; Huang, J. C.; Chen, M. Non-Isothermal Crystallization Kinetic Behavior of Alumina Nanoparticle Filled Poly (Ether Ether Ketone). *Mater. Chem. Phys*. **2006,** *99*, 258–268.

Kuo, S. W.; Chang, F. C. POSS Related Polymer Nanocomposites. *Pro. Polym. Sci*. **2011,** *36*, 1649–1696.

Kurahatti, R. V.; Surendranathan, A. O.; Kori, S. A.; Singh, N.; Kumar, A. V.; Srivastava, S. Defence Applications of Polymer Nanocomposites. *Defence Sci. J.* **2010,** *60* (5), 551–563.

Kyprianidou-Leodidou T, Caseri W, Suter, U. W. Size Variation of PbS Particles in High-Refractive-Index Nanocomposites. *J. Physi. Chem.* **1994,** *98*, 8992–8997.

Lee, J. M.; Sultan, M. T.; Kim, S. H.; Kumar, V.; Yeon, Y. K.; Lee, O. J. et al. Artificial Auricular Cartilage Using Silk Fibroin and Polyvinyl Alcohol Hydrogel. *Int. J. Mol. Sci.* **2017,** *18* (8). doi.org/10.3390/ijms18081707.

Li, G.; Yu, S.; Sun, R.; Lu, D. Clean and in-Situ Synthesis of Copper–Epoxy Nanocomposite as a Matrix for Dielectric Composites with Improved Dielectric Performance. *Composites Sci. Technol.* **2015,** *110*, 95–102.

Li, Y. Q.; Fu, S. Y.; Yang, Y.; Mai, Y. W. Facile Synthesis of Highly Transparent Polymer Nanocomposites by Introduction of Core–Shell Structured Nanoparticles. *Chem. Mater*. **2008,** *20*, 2637–2643.

Li, W.; Feng, Z.; Dai, E.; Xu, J.; Bai, G. Organic Vapour Sensing Properties of Area-Ordered and Size-Controlled Silicon Nanopillar. *Sensors* **2016,** *20*, 1880.

Lim, Y. T.; Park, O. O. Rheological Evidence for the Microstructure of Intercalated Polymer/Layered Silicate Nanocomposites. *Macromo. Rapid Commun.* **2000,** *21*, 231–235.

Lin, J. C.; Chang, L. C.; Nien, M. H.; Ho, H. L. Mechanical Behavior of Various Nanoparticle Filled Composites at Low-Velocity Impact. *Compos. Struct.* **2006,** *74*, 30–36.

Liu, H.; Huang, W.; Yang, X.; Dai, K.; Zheng, G.; Liu, C. et al. Organic Vapor Sensing Behaviors of Conductive Thermoplastic Polyurethane–Graphene Nanocomposites. *J. Mater. Chem. C.* **2016,** *4*, 4459–4469.

Lofrano, G.; Carotenuto, M.; Libralato, G.; Domingos, R. F.; Markus, A.; Dini, L. et al. Polymer Functionalized Nanocomposites for Metals Removal from Water and Wastewater: An Overview. *Water Res*. **2016,** *92*, 22–37.

Lott, J.; Xia, C.; Kosnosky, L.; Weder, C.; Shan, J. Terahertz Photonic Crystals Based on Barium Titanate/Polymer Nanocomposites. *Adv. Mater*. **2008,** *20*, 3649–3653.

Ma, P. C.; Siddiqui, N. A.; Marom, G.; Kim, J. K. Dispersion and Functionalization of Carbon Nanotubes for Polymer-Based Nanocomposites: A Review. *Compos. Part A: Appl. Sci. Manuf.* **2010,** *41*, 1345–1367.

Maiti, M.; Sadhu, S.; Bhowmick, A. K. Brominated Poly (isobutylene-co-para-methylstyrene) BIMS)-Clay Nanocomposites: Synthesis and Characterization. *J. Poly. Sci. Part B: Poly. Phys.* **2004,** *42*, 4489–4502.

Majetich, S. A. Carter, A. C. Surface Effects on the Optical Properties of Cadmium Selenide Quantum Dots. *J. Phys. Chem.* **1993,** *97*, 8727–8731.

Mallakpour, S.; Khadem, E. Recent Development in the Synthesis of Polymer Nanocomposites Based on Nano-Alumina. *Prog. Polym. Sci*. **2015,** *51*, 74–93.

Maria, H. J.; Lyczko, N.; Nzihou, A.; Joseph, K.; Mathew, C.; Thomas, S. Stress Relaxation Behavior of Organically Modified Montmorillonite Filled Natural Rubber/Nitrile Rubber Nanocomposites. *Appl. Clay Sci.* **2014,** *87*, 120–128.

Martin, C. R. Nanomaterials: A Membrane-Based Synthetic Approach. *Science* **1994,** *266*, 1961–1966.

Matei, A.; Cernica, I.; Cadar, O.; Roman, C.; Schiopu, V. Synthesis and Characterization of ZnO–Polymer Nanocomposites. *Int. J. Mat. Form.* **2008,** *1*, 767–770.

Mohan, T. P.; Kanny, K. Effects of Synthetic and Processing Methods on Dispersion Characteristics of Nanoclay in Polypropylene Polymer Matrix Composites. *Mater. Sci. Appl.* **2011,** *2*, 785–800.

Moni, G.; Abraham, J.; Kurian, C.; Joseph, A.; George, S. C. Effect of Reduced Graphene Oxide on the Solvent Transport Characteristics and Sorption Kinetics of Fluoroelastomer Nanocomposites. *Phys. Chem. Chem. Phys.* **2018a,** *20*, 17909–17917.

Moni, G.; Mayeen, A.; Abraham, J.; Jose, T.; Maya, M. G.; Bhowmik, R.; George, S. C. Flexible FKM/mRGO Nanocomposites with Excellent Thermal, Mechanical and Electrical Properties. *Arab. J. Chem.* **2018b,** *13*, 2142–2152.

Murray, C.; Norris, D. J.; Bawendi, M. G. Synthesis and Characterization of Nearly Monodisperse CdE(E= Sulfur, Selenium, Tellurium) Semiconductor Nanocrystallites. *J. Am.Chem. Soc.* **1993,** *115*, 8706–8715.

Naganuma, T.; Kagawa, Y. Effect of Particle Size on the Optically Transparent Nano Meter-Order Glass Particle-Dispersed Epoxy Matrix Composites. *Compos. Sci. Technol.* **2002,** *62*, 1187–1189.

Nalwa, H. S. *Encyclopedia of Nanoscience and Nanotechnology*; American Scientific Publishers: California, 2004; p 10.

Niihara, K. New Design Concept of Structural Ceramics. *J. Ceram. Soc. Japan.* **1991,** *99*, 974–982.

Nimesh, S.; Manchanda, R.; Kumar, R.; Saxena, A.; Chaudhary, P.; Yadav, V. et al. Preparation, Characterization and in Vitro Drug Release Studies of Novel Polymeric Nanoparticles. *Int. J. Pharm.* **2006,** *323*, 146–152.

Ning, N.; Ji, L.; Zhang, L.; Liu, J.; Lu, Y.; Wu, S. et al. High Elasticity and Conductivity of Elastomer Composites with Arrayed Carbon Nanotubes as Nanosprings. *Compos. Sci. Technol.* **2015,** *118*, 78–84.

Nisha, A.; Rajeswari, K. M.; Dhamodharan, R. Intercalative Redox Polymerization and Characterization of poly(n-vinyl-2-pyrrolidinone) in the Gallery of Vermiculite: A Novel Inorganic–Organic Hybrid Material. *J. App. Polym. Sci.* **2000,** *76*, 1825–1830.

Ock, H. G.; Ahn, K. H.; Lee, S. J.; Hyun, K. Characterization of Compatibilizing Effect of Organoclay in Poly (Lactic Acid) and Natural Rubber Blends by FT-Rheology. *Macromolecules* **2016,** *49*, 2832–2842.

Ogata, N.; Kawakage, S.; Ogihara, T. (Poly Vinyl Alcohol)-clay and (Poly Ethylene Oxide)-Clay Blends Prepared Using Water as Solvent. *J. App. Polym. Sci.* **1997,** *66*, 573–581.

Olshavsky, M. A.; Allcock, H. R. Synthesis of CdS Nanoparticles in Solution and in a Polyphosphazene Matrix. *Chem. Mater.* **1997,** *17*, 1367–1376.

Ono-Ogasawara, M.; Serita, F.; Takaya, M. Distinguishing Nanomaterial Particles from Background Airborne Particulate Matter for Quantitative Exposure Assessment. *J. Nanopart. Res.* **2009,** *11* (7), 1651–1659.

Palza, H. Antimicrobial Polymers with Metal Nanoparticles. *Int. J. Mol. Sci.* **2015,** *16*, 2099–2116.

Parthasarathy, R. V.; Martin, C. R. Synthesis of Polymeric Microcapsule Arrays and Their Use for Enzyme Immobilization. *Nature* **1994,** *369*, 298–301.

Pawar, S. P.; Gandi, M.; Bose, S. High Performance Electromagnetic Wave Absorbers Derived from PC/SAN Blends Containing Multiwall Carbon Nanotubes and Fe_3O_4 Decorated onto Graphene Oxide Sheets. *RSC Adv.* **2016,** *6*, 37633–37645.

Peigney, A.; Laurent, C.; Dobigeon, F.; Rousset, A. Carbon Nanotubes Grown in Situ by a Novel Catalytic Method. *J. Mater. Res.* **1997,** *12*, 613–615.

Peppas, N. A. Intelligent Therapeutics: Biomimetic Systems and Nanotechnology in Drug Delivery. *Adv. Drug Deliv. Revs.* **2004,** *56* (11), 1529–1531.

Pileni, M. P.; Motte, L.; Petit, C. Synthesis of Cadmium Sulfide in Situ in Reverse Micelles: Influence of the Preparation Modes on Size, Polydispersity, and Photochemical Reactions. *Chem. Mater.* **1992,** *4*, 338–345.

Pillai, V.; Kumar, P.; Hou, M. J.; Ayyub, P.; Shah, D. O. Preparation of Nanoparticles of Silver Halides, Superconductors and Magnetic Materials Using Water-in-Oil Microemulsions as Nano-Reactors. *Adv. Colloid Interf. Sci.* **1995,** *55*, 241–269.

Poothanari, M. A.; Abraham, J.; Kalarikkal, N.; Thomas, S. Excellent Electromagnetic Interference Shielding and High Electrical Conductivity of Compatibilized Polycarbonate/Polypropylene Carbon Nanotube Blend Nanocomposites. *Indust. Eng. Chem. Res.* **2018,** *57*, 4287–4297.

Prabha, G.; Raj, V. Preparation and Characterization of Polymer Nanocomposites Coated Magnetic Nanoparticles for Drug Delivery Applications. *J. Magnet. Magnet. Mater*. **2016,** *408*, 26–34.

Pramanik, M.; Srivastava, S. K.; Samantaray, B. K.; Bhowmick, A. K. Rubber–Clay Nanocomposite by Solution Blending. *J. Appl. Polym. Sci.* **2003,** *87*, 2216–2220.

Pramanik, S.; Karak, N. Polymer Nanocomposites for Adhesive, Coating, and Paint Applications. Properties and Applications of Polymer Nanocomposites: Clay and Carbon Based Polymer Nanocomposites, 2017. https://link.springer.com/chapter/10.1007/978-3-662-53517-2_8

Praveen, S.; Chattopadhyay, P. K.; Albert, P.; Dalvi, V. G.; Chakraborty, B. C.; Chattopadhyay, S. Synergistic Effect of Carbon Black and Nanoclay Fillers in Styrene Butadiene Rubber Matrix: Development of Dual Structure. *Compos. Part A: Appl. Sci. Manuf.* **2009,** *40*, 309–316.

Pu, Z.; Mark, J. E.; Jethmalani, J. M.; Ford, W. T. Mechanical Properties of a Poly (Methyl Acrylate) Nanocomposite Containing Regularly-Arranged Silica Particles. *Polym. Bull.* **1996,** *37*, 545–551.

Qu, X.; Brame, J.; Li, Q.; Alvarez, P. J.; Nanotechnology for a Safe and Sustainable Water Supply: Enabling Integrated Water Treatment and Reuse. *Acc. Chem. Res.* **2012,** *46*, 834–843.

Rajasekar, R.; Pal, K.; Heinrich, G.; Das, A.; Das, C. K. Development of Nitrile Butadiene Rubber–Nanoclay Composites with Epoxidized Natural Rubber as Compatibilizer. *Mater. Design*. **2009a,** *30*, 3839–3845.

Rajasekar, R.; Heinrich, G.; Das, A.; Das, C. K. Development of SBR-Nanoclay Composites with Epoxidized Natural Rubber as Compatibilizer. *J. Nanotechnol.***2009b,** *2009*. DOI: 10.1155/2009/405153.

Rath, T. Carbon Nanofibers: Synthesis, Properties and Applications. In *Polymer Nanocomposites Based on Inorganic and Organic Nanomaterials*; Mohanty, S., Nayak, S. K., Kaith, B. S., Kalia, S., Eds.; **2010,** *5*, 63–88.

Ravichandran, R.; Sundarrajan, S.; Venugopal, J. R.; Mukherjee, S.; Ramakrishna, S. Advances in Polymeric Systems for Tissue Engineering and Biomedical Applications. *Macromol. Biosci.* **2012,** *12*, 286–311.

Sabir, B. B. Strength and Fracture Toughness of Silica Fume Concrete. *Mag. Concr. Res.* **1997,** *49*, 139–146.

Sadasivuni, K. K.; Ponnamma, D.; Thomas, S.; Grohens, Y. Evolution from Graphite to Graphene Elastomer Composites. *Prog. Polym. Sci.* **2014,** *39*, 749–780.

Sambhy, V.; MacBride, M. M.; Peterson, B. R.; Sen, A. Silver Bromide Nanoparticle/Polymer Composites: Dual Action Tunable Antimicrobial Materials. *J. Am. Chem. Soc.* **2006,** *128*, 9798–9808.

Sankaran, V.; Yue, J.; Cohen, R. E.; Schrock, R. R.; Silbey, R. J. Synthesis of Zinc Sulfide Clusters and Zinc Particles Within Microphase-Separated Domains of Organometallic Block Copolymers. *Chem. Mater.* **1993,** *5*, 1133–1142.

Shaheen, S.; Olson, D.; White, M.; Mitchell, W.; Miedaner, A.; Curtis, C. et al. Polymer Based Nanocomposites for Solar Energy Conversion. National Renewable Energy Laboratory (NREL), Golden, CO, 2005.

Shao, X.; Tian, J.; Liu, W.; Xue, Q.; Ma, C. Tribological Properties of SiO_2 Nanoparticle Filled–Phthalazine Ether Sulfone/Phthalazine Ether Ketone (50/50 mol%) Copolymer Composites. *J. Appl. Poly. Sci.* **2002,** *85*, 2136–2144.

Sharma, M.; Sharma, S.; Abraham, J.; Thomas, S.; Madras, G.; Bose, S. Flexible EMI Shielding Materials Derived by Melt Blending PVDF and Ionic Liquid Modified MWNTs. *Mat. Res. Express.* **2014**. DOI: 10.1088/2053–1591/1/3/035003.

Shrivastava, S.; Jadon, N.; Jain, R. Next-Generation Polymer Nanocomposite-Based Electrochemical Sensors and Biosensors: A Review. *TrAC Trends Anal. Chem.* **2016,** *82*, 55–67.

Singh, N. B.; Susan, A. B. Polymer Nanocomposites for Water Treatments. (In) Polymer-Based Nanocomposites for Energy and Environmental Applications **2018,** 569–595.

Singh, R. P.; Zhang, M.; Chan, D. Toughening of a Brittle Thermosetting Polymer: Effects of Reinforcement Particle Size and Volume Fraction. *J. Mater. Sci.* **2002,** *37*, 781–788.

Soppimath, K. S.; Aminabhavi, T. M.; Kulkarni, A. R.; Rudzinski, W. E. Biodegradable Polymeric Nanoparticles as Drug Delivery Devices. *J. Contr. Rel.* **2001,** *70*, 1–20.

Sridhar, V.; Tripathy, D. K. Barrier Properties of Chlorobutylnanoclay Composites. *J. Appl. Poly. Sci.* **2006,** *101*, 3630–3637.

Strawhecker, K. E.; Manias, E. Structure and Properties of Poly (Vinyl Alcohol)/Na+ Montmorillonite Nanocomposites. *Chem. Mater.* **2000,** *12*, 2943–2949.

Suslick, K. S.; Didenko, Y.; Fang, M. M.; Hyeon, T.; Kolbeck, K. J.; McNamara, W. B. M.; Mdleleni, M.; Wong, M. Acoustic Cavitation and Its Chemical Consequences. *Phil. Trans. Royal Soc. Lond. A: Math. Phys. Eng. Sci.* **1999,** *357*, 335–353.

Suslick, K. S.; Fang, M. M.; Hyeon, T.; Mdleleni, M. M. Applications of Sonochemistry to Materials Synthesis. In *Sonochemistry and Sonoluminescence*; Springer: Dordrecht, 1999b; pp 291–320.

Tang, H.; Lin, Y.; Andrews, C.; Sodano, H. A. Nanocomposites with Increased Energy Density through High Aspect Ratio PZT Nanowires. *Nanotechnology* **2010**. DOI: 10.1088/0957-4484/22/1/015702.

Terzopoulou, Z.; Bikiaris, D. N.; Triantafyllidis, K. S.; Potsi, G.; Gournis, D.; Papageorgiou, G. Z.; Rudolf, P. Mechanical, Thermal and Decomposition Behavior of Poly ε-caprolactone) Nanocomposites with Clay-Supported Carbon Nanotube Hybrids. *Thermochim. Acta.* **2016,** *642*, 67–80.

Thostenson, E. T.; Li, W. Z.; Wang, D. Z.; Ren, Z. F.; Chou, T. W. Carbon Nanotube/Carbon Fiber Hybrid Multiscale Composites. *J. App. Phys.* **2002,** *91*, 6034–6037.

Tseng, C. R.; Wu, J. Y.; Lee, H. Y.; Chang, F. C. Preparation and Crystallization Behavior of Syndiotactic Polystyrene–Clay Nanocomposites. *Polymer* **2001,** *42*, 10063–10070.

Tung, T. T.; Robert, C.; Castro, M.; Feller, J. F.; Kim, T. Y.; Suh, K. S. Enhancing the Sensitivity of Graphene/Polyurethane Nanocomposite Flexible Piezo-Resistive Pressure Sensors with Magnetite Nano-Spacers. *Carbon*. **2016,** *108*, 450–460.

Unuabonah, E. I.; Taubert, A. Clay–Polymer Nanocomposites (CPNs): Adsorbents of the Future for Water Treatment. *Appl. Clay Sci.* **2014,** *99*, 83–92.

Venkatesan, J.; Bhatnagar, I.; Manivasagan, P.; Kang, K. H.; Kim, S. K. Alginate Composites for Bone Tissue Engineering: A Review. *Int. J. Biol. Macromol.* **2015,** *72*, 269–281.

Virtanen, S.; Krentz, T. M.; Nelson, J. K.; Schadler, L. S.; Bell, M.; Benicewicz, B.; Hillborg, H.; Zhao, S. Dielectric Breakdown Strength of Epoxy Bimodal-Polymer-Brush-Grafted Core Functionalized Silica Nanocomposites. *IEEE Trans. Dielectr. Electr. Insul.* **2014,** *21*, 563–570.

Wang, C. C.; Zhang, Z.; Ying, J. Y. Photocatalytic Decomposition of Halogenated Organics over Nanocrystallinetitania. *Nanostruct. Mater.* **1997,** *9*, 583–586.

Wang, J.; Hu, J.; Yang, L.; Zhu, K.; Li, B. W.; Sun, Q.; Li, Y.; Qiu, J. High Discharged Energy Density of Polymer Nanocomposites Induced by Nd-doped $BaTiO_3$ Nanoparticles. *J. Materiomics.* **2018,** *4*, 44–50.

Wang, L.; Wu, X. L.; Xu, W. H.; Huang, X. J.; Liu, J. H.; Xu, A. Stable Organic–Inorganic Hybrid of Polyaniline/α-Zirconium Phosphate for Efficient Removal of Organic Pollutants in Water Environment. *ACS Appl. Mater. Interf.* **2012,** *4*, 2686–2692.

Weimer, M. W.; Chen, H.; Giannelis, E. P.; Sogah, D. Y. Direct Synthesis of Dispersed Nanocomposites by in Situ Living Free Radical Polymerization Using a Silicate-Anchored Initiator. *J. Am. Chem. Soc.* **1999,** *121*, 615–616.

Whitesides, G. M.; Grzybowski, B. Self-Assembly at All Scales. *Science* **2002,** *295*, 2418–2421.

Wu, H. D.; Tseng, C. R.; Chang, F. C. Chain Conformation and Crystallization Behavior of the Syndiotactic Polystyrene Nanocomposites Studied Using Fourier Transform Infrared Analysis. *Macromolecules* **2001,** *34*, 2992–2999.

Xia, X.; Yih, J.; D'Souza, N. A.; Hu, Z. Swelling and Mechanical Behavior of Poly N-isopropylacrylamide)/Na-Montmorillonite Layered Silicates Composite Gels. *Polymer* **2003,** *44*, 3389–3393.

Xiao, P.; Xiao, M.; Gong, K. Preparation of Exfoliated Graphite/Polystyrene Composite by Polymerization-Filling Technique. *Polymer* **2001,** *42*, 4813–4816.

Yang, K.; Huang, X.; He, J.; Jiang, P. Strawberry-like Core–Shell Ag@ Polydopamine@ BaTiO3 Hybrid Nanoparticles for High-k Polymer Nanocomposites with High Energy Density and Low Dielectric Loss. *Adv. Mater. Interf.* **2015**. DOI: 10.1002/admi.201500361.

Yang, Y.; Gupta, M. C.; Dudley, K. L.; Lawrence, R. W. A Comparative Study of EMI Shielding Properties of Carbon Nanofiber and Multi-Walled Carbon Nanotube Filled Polymer Composites. *J. Nanosci. Nanotechnol*. **2005a,** *5*, 927–931.

Yang, Y.; Gupta, M. C.; Dudley, K. L.; Lawrence, R. W. Conductive Carbon Nanofiber–Polymer Foam Structures. *Adv. Mater.* **2005b,** *17*, 1999–2003.

Yaragalla, S.; Chandran, C. S.; Kalarikkal, N.; Subban, R. H.; Chan, C. H.; Thomas, S. Effect of Reinforcement on the Barrier and Dielectric Properties of Epoxidized Natural Rubber–Graphene Nanocomposites. *Poly. Eng. Sci*. **2015a,** *55*, 2439–2447.

Yaragalla, S.; Meera, A. P.; Kalarikkal, N.; Thomas, S. Chemistry Associated with Natural Rubber–Graphene Nanocomposites and Its Effect on Physical and Structural Properties. *Industr. Crops Prod.* **2015b,** *74*, 792–802.

Yaragalla, S.; Sindam, B.; Abraham, J.; Raju, K. J.; Kalarikkal, N.; Thomas, S. Fabrication of Graphite-Graphene-Ionic Liquid Modified Carbon Nanotubes Filled Natural Rubber Thin Films for Microwave and Energy Storage Applications. *J. Poly. Res.* **2015c,** *22*. DOI: 10.1007/s10965-015-0776-5.

Ying, J. Y.; Sun, T. Research Needs Assessment on Nanostructured Catalysts. *J. Electroceram.* **1991,** *1*, 219–238.

Yu, X.; Wang, J.; Du, Y.; Ma, Z.; He, W. Nanotechnology for Tissue Engineering Applications. *J. Nanomater*. **2011**. doi.org/10.1155/2011/506574.

Yuan, Y.; Fendler, J. H.; Cabasso, I. Photoelectron Transfer Mediated by Size-Quantized Cadmium Sulfide Particles in Polymer-Blend Membranes. *Chem. Mater.* **1992,** *4*, 312–318.

Zeng, Q. H.; Wang, D. Z.; Yu, A. B.; Lu, G. Q. Synthesis of Polymer–Montmorillonite Nanocomposites by in Situ Intercalative Polymerization. *Nanotechnology* **2002,** *13*. DOI: 10.1088/0957-4484/13/5/301/meta.

Zhang, C.; Chi, Q.; Dong, J.; Cui, Y.; Wang, X.; Liu, L.; Lei, Q. Enhanced Dielectric Properties of Poly Vinylidene Fluoride) Composites Filled with Nano Iron Oxide-Deposited Barium Titanate Hybrid Particles. *Sci. Rep*. **2016**. DOI: 10.1038/srep33508.

Zhang, D.; Tong, J.; Xia, B. Humidity-Sensing Properties of Chemically Reduced Graphene Oxide/Polymer Nanocomposite Film Sensor Based on Layer-by-Layer Nano Self-Assembly. *Sens. Actuat. B: Chem.* **2014,** *197*, 66–72.

Zhang, L.; Manthiram, A. Chains Composed of Nanosize Metal Particles and Identifying the Factors Driving Their Formation. *App. Phys. Lett.* **1997,** *70*, 2469–2471.

Zhao, X.; Urano, K.; Ogasawara, S. Adsorption of Polyethylene Glycol from Aqueous Solution on Montmorillonite Clays. *Colloid Polym. Sci.* **1989,** *267*, 899–906.

Zhu, J.; Morgan, A. B.; Lamelas, F. J.; Wilkie, C. A. Fire Properties of Polystyrene– Clay Nanocomposites. *Chem. Mater.* **2001,** *13*, 3774–3780.

Index